325

Topics in Current Chemistry

Topics in Current Chemistry
Recently Published and Forthcoming Volumes

Molecular Imprinting

Volume Editor: Karsten Haupt

With Contributions by

C. Ayela · M.E. Benito-Peña · A. Biffis · M. Bompart · D. Carboni ·
P.J. Cywinski · G. Dvorakova · A. Falcimaigne-Cordin · K. Flavin ·
K. Haupt · G. Horvai · W. Kutner · A.V. Linares · G.J. Mohr ·
M.C. Moreno-Bondi · A.J. Moro · G. Orellana · M. Resmini ·
S. Suriyanarayanan · B. Tse Sum Bui · B. Tóth · J.L. Urraca

Springer

Editor
Prof. Karsten Haupt
Compiègne University of Technology
UMR CNRS 6022
Compiègne
France

ISSN 0340-1022   e-ISSN 1436-5049
ISBN 978-3-642-28420-5   e-ISBN 978-3-642-28421-2
DOI 10.1007/978-3-642-28421-2
Springer Heidelberg Dordrecht London New York

Library of Congress Control Number: 2012932901

Printed on acid-free paper

Springer is part of Springer Science+Business Media (www.springer.com)

Topics in Current Chemistry
Also Available Electronically

Topics in Current Chemistry is included in Springer's eBook package *Chemistry and Materials Science*. If a library does not opt for the whole package the book series may be bought on a subscription basis. Also, all back volumes are available electronically.

For all customers with a print standing order we offer free access to the electronic volumes of the series published in the current year.

If you do not have access, you can still view the table of contents of each volume and the abstract of each article by going to the SpringerLink homepage, clicking on "Chemistry and Materials Science," under Subject Collection, then "Book Series," under Content Type and finally by selecting *Topics in Current Chemistry*.

You will find information about the

– Editorial Board
– Aims and Scope
– Instructions for Authors
– Sample Contribution

at springer.com using the search function by typing in *Topics in Current Chemistry*.

Color figures are published in full color in the electronic version on SpringerLink.

Aims and Scope

The series *Topics in Current Chemistry* presents critical reviews of the present and future trends in modern chemical research. The scope includes all areas of chemical science, including the interfaces with related disciplines such as biology, medicine, and materials science.

The objective of each thematic volume is to give the non-specialist reader, whether at the university or in industry, a comprehensive overview of an area where new insights of interest to a larger scientific audience are emerging.

Thus each review within the volume critically surveys one aspect of that topic and places it within the context of the volume as a whole. The most significant developments of the last 5–10 years are presented, using selected examples to illustrate the principles discussed. A description of the laboratory procedures involved is often useful to the reader. The coverage is not exhaustive in data, but rather conceptual, concentrating on the methodological thinking that will allow the non-specialist reader to understand the information presented.

Discussion of possible future research directions in the area is welcome.

Review articles for the individual volumes are invited by the volume editors.

In references *Topics in Current Chemistry* is abbreviated *Top Curr Chem* and is cited as a journal.

Impact Factor 2010: 2.067; Section "Chemistry, Multidisciplinary": Rank 44 of 144

Preface

Like a key in a lock, antibodies fit perfectly with their target antigens and are able to recognise and bind them with high affinity and selectivity. They have, therefore, found numerous applications in medicine, both for diagnostics and treatment, and in biotechnology. For example, enzyme-linked immunosorbent assays are widely used to detect and quantify small and large targets via a specific biological recognition mechanism. On the other hand, antibodies are also used to treat certain infections and other diseases, and have saved many lives. More recently, new cancer therapies have been developed based on antibodies. However, antibodies are not always perfect for these applications because they are unstable out of their native environment, may be degraded by proteases, and tend to be difficult to integrate into standard industrial fabrication processes. In addition, an antibody for the particular target of interest, in particular for small molecules, can sometimes be difficult to obtain. It has therefore been a long-term dream of researchers to be able to obtain such structures synthetically – creating tailor-made receptors for a given molecular target. One surprisingly simple way of achieving this is through the molecular imprinting of synthetic polymers.

Molecular imprinting is a process where interacting and cross-linking monomers are arranged around a molecular template, followed by polymerisation to form a cast-like shell. The template is usually the target molecule to be recognised by the synthetic antibody, or a derivative thereof. Initially, the monomers form a complex with the template through covalent or non-covalent interactions. After polymerisation and removal of the template, binding sites are exposed that are complementary to the target molecule in size, shape, and position of functional groups, which are held in place by the cross-linked polymer matrix. In essence, a molecular memory is imprinted in the polymer, which is now capable of selectively rebinding the target. Thus, molecularly imprinted polymers (MIPs) possess two of the most important features of biological antibodies – the ability to recognise and bind specific target molecules.

When these MIPs were first described in the 1970s and early 1980s by Wulff [1] and Mosbach [2], they were merely used as specific separation materials, for example, for the chromatographic separation of enantiomers. It was not until 1993 and Mosbach's seminal paper in Nature [3] that the great potential of MIPs as synthetic antibody mimics was recognised. This resulted in a nearly exponential increase in the number of publications in the area, with several hundreds per annum over the recent years. There are a number of potential application areas that have been identified for MIPs, all based on their capability to specifically recognise molecular targets: affinity separation, chemical sensors and assays, directed synthesis and enzyme-like catalysis, and biomedical applications like drug delivery. To date, the main application area is analytical chemistry, and during the past decade, the only commercially available MIPs have been solid-phase extraction matrices for sample preparation and analyte pre-concentration, mainly for biomedical and food analyses. These are commercialised by the Swedish company *Biotage* and by the French *PolyIntell*. However, apart from separation, other promising commercial applications of MIPs are sensors and assay systems. Indeed, MIP-coated wipes for the detection of explosives are more recent products commercialised by the US company *Raptor*.

An exciting recent trend goes towards the use of MIPs for medical treatment, in particular for drug delivery. One example is the use of MIPs in contact lenses for drug delivery to the eye. In fact, there is a considerable need for more efficient delivery of ocular therapeutics. This can be achieved by using the molecular selectivity and affinity of an MIP to extend and control the residence time of drugs on the eye surface and thereby limiting drug loss by lacrimation, drainage, and non-productive absorption [4]. On the other hand, it is also conceivable to use MIPs for the removal of unwanted molecules from our body. While there have been no reports in the literature on real applications with living organisms, there are a few examples on the removal of toxic substances, for example bilirubin and metal ions, from biological fluids using extracorporeal devices [5], and the Israeli–US company *Semorex* lists an MIP for phosphate removal from the gastrointestinal system as one of their products.

In a similar direction, that is, using MIPs directly as drugs, goes the work by Cutivet et al. [6], who have developed water-compatible MIP microgels that strongly and selectively inhibit the protease trypsin, enzyme inhibitors being potential drug candidates. The inhibitory power of these MIPs was three orders of magnitude higher than that of small-molecule inhibitors like benzamidine. Very recently, Shea and colleagues [7] have made an exciting new contribution to the MIP field. They created molecular imprints for the cytotoxic peptide melittin, the main component of bee venom, in the surface of polymer nanoparticles, obtaining, as a result, an artificial antibody that could be used for the in vivo capture and neutralisation of melittin in mice. The authors for the first time demonstrated the possibility of in vivo application of the imprinted nanoparticles in mice. Normally, when mice are injected into the blood stream with a certain dose of melittin, they die within less than an hour due to the cytolytic activity of this peptide. However, when the MIP nanoparticles were injected shortly after the peptide, the survival time and rate of the mice increased considerably.

From a materials point of view, there is still much room of improvement for MIPs. Indeed, during the past 10 years, the development in the area has taken a few different directions aiming at making these improvements possible. For example, it has been suggested to apply controlled polymerisation techniques for MIP synthesis, in order to improve their inner morphology. Another trend is the combination of molecular imprinting and nanostructures, yielding materials with interesting additional properties. Examples are MIP photonic crystals, which can be used as optical sensors [8], or layers of surface-bound MIP nanofilaments, which allow us to tune the surface properties of a MIP film towards superhydrophobicity [9]. An important development is the systematic decrease of particle size, resulting in nanogels with sizes in the lower nanometre range [10], which seems to convey to MIP's properties closer to those of biological antibodies, such as a quasi solubility, very few, or even one, binding site per particle, and a more homogeneous affinity distribution, which can be even further improved by fractioning the particles by affinity chromatography [11]. Resmini and colleagues [12] have shown that these particles, when imprinted with a transition state analogue, can be efficient enzyme mimics. Others have worked on improving the compatibility of MIPs with aqueous solvents, by using monomers that interact more strongly with the molecular template [13]. The development of MIPs that can recognise proteins has also been a long-time dream of many researchers working in the area, which now seems to come true, an example being MIPs that very specifically recognise peptide epitopes of proteins [14].

It appears now that MIPs will finally find their own applications, rather than trying to do what antibodies already can do, and better. Industry is currently evaluating the potential application and commercial opportunities for MIPs. Companies need to investigate the selectivity for MIPs for their targets not in pure solvents but in the environment in which they are to be used, including complex ones like biological fluids. Criteria like the ready integration of molecular imprinting within existing industrial fabrication processes, yields, cost, and the competitiveness of MIPs with existing affinity materials also need to be examined. Certainly, a simple proof of principle will not be sufficient in this case. In the different chapters of this book, the reader will be able to learn about the latest developments in the area of molecular imprinting, about inspiring new concepts leading to considerable improvements of these materials, and also about many remaining challenges for their application in the Real World.

Karsten Haupt

References

1. Wulff G, Sarhan A (1972) Angew Chem Int Ed 11:341
2. Arshady R, Mosbach K (1981), Makromol Chem 182:687
3. Vlatakis G, Andersson LI, Müller R, Mosbach K (1993) Nature 361:645
4. White CJ, Byrne ME (2010) Expert Opin Drug Deliv 7:765
5. Baydemir G, Andacü M, Bereli N, Say R, Denizli A (2007) Ind Eng Chem Res 46:2843

6. Cutivet A, Schembri C, Kovensky J, Haupt K (2009) J Am Chem Soc 131:14699
7. Hoshino Y, Koide H, Urakami T, Kanazawa H, Kodama T, Oku N, Shea KJ (2010) J Am Chem
     Soc 132:6644
8. Hu XB, An Q, Li GT, Tao SY, Liu J (2006) Angew Chem Int Ed 45:8145
9. Vandevelde F, Belmont A-S, Pantigny J, Haupt K (2007) Adv Mater 19:3717
10. Wulff G, Chong B-O, Kolb U (2006) Angew Chem Int Ed 45:2955
11. Guerreiro AR, Chianella I, Piletska E, Whitcombe MJ, Piletsky SA (2009) Biosens Bioelectron
     24:2740
12. Carboni D, Flavin K, Servant A, Gouverneur V, Resmini M (2008) Chem Eur J 14:7059
13. Urraca JL, Moreno-Bondi MC, Hall AJ, Sellergren B (2007) Anal Chem 79:695
14. Nishino H, Huang C-S, Shea KJ (2006) Angew Chem Int Ed 45:2392

Contents

Top Curr Chem (2012) 325: 1–28
DOI: 10.1007/128_2011_307
© Springer-Verlag Berlin Heidelberg 2011
Published online: 20 December 2011

Molecularly Imprinted Polymers

**Karsten Haupt, Ana V. Linares, Marc Bompart,
and Bernadette Tse Sum Bui**

Abstract Molecular imprinting is a process that allows for the synthesis of artificial receptors for a given target molecule based on synthetic polymers. The target molecule acts as a template around which interacting and cross-linking monomers are arranged and co-polymerized to form a cast-like shell. In essence, a molecular memory is imprinted in the polymer, which is now capable of selectively binding the target. Molecularly imprinted polymers (MIPs) thus possess the most important feature of biological antibodies - specific molecular recognition. They can thus be used in applications where selective binding events are of importance, such as immunoassays, affinity separation, biosensors, and directed synthesis and catalysis. Since its beginnings in the 1970s, the technique of molecular imprinting has greatly diversified during the last decade both from a materials point of view and from an application point of view. Still, there is much room for further improvement. The key challenges, in particular the binding site homogeneity and water compatibility of MIPs, and the possibility of synthesizing MIPs specific for proteins, are actively addressed by research groups over the World. Other important points are the conception of composite materials based on MIPs, in order to include additional interesting properties into the material, and the synthesis of very small and quasi-soluble MIPs, close in size to proteins.

Keywords Artificial receptors · Molecularly imprinted polymers · Plastic antibodies · Protein imprinting · Water compatible MIP · Controlled/living radical polymerization

K. Haupt (✉), A.V. Linares, M. Bompart, and B. Tse Sum Bui
Compiègne University of Technology, UMR CNRS 6022, BP 20529, Compiègne 60205,
France
e-mail: karsten.haupt@utc.fr

Contents

1   Introduction: Polymers as Artificial Receptors

Molecular recognition is the underlying principle of many biological processes. Specialized structures fit perfectly with their natural targets, antibodies with their antigens, enzymes with their substrates, or hormone receptors with their hormones. These biomacromolecules are therefore used as recognition systems in affinity technology with applications for example in the biomedical and analytical chemistry fields. However, they are far from perfect tools – they are unstable out of their native environment, and therefore difficult to integrate in standard industrial fabrication processes. Also, they are often low in abundance, and a natural receptor for the particular target molecule of interest may not exist. So, it has been a long-term dream of researchers to be able to build such structures artificially – creating tailor-made receptors that are capable of recognizing and binding the desired molecular target with a high affinity and selectivity. One surprisingly simple way of generating artificial macromolecular receptors is through the molecular imprinting of synthetic polymers.

Molecular imprinting is a process where the target molecule (or a derivative thereof) acts as a template around which interacting and cross-linking monomers are arranged and copolymerized to form a cast-like shell (Fig. 1) [1–3]. Initially, the monomers form a complex with the template through covalent or non-covalent interactions. After polymerization and removal of the template, binding sites are exposed that are complementary to the template in size, shape, and position of the functional groups, which are held in place by the cross-linked structure. In essence, a molecular memory is imprinted in the polymer, which is now capable of rebinding

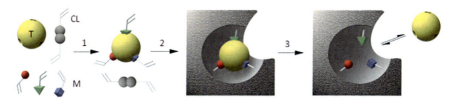

Fig. 1 General principle of molecular imprinting. A molecular template (*T*) is mixed with functional monomers (*M*) and a cross-linker (*CL*) resulting in the formation of a self-assembled complex (*1*). The polymerization of the resulting system produces a rigid structure bearing imprinted sites (*2*). Finally removal of the template liberates cavities that can specifically recognize and bind the target molecule (*3*). Adapted with permission from [3]. Copyright 2003 American Chemical Society

the template selectively. Thus, molecularly imprinted polymers (MIPs) possess the most important feature of biological receptors – the ability to recognize and bind specific target molecules. From a structural point, however, MIPs are very different from biological receptors. While in proteins the monomers (amino acids) are arranged according to a sequence and the polypeptide chain folds into a well-defined secondary and tertiary structure, in MIPs the monomers are incorporated randomly, the chain length is normally not controlled, and the high degree of cross-linking adds even more to the chaotic structure of the network. Although specific binding sites are created by the imprinting process, the population of these sites is often heterogeneous, because of the influence of the equilibria that govern the template–monomer complex formation, and the dynamics of the growing polymer chains prior to complete polymerization. The often heterogeneous pore size distribution in the final material, and the fact that the binding sites are contained within the bulk, tend to make mass transfer slow. Depending on their size, MIPs can bear thousands or millions of binding sites, whereas biological receptors have a few binding sites, or just one. Although not always problematic, these characteristics can prevent MIPs from being substituted, for example, for antibodies in certain applications, and part of the current MIP research is focusing on finding solutions or workarounds for these potential shortcomings. One of the attractive features of the molecular imprinting technique is that it can be applied to a wide range of target molecules. The imprinting of small, organic molecules (e.g. pharmaceuticals, pesticides, amino acids and peptides, nucleotides, steroids, and sugars) is now well established and considered almost routine. Moreover, MIPs can be synthesized in a variety of physical forms, such as, porous microspheres, nanospheres, nanowires, thin films, nanostructured films, nanocomposites, and quasi-soluble nanogels.

Similarly to their natural counterparts (enzymes, antibodies, and hormone receptors), MIPs have found numerous applications in various areas. They have been used as antibody mimics in immunoassays and sensors and biochips as affinity separation materials and for chemical and bioanalysis, for directed synthesis and enzyme-like catalysis, and for biomedical applications. Concerning their commercialization, there has been great progress during the past decade, in particular in the analytical chemistry and biochemistry fields (sample clean-up and pre-concentration), an application where the clear advantages of MIPs over biomacromolecules are easily visible.

2 Imprinting Matrices

At the present time, the majority of reports on MIPs describe organic polymers synthesized by radical polymerization of functional and cross-linking monomers having vinyl groups, and using non-covalent interactions with the template (see http://www.mipdatabase.com/Database.html for a complete collection of references in the MIP area). This can be attributed to the rather straightforward synthesis of these materials, and to the vast choice of commercially available monomers. Just like the 20 amino acids in proteins, monomers can be chosen for MIPs to play a specific role for structure or function. These can carry basic (e.g. vinylpyridine), acid (e.g. methacrylic acid), hydrogen bonding (e.g. methacrylamide), or hydrophobic (e.g. styrene), groups.

Nevertheless, other materials have started to appear during recent years that are either better suited for a given application or easier to synthesize in the desired form. For example, other organic polymers such as polyurethanes [4] have been used. Compared to polymers based on vinylic monomers, the use of the above-mentioned polymers seems to be somewhat restricted due to the limited choice of functional monomers. Sol–gels such as silica and titanium dioxide are now gaining in importance as imprinting matrices, although they were introduced years ago. Katz and Davis have reported the molecular imprinting of bulk amorphous silica with single aromatic molecules using a covalent monomer template complex, creating shape-selective catalysts [5]. It has been shown that for some target molecules, sol–gel matrices seem to yield better results than the common polymethacrylate polymers [6]. Finally, during the last few years, composite materials have started to appear that combine the recognition properties of MIPs with the specific mechanical, optical, electrical, or other functional properties of a second material [7–10].

In the following, a number of important aspects and challenges concerning MIPs in general and MIPs synthesized by radical polymerization in particular will be discussed in more detail. These include improving the binding site homogeneity of MIPs, as well as their performance in aqueous and other polar solvents, and adapting the technique to biomacromolecular targets. In fact, while the development of MIPs for small targets is considered merely routine, the synthesis of MIPs for proteins and other biomacromolecules has remained a real challenge to date, but new approaches and workarounds are constantly being proposed by researchers in the field.

3 MIPs Synthesized by Free Radical Polymerization

In its simplest form, a typical MIP synthesis protocol contains the template, one or more functional monomers, a cross-linker, a polymerization initiator, and a solvent. The solvent acts at the same time as a porogen, generating a porous material for better access of the imprinting binding sites. The functional monomer that interacts with the template during molecular imprinting provides the functional groups in the binding sites of the MIP. It can be selected according to two different philosophies. Mosbach [1] has developed a "biochemist's" approach based on the well-known fact that most molecular recognition phenomena in nature are based on weak non-covalent bonds. This so-called non-covalent or self-assembly protocol is today the most commonly employed, since it is relatively easy to put into practice, and rather flexible since a large number of functional monomers able to interact with almost any kind of target molecule are available. In this approach, the complex between the template and the functional monomer is formed by interactions such as hydrogen bonds, ionic bonds, van der Waals forces, and the hydrophobic effect. Association between the monomer and template is governed by an equilibrium and the functional monomers normally have to be added in excess relative to the number of moles of template to favor the formation of the complex (template to functional monomer ratio of 1:4 or more). Consequently, this gives rise to a number of different configurations of the template–functional monomer complex, leading to a heterogeneous binding site distribution in the final MIP with a range of affinity constants [11]. Besides, the remaining uncomplexed monomers are randomly incorporated in the polymer matrix, resulting in non-imprinted binding sites (Fig. 2). Such MIPs are hence analogous to polyclonal antibodies and are sometimes "group-specific," that is, selective not only towards the original template but also to other structurally related compounds, which can be quite useful, for example, for the separation of a class of analytes [13]. After polymerization, the

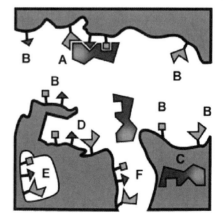

Fig. 2 Representation of binding sites' heterogeneity: high affinity site in macropore (**A**) and micropore (**F**), and lower affinity sites (**B**) in macropore, (**C**) trapped template, (**E**) embedded site, (**D**) highest affinity site with shape selectivity from polymer. Reproduced from [12] with permission of The royal Society of Chemistry

imprinted molecule can be removed by simple solvent extraction. Binding of the analyte or target molecule to such an MIP is by the same non-covalent interactions. The yield of imprinted sites is typically quite low in this type of protocol [14]. This is due to the fact that, during the synthesis of the cross-linked polymer, stress is built up within the polymer network, some of which is continuously released owing to the reversible nature of the template–monomer or template–oligomer bonds. This leads to the relaxation of the polymer chains and to the destruction of a large percentage of the binding sites before the system is finally frozen in space.

In contrast, Wulff [2] and Shea [15] have developed a "chemist's" approach based on covalent bonds between the template and the monomers. The complex is formed by the reversible covalent linkage of a functional monomer to the template prior to polymerization, thus requiring some chemical synthesis prior to the synthesis of the MIP itself. After polymerization, the imprint molecule is removed by chemical cleavage. Binding of the analyte occurs via the same covalent interactions. The obvious advantage of this technique is that the mono-mer/template complex is stoichiometric and hence results in a more homogeneous population of binding sites within the polymer. However, this approach may be less straightforward for non-chemists, and from a general point of view it is somewhat less versatile since not all targets of interest can be easily derivatized with one or more polymerizable moieties for molecular imprinting. Moreover, the need to re-establish covalent bonds upon binding of the target to the MIP may render the binding kinetics slow. To solve the latter problem, Whitcombe and co-workers [16] have suggested using covalent bonds only during imprint formation, while binding of the target to the MIP could be by non-covalent interactions. This approach seems to combine the advantages of both the covalent and the non-covalent protocols. Whitcombe and co-workers have demonstrated its feasibility on the example of a MIP specific for cholesterol. A covalent template–functional monomer complex, cholesteryl (4-vinyl)phenyl carbonate ester, was imprinted. The carbonate group was then eliminated by chemical cleavage, creating some space in the binding site and leaving behind a phenolic hydroxyl group, to which cholesterol could bind via a hydrogen bond. Whether covalent imprinting protocols present a real advantage with respect to a more homogeneous distribution of binding site affinities and a higher imprinting efficiency will depend on the particular case. In fact, stress is built up in the cross-linked polymer network during its synthesis. In covalent imprinting protocols, this stress may in part be released upon template removal, causing conformational changes in the polymer, and thus changes in binding site conformation.

4   Choosing the Right Monomers

Commonly, new "non-covalent" MIPs are designed using a generic approach where the functional groups on the binding monomers are chosen according to their complementarity with the chemical groups of the template. In order to speed up the optimization procedure and to obtain MIPs with the best recognition properties, combinatorial synthesis accompanied with high throughput screening, and molecular modeling, can be used. The latter technique uses molecular modeling software to screen a virtual library of monomers for a given template. Monomers that form the most stable complex with the template, sometimes taking into account the solvent used, are identified, and the corresponding MIPs are synthesized for experimental confirmation [17–19]. The use of NMR titration or other spectroscopic methods as standalone technique or for the confirmation of modeling results [20–23] may seem obvious, but is not routinely done as yet.

The combinatorial MIP optimization [24, 25] is based on the preparation of a library of different MIPs, from which the best is selected for strong and selective target binding and low non-specific binding. In one example, to speed up the process, the MIPs for the herbicide terbutylazine were synthesized in small quantities ("miniMIPs" – 55 mg) in HPLC autosampler vials [24], allowing for automated processing. During a first round of screening, the selection criteria were based on the amount of template released upon washing with the solvent used as porogen during polymerization. Thus, for a particular MIP, a quantitative release of the template indicates that target binding will be weak and the MIP will thus be discarded. In the second screening round, maximizing the imprinting effect was the goal; thus, the rebinding of the template to the miniMIPs and to non-imprinted polymers was compared. An improved version of this approach was developed using 96-well microtiter plates [26, 27]. A library of 80 polymers could thereby be synthesized and evaluated in only 1 week [27]. Similarly, Ceolin et al. have recently proposed a novel technique for the synthesis and testing of large numbers of molecularly imprinted polymers, apparently requiring even less time than the "miniMIP" approach. Instead of vials, the polymers are synthesized on the surface of microfiltration membranes in multiwell filterplates. The thin polymeric films enable accelerated template removal. The performance of the system was demonstrated by creating a combinatorial library of MIPs selective for cimetidine, an antiulcer drug. An experimental design combined with a multivariate analysis (i.e., response surface modeling) was used to minimize the number of experiments in the optimization process. The highest imprinting factor was obtained using an MAA/EDMA/template molar ratio of 3.5:19.5:1 [28]. This combinatorial approach should be a good complement or an alternative to the computational approach and to spectroscopic titration experiments, and might, as a standalone technique, even yield better results, at least until the molecular imprinting process with all the intermediate states of polymerization is fully understood. This is because it is based on real binding data between template and polymer, automatically comprising the possible contribution of the entropy term, rather than being restricted to the template–monomer interactions.

Combinatorial MIP development can be speeded up considerably by using experimental design approaches [27–31]. These allow one to identify the most significant factors determining the performance of imprinted polymers (for example the degree of cross-linking, the monomer:template ratio, the initiator

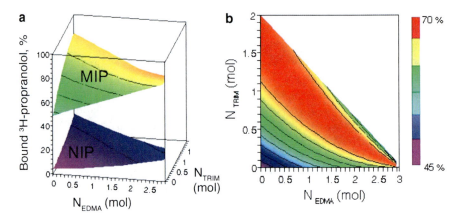

Fig. 3 Application of the Doehlert experimental design to optimize a MIP for propranolol with respect to the type of cross-linker (*EDMA* or *TRIM*) and the degree of cross-linking. (**a**) Three-dimensional representation of response surfaces for the percentage of bound [^{3}H]propanolol to the molecularly imprinted polymer (*MIP*) and the corresponding non-imprinted control polymer (*NIP*). (**b**) Contour plot of the function describing binding of [^{3}H]propanolol to MIPs relative to the degree and the kind (bi or trifunctional) cross-linking. The values were corrected for non-specific binding to the non-imprinted control polymer. Adapted from [31] with kind permission from Springer Science + Business Media

concentration), to detect links between these factors, and to optimize the composition of the MIP accordingly, while minimizing the number of necessary experiments (MIP syntheses, binding assays). For example, we recently proposed the use of Doehlert's experimental design, a second-order uniform shell design, for the optimization of MIPs. The influence of the kind and degree of cross-linking on specific target (S-propranolol) recognition was studied. With very few experiments, the evolution of the binding capacity of a MIP as a function of the different parameters was studied (Fig. 3). It is believed that such chemometric tools can significantly accelerate the development of new MIPs, particularly in the context of a given application.

5 MIP Synthesis by Controlled Radical Polymerization

5.1 *Improving the Inner Morphology of MIPs*

There have been a number of reports aiming at improving the morphology of MIPs and thus improving the imprinting efficiency. One interesting proposal has been made by Steinke and co-workers [32, 33]. Their argument is that the statistically and kinetically-driven nature of the network-forming process in conventional free radical polymerization (FRP) makes it impossible to achieve a homogeneous distribution of binding site affinities in MIPs. As a remedy, they suggest a thermodynamically controlled process via ring-opening metathesis polymerization (ROMP). By using a covalent imprinting approach (the template L-menthol, was covalently bound to polymerizable moiety, and cleaved off chemically after polymerization), Steinke and co-workers were able to demonstrate that a MIP with some selectivity for L-menthol over D-menthol (20% enantiomeric excess) could be synthesized by this method [33]. While it is difficult to evaluate the usefulness of this approach, as the authors have not made a comparison with a MIP synthesized by the conventional FRP method with respect to polymer morphology and site homogeneity, the basic idea seems very sound: to improve over the chaotic inner morphology of highly cross-linked MIPs synthesized by free radical polymerization. Another approach that might allow for this is the use of controlled radical polymerization (CRP) for the synthesis of MIPs.

5.2 Controlled Radical Polymerization of MIPs

It is rather surprising that modern methods of controlled/living radical polymerization are still very little used in molecular imprinting, as these methods should have a great potential for generating a controlled morphology, and, in addition, they give access to thin films and nanostructures with controlled dimensions and complex architectures. Among all polymerization methods, FRP is the most flexible in terms of reagent purity, experimental conditions, choice of monomers, etc. This makes it accessible even for researchers who are not experts in polymer chemistry. In the context of molecular imprinting, this method is very attractive since it is compatible with a wide range of monomers carrying different functional groups, and because it usually tolerates the presence of additional chemicals, most importantly the imprint molecule. It is therefore not surprising that this polymerization method has been widely adopted by the molecular imprinting community, and is principally seen in the majority of the publications in the field.

FRP proceeds in three steps:

1. The polymerization is *initiated* (1) by generating free radicals $R^{\bullet}$ using an initiator, a molecule that decomposes under heating or under UV or visible irradiation. A radical attacks the double bond of a monomer, resulting in the formation of an intermediate radical $M_1^{\bullet}$. This is the rate-limiting step of the process.
2. The *propagation* (2) is the main step of radical polymerization. Here, the macromolecular chain is formed by successive additions of monomers on the growing macroradical $M_n^{\bullet}$. The result is a mixture of polymer chains with high molecular weights.

3. The *termination* of the reaction can take place in two ways: recombination (3) of two macroradicals $M_n^\bullet$ and $M_p^\bullet$ forming a macrochain M_{n+p} or disproportionation (4) yielding a double bond $M_n =$ and a C–H bond at the chain terminus M_pH.

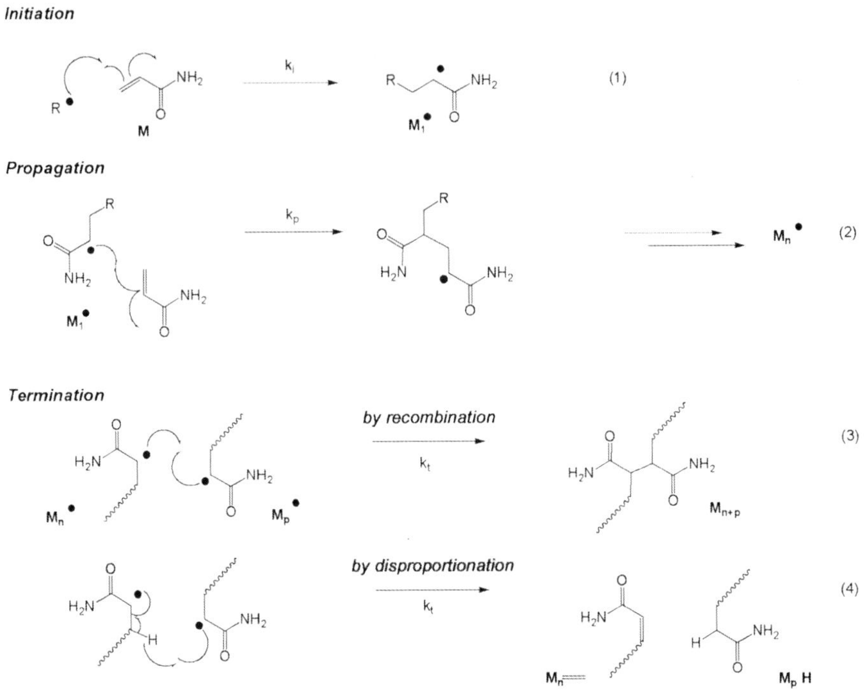

Initiation

(1)

Propagation

(2)

Termination

by recombination

k_t

(3)

by disproportionation

k_t

(4)

In addition, side reactions with impurities (oxygen, inhibitors) or chain transfer reactions (to solvent molecules, to the polymer, or, in the case of the MIP, to the template) can occur, preventing further growth of the chain and resulting in shorter polymer chains. Owing to these characteristics, FRP has a major drawback. It does not allow one to control the size, architecture, and number of the macromolecules synthesized, due to the high reactivity of alkyl radicals produced constantly during the polymerization process, which favor irreversible termination reactions by recombination and disproportionation. Thus, the molecular weight of the polymer cannot be controlled or predicted, and block copolymers and other polymers of complex architecture are totally inaccessible. It is therefore of great interest to be able to control the growing and/or the termination steps of the reaction, and, thus, to use controlled/living radical polymerization methods. For example, controlling chain growth and termination gives access to the design of polymeric core-(multiple) shell particles, with defined layer thicknesses and where each layer has a different function. In addition, the possibility of synthesizing block copolymers from the surface of such particles enables one, for example, to form a biocompatible outer layer, to grow polymer brushes for restricted access by steric hindrance, to fine-tune the hydrophilicity/hydrophobicity of the particles and thus their stability in a given solvent, and so on [34].

A number of controlled/living radical polymerization methods have been developed. These include:

– *Reversible addition-fragmentation chain transfer polymerization (RAFT)* [35, 36]. This method involves reversible addition-fragmentation sequences in which the transfer of a dithioester moiety between active and dormant chains serves to maintain the living character of the polymerization. Initiation of polymerization is by a classical FRP initiator. Growing chains (Pn˙ or Pm˙) are added up successively to the chain transfer agent (CTA). Addition of this macroradical on CTA gives rise to an unstable radical which fragments into either a new radical R˙ which can polymerize a new polymer chain or Pn˙ (or Pm˙) that will propagate until they encounter another CTA [37–42]:

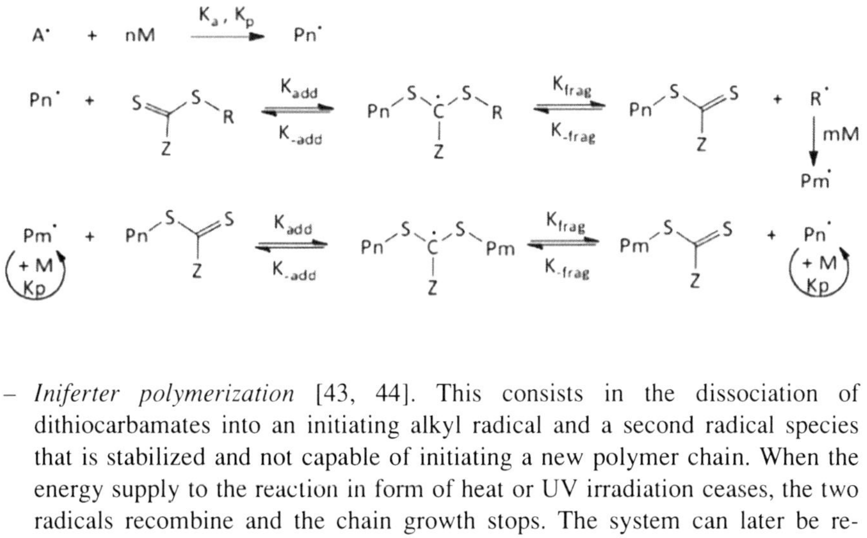

– *Iniferter polymerization* [43, 44]. This consists in the dissociation of dithiocarbamates into an initiating alkyl radical and a second radical species that is stabilized and not capable of initiating a new polymer chain. When the energy supply to the reaction in form of heat or UV irradiation ceases, the two radicals recombine and the chain growth stops. The system can later be reinitiated with the same or other monomers, thus providing some degree of living character (for MIP: [45–51]).

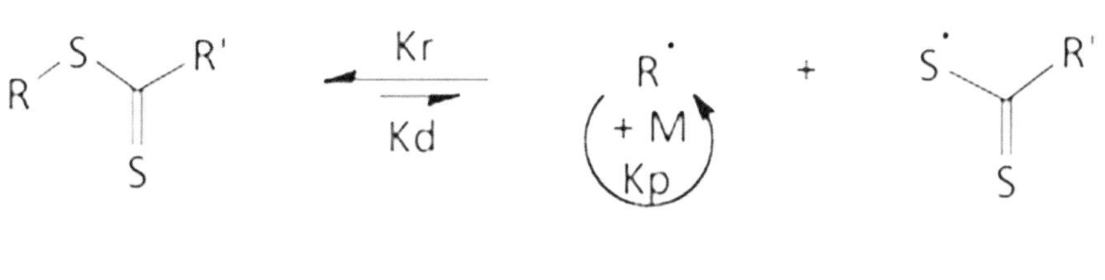

– *Atom transfer radical polymerization (ATRP)* [52–55]. Active species are produced by a reversible redox reaction, catalyzed by a transition metal/ligand
   complex (M_t^n–Y/L_x). This catalyst is oxidized via the halogen atom transfer
   from the dormant species (Pn–X) to form an active species (Pn$^{\bullet}$) and the
   complex at a higher oxidation state (X-M_t^{n+1}–Y/L_x).

$$Pn - X \; + \; M_t^n - Y/L_x \quad \underset{K_{deact}}{\overset{K_{act}}{\rightleftharpoons}} \quad Pn^{\bullet} \; + \; X - M_t^{n+1} - Y/L_x$$

$$\left(\begin{array}{c} + \, M \\ Kp \end{array}\right)$$

– *Nitroxide mediated polymerization (NMP)* [56, 57]. This consists in a thermally
   reversible termination reaction by a homolytic cleavage of a C–ON bond of an
   alkoxyamine, giving rise to an initiating alkyl radical (active species) and a
   nitroxyl radical, which brings control to the reaction [58].

$$R-(M)n-O \overset{R'}{\underset{}{\overset{|}{N}}} R'' \quad \underset{K_c}{\overset{K_d}{\rightleftharpoons}} \quad R-(M)n\text{-}1-M^{\bullet} \; + \; {}^{\bullet}O-N\overset{R'}{\underset{R''}{}}$$

Dormant species   Nitroxyl radical

$$\left(\begin{array}{c} + \, M \\ Kp \end{array}\right)$$

It should be noted that the presence of cross-links results in the partial or
complete loss of control over the size of the polymer molecules, even if the living
character of the polymerization can sometimes be preserved. Incidently, one of the
characteristics of MIPs is that they are cross-linked polymers. This cross-linking is
necessary in order to maintain the conformation of the three-dimensional binding
sites obtained through the molecular imprinting process, and thus the ability of the
polymer to recognize specifically and selectively its target molecule. Nevertheless,
even with cross-linked polymers, the use of CRP methods may be beneficial, as it
can, up to a certain point, improve the structure of the polymer matrix. Indeed, all of
the above CRP methods have been applied to MIPs.

Concerning the use of ATRP with MIPs, the major limitation for this technique
in the context of MIP synthesis is the small choice of monomers with suitable
functional groups. Typical monomers used for molecular imprinting such as
methacrylic acid (MAA) are incompatible, as they inhibit the metal–ligand complex involved in ATRP. With other monomers like methacrylamide [59] and
vinylpyridine [60] it is difficult to achieve high monomer conversion. Template
molecules also often carry functional groups that may inhibit the catalyst. All this
seems to make ATRP not the best choice for molecular imprinting. Nevertheless,

it has been used on a number of occasions [61–66]. For example, Takeuchi's group used ATRP for the imprinting of bisphenol A through a covalent approach. They showed that the swelling degrees of ATRP MIPs and non-imprinted controls were approximately twice as high as those of FRP polymers, indicating a lower cross-linking density for the former. More template could be extracted compared to the FRP MIP, and capacity, selectivity, and imprinting factor were improved [65]. Zu et al. used ATRP combined with precipitation polymerization. ATRP provided larger spherical particles (2–5 μm vs 200–430 nm) compared to FRP. The loading capacities are higher with ATRP than with FRP, although the imprinting factors obtained were similar [64].

In the case of NMP, polymerization generally requires high temperatures for the activation–deactivation of alkoxyamines, though high temperatures are not compatible for MIPs synthesized with non-covalent monomer-template complexes, but if the so-called covalent approach is used, this should be of lesser concern, since covalent bonds are more stable even at higher temperatures. Indeed, Ye and co-workers [58] employed NMP to synthesize a MIP specific for cholesterol through the covalent approach. They included a sacrificial spacer between the template and the functional monomer, and the binding of the target molecule to the MIP is via non-covalent hydrogen bonding interactions, a method that had been introduced earlier by Whitcombe's group [16]. They could show a more efficient template cleavage and superior imprinting effect with the NMP MIP compared to the FRP MIP, which can be attributed to the more ordered structure of the polymer network of the former.

Iniferter-initiated polymerization, in the context of MIPs, has the advantage of being compatible with the majority of functional monomers commonly used, and seems to be the method best compatible with photopolymerization. Iniferters were mainly used to achieve the synthesis of thin MIP films on supports by surface-initiation. Kobayashi and co-workers [45] were the first to use this CRP method for molecular imprinting. They reported photografting a MIP layer on a polyacrylonitrile membrane modified with a diethyldithiocarbamate iniferter. Later, Sellergren and co-workers [46] modified different silica beads and polystyrene beads with iniferter groups for grafting a MIP layer imprinted with D- or L-phenylalanine anilide for use in chiral chromatography. The composite materials exhibited enantioselectivity in chromatography mode similar to the system based on immobilized azoinitiators [67], with the advantage of no or minimal propagation occurring in solution. An additional benefit is the possibility of grafting multiple consecutive polymer layers [47]. Pérez-Moral and Mayes [49] used a dithiocarbamate iniferter to synthesize molecularly imprinted core-shell particles. Polystyrene nanoparticles obtained by emulsion polymerization were modified with the iniferter, and multiple shells of polymer were then sequentially added by UV-initiated polymerization in an organic solvent. The imprinted sites (specific for the beta-blocking drug propranolol) created in the first shell, an ethylene glycol dimethacrylate-methacrylic acid copolymer of around 20 nm thickness, were still accessible and maintained their ability to bind the target specifically even after two more layers of approximately 20 nm each were added. While the above examples demonstrate the usefulness of iniferter-based CRP for modifying solid support materials with thin MIP films by surface-initiated polymerization, Byrne and co-workers [50] explored possible additional benefits of the method, namely, an improved fractional double bond conversion. This was hoped to be associated with better target binding properties of MIPs specific for ethyladenine-9-acetate, synthesized in a bulk format. The degree of double bond conversion can be modulated by varying initiator type and concentration, cross-linking monomer length, temperature, and the amount of added solvent. The authors showed that, using an iniferter initiator, the number of binding sites in a poly(methacrylic acid-co-ethylene glycol dimethacrylate) MIP was increased by 63% compared to a similar MIP synthesized by FRP using azobisisobutyronitrile as a photoinitiator, the binding affinity remaining roughly the same. This was attributed to an increased fractional double bond conversion. The same authors later used iniferter polymerization for the synthesis of MIPs for controlled drug release. With CRP MIPs they observed a substantially increased drug load capacity and a delayed template release with respect to FRP MIPs [68].

RAFT has also been used for MIP synthesis [37, 39, 41, 69]. Titirici and Sellergren [37] reported the modification of mesoporous silica beads by grafting of very thin (a few nanometers) cross-linked MIPs with RAFT polymerization. RAFT apparently prevented the gel formation in solution and improved the composite morphology when compared with an FRP MIP. When the silica spheres were used as stationary phase in HPLC, efficient enantioseparation of the chiral template molecule, phenylalanine anilide, was obtained. Unfortunately, the use of RAFT polymerization did not seem to improve the peak shape of the separation, as the same strong peak broadening for the more strongly retained enantiomer was observed that is typical for MIPs. Pan et al. synthesized a MIP by RAFT precipitation polymerization. They observed a higher capacity, a better binding constant, and an increased density of high-affinity sites compared to the FRP MIP [70]. Others have used the living character of RAFT in order to change the surface properties of MIP spheres [42, 71]. For example, the living ends were used to modify MIPs with poly-HEMA brushes and hydrogels for improved water compatibility [71].

6   Synthesis and Use of MIPs in Aqueous Solvents

The majority of molecular imprinting processes involving non-covalent synthesis are performed in aprotic and low polarity organic solvents like chloroform, dichloromethane, and toluene [72]. The recognition between template and functional monomer relies principally on electrostatic interactions and hydrogen bonding. However, to overcome solubility problems, more polar solvents like acetonitrile, dimethyl sulfoxide, and dimethyl formamide are often employed. Though electrostatic interactions are lower in these solvents, they prove suitable for the imprinting of molecules which bind predominantly by hydrogen bonding. It remains that the overall general rule is to exclude polar protic solvents such as alcohols and water, due to their competition for forming interactions with both monomers and template. As outlined below, some of the MIPs synthetized in organic solvents can then be used in aqueous solvents, although normally with a somewhat altered selectivity [73, 74]. Nevertheless, a few exceptions of imprinting in mixed water systems and even in pure water do exist.

6.1 Interactions Driven by the Synergistic Effect Brought by Both Solvent and Template

Imprinting in a mixture of water-containing solvents as porogen has been done either for solubility reasons, when the template is a polar compound and is not soluble in organic solvents, or when the use of a polar solvent favors the interactions of the template with the monomers. For instance, 2,4-dichlorophenoxyacetic acid (2,4-D) [75] and 1-naphthalene sulfonic acid (1-NS) [76] were imprinted in a mixture of methanol/water. 2,4-D and 1-NS bear an acidic group and an aromatic ring (Fig. 4) and both were imprinted using the basic functional monomer 4-vinylpyridine. The specificity is driven by a combination of ionic and $\pi-\pi$ stacking interactions, which are strong enough to allow for complex formation in a polar environment, as confirmed by NMR titration experiments for 2,4-D [20]. The high affinity and specificity of this MIP in aqueous systems were independently demonstrated by rebinding studies at equilibrium [75, 77] and by chromatography [78].

Other successful imprintings in pure methanol or methanol–water mixtures were observed with piperazine-based fluoroquinolones like, for instance, ciprofloxacin [79] and ofloxacin [80], respectively. In both cases, the functional monomer is MAA and the interaction is pH-dependent and due to strong ion pairs formed between the piperazine ring of the quinolones and the carboxylate group of MAA.

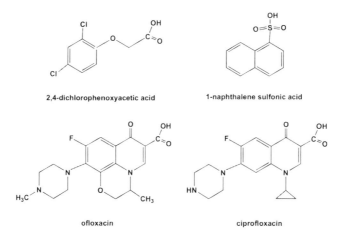

2,4-dichlorophenoxyacetic acid

1-naphthalene sulfonic acid

ofloxacin

ciprofloxacin

Fig. 4 Chemical structures of molecular templates for which MIPs were synthesized in mixed organic/aqueous solvents

6.2   Stoichiometric Non-Covalent Template–Monomer Complexes

Electrostatic effect-based recognition. As mentioned above, in aprotic solvents of low polarity, recognition between the template molecule and the functional mono-mer relied on electrostatic interactions in addition to hydrogen bonding. For example, if the template has an acidic functional group (carboxylate, phosphonate), basic functional monomers, available commercially, like vinylpyridine and *N,N*-diethylaminoethyl methacrylate, can be employed. However, these interactions are weak and, hence, a large excess of functional monomer (at least fourfold) is used in order to ensure a sufficiently high degree of complexation with functional groups of the template for effective imprinting to occur. This leads to a substantial number of non-specific binding sites. But, if the association constant between template and functional monomer is high enough ($K_a \geq 10^3$ M^{-1}; for comparison, the K_a of hydrogen bonds, electrostatic interactions between carboxylic acids and basic nitrogen and electrostatic interactions between carboxylic acids and basic nitrogen with additional hydrogen bonds are respectively around 1.7, 3.3, and 30 M^{-1} in acetonitrile) [81], they will completely bind to each other in a 1:1 molar ratio. With this procedure, no non-specific binding sites are produced in the polymer. For this purpose, designed monomers with superior association constants have been devel-oped by several groups. Wulff and co-workers were the first to develop a series of host monomers bearing an amidine group [82], such as *N,N'*-diethyl-4-vinyl-benzamidine **1** (Fig. 5a) which can form strong electrostatic interactions with

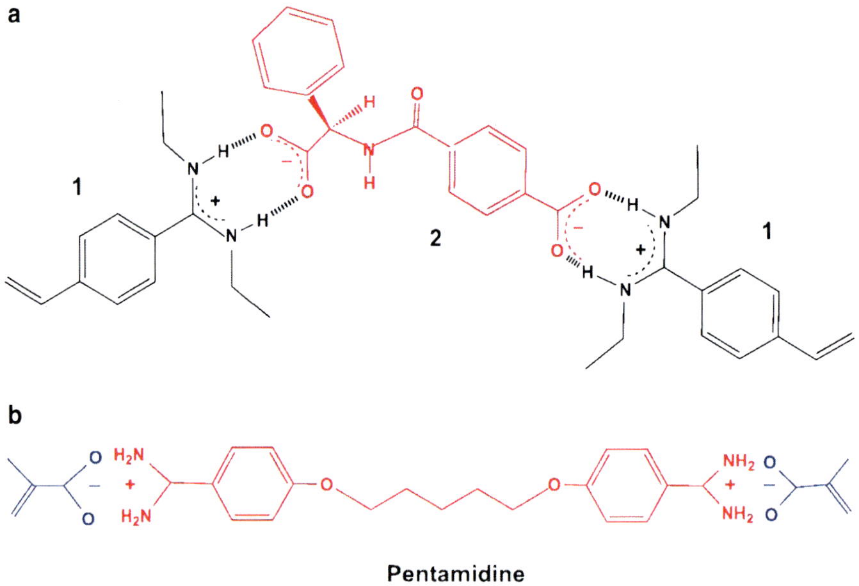

Fig. 5 Complexes formed (**a**) between amidine groups in the functional monomer *N,N'*-diethyl-4-vinyl-benzamidine and carboxyl groups in the imprinting template *N*-terephthaloyl-D-phenylglycine [82] and (**b**) between amidine groups in the imprinting template *pentamidine* and carboxyl groups in the functional monomer methacrylic acid [83]

carboxylates, phosphonates and phosphates ($5 \times 10^3 \text{ M}^{-1} < K_a < 10^6 \text{ M}^{-1}$). For example, enantioselective MIPs were prepared by targeting the oxyanions of N-terephthaloyl-D-phenylglycine **2** with monomer **1** in THF. The binding here was strong enough to provide quantitative rebinding in methanol [14]. Though its rebinding behavior in water is not mentioned, it is very likely that strong specific interactions would also prevail in this medium, since there has been a precedent involving amidine moieties from the template pentamidine and carboxyl moieties from the monomer MAA that interact very strongly and specifically in aqueous solutions (Fig. 5b) [83].

Later, other monomers using this concept of host–guest interaction were developed and their selectivities in water were demonstrated. Lübke et al. [84] synthesized two functional monomers, one is a derivative of a boron-containing receptor **3** and the other is a quinone **4** to react respectively with the carboxylate and the amino groups of the antibiotic ampicillin **5** (Fig. 6). The association constant of the polymerizable boronic acid-containing receptor with ampicillin carboxylate in a 1:1 complex was determined to be 2,800 M^{-1} (in d_3-acetonitrile). Binding of the amino group in ampicillin with the electron-deficient quinone occured through N–π interactions and the K_a of a 2:1 complex was estimated to be >30,000 M^{-1} (in d_6-DMSO). The polymers prepared with ampicillin carboxylate and these monomers in a 1:1:1 ratio in DMSO afforded efficient binding of ampicillin, as compared to the non-imprinted polymer, in aqueous buffer solutions. These two new functional monomers demonstrate the high potential of imprinting of a target carrying carboxyl and amino groups which are common to many other antibiotics, amino acids, peptides, nucleotides, and alkaloids, and therefore could be generalized to the imprinting of these bioactive compounds and give rise to polymers which would be very specific in aqueous media.

Similarly, a series of urea-based vinyl monomers were synthesized for stoichiometric oxyanion recognition [22]. One of these urea-based monomers,

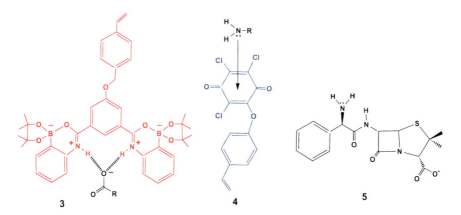

Fig. 6 Monomers (**3** and **4**) for the stoichiometric complexation of ampicillin (**5**) [84]

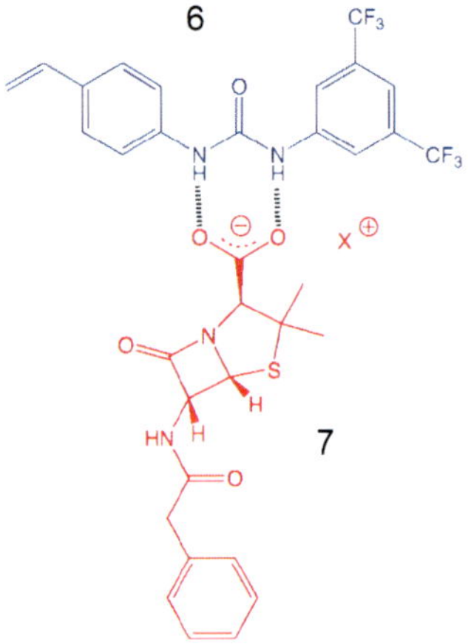

Fig. 7 Urea-based functional
monomer (**6**) for the
complexation of penicillin G
(**7**) [22]

Fig. 7 Urea-based functional monomer (**6**) for the complexation of penicillin G (**7**) [22]

1-(4-vinylphenyl)-3-(3,5-bis(trifluoromethyl)phenyl)urea **6**, was employed stoichio-metrically with the template penicillin G **7** (Fig. 7) for the preparation of a MIP to extract penicillin G and its beta-lactam derivatives from aqueous samples. The K_a between this monomer and tetrabutylammonium benzoate is 8,820 M^{-1} (in d_6-DMSO). The MIP was synthesized in acetonitrile as porogen and the loading of the antibiotics was done in Hepes buffer where the development of strong stoichiometric electrostatic interactions between the carboxylate groups of the antibiotics and the urea moiety of the monomer allowed for retention. The clean-up was achieved simply by percolating the loading buffer containing 10 vol.% CH_3CN: all non-specific interactions were eliminated, as monitored on the NIP, leaving the specific interactions untouched as judged by the high recoveries of the analytes during elution [85].

6.3 Hydrophobic Effect-Based Recognition

Hydrophobic effect-driven recognition in aqueous environments has made use of polymerizable β-cyclodextrin and its derivatives as functional monomers for molecular imprinting. Cyclodextrins are cyclic oligosaccharides with a hydrophilic exterior and an internal hydrophobic pocket, which is utilized to bind analytes containing hydrophobic moieties. Using this methodology, it is possible to do the imprinting and the recognition in both high polarity organic solvents or pure water or a mixture of both. The association constant between β-cyclodextrin and, for example, the steroid estradiol in methanol–water, using a chromatographic method, is 6,830 M^{-1} [86]. This type of recognition has been applied to the imprinting of amino acids [87], oligopeptides [88], or steroids [89]. In the case of large templates like steroids, two or more cyclodextrin molecules with each binding part of the template, can be cross-linked together to produce an ordered assembly to recognize the template very selectively (Fig. 8).

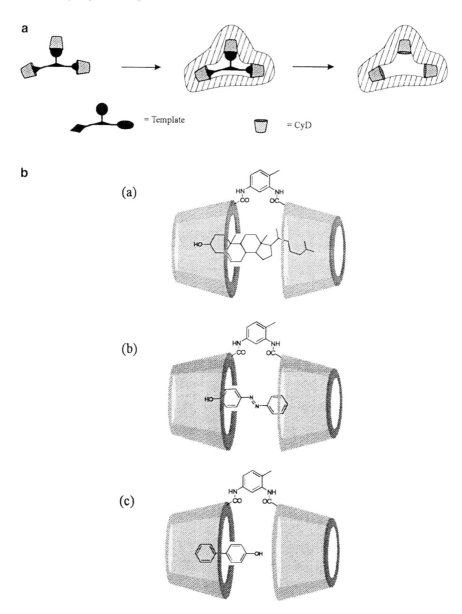

Fig. 8 (a) Schematic representation of an ordered assembly of cyclodextrin molecules with each one binding a part of the template. (b) Binding modes of (a) cholesterol, (b) 4-hydroxyazobenzene and (c) 4-phenylphenol to the guest-binding sites in the cholesterol-imprinted beta-cyclodextrin polymer. Reprinted with permission from [89]. Copyright 1999 American Chemical Society

6.4   Metal Coordination-Based Recognition

MIPs containing metal chelates have been synthesized, for instance, for the selective recognition of histidine-containing short peptides in aqueous environment. The binding relies on strong coordination between an N-terminal His residue of the template and a Ni(II) complex bound to the polymer. This approach is adapted from the work of Hochuli et al. [90], who introduced an adsorbent based on a Ni(II)-nitrilotriacetic acid (NTA) complex to bind selectively genetically engineered N-terminal histidine sequences during protein purification. The association constant, K_a, of the complexation of histidine with Ni(II)-NTA is 10^5 M^{-1} [91]; in addition histidine, Ni(II), and NTA form a 1:1:1 ternary-complex together. For creating the peptide receptor, a polymerizable functionalized NTA monomer was first synthesized. A solution of $NiSO_4$ was then added to form the methacrylamide–NTA–Ni^{2+} mixed complex and the pre-polymerization mixture was obtained by incorporating the template peptide, His-Ala. The polymer was prepared by copolymerization of this complex with acrylamide and N,N'-ethylenebisacrylamide (Fig. 9). Binding studies showed that the His-Ala-imprinted polymer has a significantly higher capacity for the template peptide over the other sequences (His-Phe, His-Ala-Phe, Ala-Phe) examined and that non-histidine-containing peptides have almost no affinity for the polymer. In this example, the imprinting and the rebinding experiments were all done in water [92].

Template–metal coordinative bonds have also been used in the imprinting and recognition of carbohydrates in aqueous media. A MIP, templated with glucose and using 4-(N-vinylbenzyl)diethylenetriamine)Cu(II)]diformate as

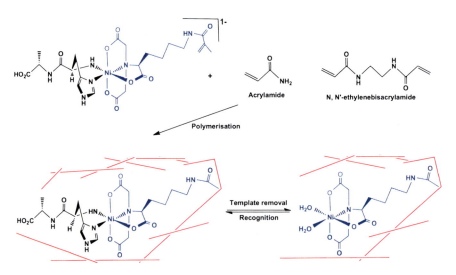

Fig. 9 Copolymerization of the (His-Ala)-Ni–NTA complex with acrylamide and N,N'-ethylene-bisacrylamide. Extraction of the peptide provides a polymer containing Ni–NTA complexes capable of rebinding the template peptide. Adapted with permission from [92]. Copyright 2001 American Chemical Society

functional monomer and pentaerythritol tetraacrylate as cross-linker was synthesized in methanol–water (3:1). Binding studies showed that a 1:1 complex of carbohydrate and the Cu(II)-chelating monomer was formed. The glucose-imprinted polymer exhibits high selectivity for glucose over the closely related *cis*-1,2-diols like galactose and mannose and the non-imprinted control polymer in water [93, 94].

6.5 Utilization in Aqueous Environment of MIPs Synthesized in Organic Solvents

As mentioned above, most MIPs are synthesized in organic solvents to preserve the hydrogen and electrostatic interactions between template and monomer. However, for the application to solid-phase extraction (SPE) where the target is most of the time in water samples or in biological fluids, a lot of studies have been carried out to examine the influence of binding media parameters (solvent polarity and composition, buffer pH, concentration, ionic strength, etc.) with the aim of attenuating non-specific adsorption of the analyte due to hydrophobic interactions which predominate in such media. For a recent review, see Tse Sum Bui and Haupt [95].

A means to avoid such tedious optimization can be envisaged by employing stoichiometric monomers to develop strong interactions with the template as mentioned above. The other way is to incorporate hydrophilic comonomers (2-hydroxyethyl methacrylate (HEMA), acrylamide) or cross-linkers (pentaerythritoltriacrylate, methylene bisacrylamide) in the polymer. This results in an increase of the hydrophilicity of the polymer. Indeed, the use of HEMA for a MIP directed towards the anesthetic bupivacaine resulted in high imprinting factors due to reduced non-specific hydrophobic adsorption in aqueous buffer. This was not the case when HEMA was omitted from the polymerization mixture [27]. These conditions were exploited for the direct and selective extraction of bupivacaine from blood plasma samples.

7 Imprints of Proteins

As mentioned above, creating a MIP against a small molecule is often straightforward, and even short peptides like the neurotransmitter enkephaline can still be imprinted with the conventional protocol [96]. However, the imprinting of larger molecules like proteins is still a challenge, due to their complex chemical nature, their large size and conformational flexibility, and their insolubility in most organic solvents. Therefore, with protein templates, there is a particular danger of artifacts [97, 98], and special care must be taken to perform the necessary controls. Early developments for the imprinting of proteins were performed in polyacrylamide gels similar to those used in electrophoresis [99]. During the last 5 years, considerable progress has been made and protein specific MIPs have now become more accessible

[100–102]. Three distinct approaches have been developed in this context: (1) the imprinting on a surface using immobilized protein template, (2) the imprinting in hydrogels including microgels (nanoparticles), and (3) the imprinting of a structural epitope (a peptide sequence accessible at the surface of the protein).

7.1 Surface Imprinting of Proteins

Ratner and co-workers were among the first to create imprints of proteins in a surface-bound polymer film [103]. The protein is first adsorbed onto an atomically flat mica surface, and then spin-coated with a disaccharide solution that, upon drying, forms a thin layer (1–5 nm) on the protein. This disaccharide shell is covered with a fluoropolymer layer via glow-discharge plasma deposition, which covalently incorporates and cross-links the sugar molecules. Finally, the polymer layer is attached to a glass substrate using an epoxy glue. After peeling off the mica, the protein is removed by treatment with aqueous NaOH/NaClO, leaving nanocavities as revealed by atomic force microscopy (AFM). The authors reported that the cavities were complementary in size, and to some extent also in functionality, to the template protein. They showed that a surface imprinted with bovine serum albumin (BSA) preferentially adsorbed the template protein from a binary mixture with IgG. Haupt and co-workers later used a similar, though technically simpler approach, where cytochrome c (Cyt c) was coupled to mica and then molded into a thin hydrogel layer using a sandwich technique (Fig. 10). In that way, surface imprints were created and their affinity and selectivity for the target protein was studied using fluorescein isothiocyanate (FITC)-labeled Cyt c and fluorescence detection. The authors were even able to detect directly the imprinted binding sites in the polymer film and measure the force of interaction by chemical force spectroscopy (Fig. 10b), with Cyt c immobilized onto an AFM cantilever [104].

    In an attempt to increase the surface area of the protein-imprinted polymer and the accessibility of the binding sites, the generation of surface-bound nanofilaments has been suggested. The MIP is nanomolded on a porous alumina substrate carrying the immobilized template protein. After the dissolution of the alumina, layers of parallel nanofilaments with a high aspect ratio appear that carry surface imprints and can adsorb specifically myoglobin, the target protein [105].

7.2 Imprinting of Proteins in Hydrogels and Microgels

Hydrogels are used as molecular imprinting matrix for proteins for their loose networks enabling the diffusion of the protein and for their aqueous environment and thus compatibility with biomacromolecules [106, 107]. If the MIP precursors mixture contains a very low concentration of monomers, nanometer-sized microgels are sometimes formed upon polymerization rather than macrogels. In order to generate

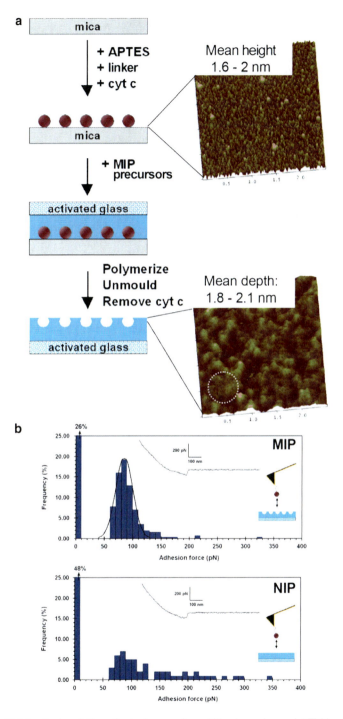

Fig. 10 (**a**) Molecular imprinting of cytochrome c immobilized on mica, and AFM images of the surfaces. (**b**) Molecular force spectroscopy with a cytochrome c-derivatized AFM cantilever of the cytochrome c-imprinted polymer (*MIP*) and the non-imprinted control polymer (*NIP*). Adapted from [104] with permission from Elsevier

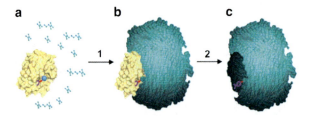

Fig. 11 Schematic representation of the molecular imprinting of trypsin using a polymerizable inhibitor as an anchoring monomer. The enzyme is put into contact with the anchoring monomer and co-monomers (**a**); polymerization is conducted (*1*); a cross-linked polymer is molded around the substrate binding site (**b**); the enzyme is removed (*2*), revealing a specific recognition site with inhibitory properties (**c**). Reproduced with permission from [108]. Copyright 2009 American Chemical Society

imprints in polymer microgels, Haupt and co-workers [108] have developed an approach based on a specific anchoring monomer (Fig. 11). The enzyme trypsin was imprinted using a polymerizable inhibitor, methacrylamidobenzamidine, together with methacrylamide or hydroxyethylmethacrylate as co-monomers, and a water-soluble cross-linker, *N,N*-ethylene bisacrylamide. The authors were able to show that the resulting microgel particles bound trypsin with a good selectivity over other proteins of similar molecular weights and isoelectric points. Moreover, this MIP was shown to inhibit competitively trypsin activity, the inhibition being almost three orders of magnitude more effective than that of the small molecule inhibitor benzamidine (inhibition constants K_i of 44 nM and 18.9 μM, respectively).

Shea and colleagues [109–111] added an exciting contribution to this field: They created molecular imprints for the peptide melittin, the main component of bee venom, in polymer nanoparticles, resulting in artificial antibody mimics that can be used for the in vivo capture and neutralization of melittin. Melittin is a peptide comprising 26 amino acids which is toxic because of its cytolytic activity. Shea and colleagues' strategy was to synthesize cross-linked, acrylamide-based MIP nanoparticles by a process based on precipitation polymerization using a small amount of surfactant. To maximize the specificity and the affinity for melittin, a number of hydrophilic monomers were screened for complementarity with the template. The imprinted nanoparticles were able to bind selectively the peptide with an apparent dissociation constant of $K_{Dapp} > 1$ nM [109].

7.3 *Imprinting of Peptide Epitopes*

Since antibodies against a protein recognize their target via certain sequential or conformational epitopes, it has been suggested to use an epitope approach also with MIPs [112, 113]. Only part of the protein, for example a surface-accessible oligopeptide, is imprinted, and the MIP can then recognize the target protein via this epitope. This has been nicely demonstrated by Shea and colleagues [114] on the

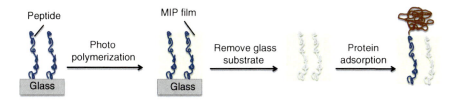

Fig. 12 Illustration outlining MIP film fabrication. The C-terminus nonapeptide epitope is attached through a tether to a glass or oxidized silicon surface by the N-terminal amino acid of the peptide. Monomers are photochemically cross-linked while remaining in contact with the peptide modified surface. Following polymerization, the glass substrate is removed. The protein can now bind to the MIP via its C-terminus nonapeptide epitope. Modified from [114]

examples of Cyt c, BSA, and alcohol dehydrogenase (ADH). Nonapeptides were chosen as the sequence length, as short epitopes focus on developing capture agents for the primary structure of the peptide rather than the more complex secondary and tertiary structures of a target protein. Three protein targets were chosen, Cyt c, ADH, and BSA (Fig. 12). The peptide sequences corresponding to the C-terminus domains of these proteins are respectively AYLKKATNE, GRYVVDTSK, and VVSTQTALA. The peptides were attached to glass substrates treated with 7-octenyltrichlorosilane. Synthesis of the imprinted polymers was achieved by exposing the functionalized surfaces to a monomer solution consisting of acrylamide, N,N-ethylene-bis-acrylamide, and polyethylene glycol 200-diacrylate in buffer solution containing a photoinitiator. The monomer solution was poured between the peptide containing substrate and a cover glass that were separated by a thin Teflon spacer. The monomers were photochemically polymerized and the glass substrate separated, revealing MIP films of a thickness of approximately 520 mm. Equilibrium binding assays yielded an average K_D for Cyt c of 72.3 nM. The MIPs were very specific; in the case of BSA, the selectivity of the imprinted MIP film was diminished by single amino acid substitutions.

8 Conclusions

In conclusion, the technique of molecular imprinting has greatly diversified during the last decade both from a materials point of view and from an application point of view. Still, there is much room for further improvement. The aspects presently considered key challenges, in particular the binding site homogeneity and water compatibility of MIPs, and the possibility of synthesizing MIPs specific for proteins, are actively addressed by research groups all over the world. Other important points are the conception of composite materials based on MIPs in order to include additional interesting properties into the material, and the synthesis of very small and quasi-soluble MIPs, close in size to proteins. For several of these aspects, the reader is directed to the other chapters of this book.

References

1. Arshady R, Mosbach K (1981) Makromol Chem 182:687
2. Wulff G, Sarhan A (1972) Angew Chem Int Ed 11:341
3. Haupt K (2003) Anal Chem 75:376A
4. Dickert FL, Hayden O, Halikias KP (2001) Analyst 126:766
5. Katz A, Davis ME (2000) Nature 403:286
6. Lordel S, Chapuis-Hugon F, Eudes V, Pichon V (2010) J Chromatogr A 1217:6674
7. Ansell RJ, Mosbach K (1998) Analyst 123:1611
8. Pérez N, Whitcombe MJ, Vulfson EN (2000) J Appl Polym Sci 77:1851
9. Bompart M, De Wilde Y, Haupt K (2009) Adv Mater 22:2343
10. Lakshmi D, Bossi A, Whitcombe MJ, Chianella I, Fowler SA, Subrahmanyam S, Piletska EV, Piletsky SA (2009) Anal Chem 81:3576
11. Umpleby RJ, Baxter SC, Chen Y, Shah RN, Shimizu KD (2001) Anal Chem 73:4584
12. Zimmerman SC, Lemcoff NG (2004) Chem Commun 1:5
13. Möller K, Crescenzi C, Nilsson U (2004) Anal Bioanal Chem 378:197
14. Wulff G, Knorr K (2002) Bioseparation 10:257
15. Shea KJ, Thompson EA (1978) J Org Chem 43:4253
16. Whitcombe MJ, Rodriguez ME, Villar P, Vulfson EN (1995) J Am Chem Soc 117:7105
17. Chianella I, Lotierzo M, Piletsky SA, Tothill IE, Chen B, Karim K, Turner APF (2002) Anal Chem 74:1288
18. Wu L, Li Y (2004) J Mol Recognit 17:567
19. Yañez F, Chianella I, Piletsky SA, Concheiro A, Alvarez-Lorenzo C (2010) Anal Chim Acta 659:178
20. O'Mahony J, Molinelli A, Nolan K, Smyth MR, Mizaikoff B (2005) Biosens Bioelectron 20:1884
21. Molinelli A, O'Mahony J, Nolan K, Smyth MR, Jakusch M, Mizaikoff B (2005) Anal Chem 77:5196
22. Hall AJ, Manesiotis P, Emgenbroich M, Quaglia M, De Lorenzi E, Sellergren B (2005) J Org Chem 70:1732
23. Ansell RJ, Wang DY, Kuah JKL (2008) Analyst 133:1673
24. Lanza F, Sellergren B (1999) Anal Chem 71:2092
25. Takeuchi T, Fukuma D, Matsui J (1999) Anal Chem 71:285
26. Takeuchi T, Seko A, Matsui J, Mukawa T (2001) Instrum Sci Technol 29:1
27. Dirion B, Cobb Z, Schillinger E, Andersson LI, Sellergren B (2003) J Am Chem Soc 125:15101
28. Ceolin G, Navarro-Villoslada F, Moreno-Bondi MC, Horvai G, Horvath V (2009) J Comb Chem 11:645
29. Navarro-Villoslada F, San Vicente B, Moreno-Bondi MC (2004) Anal Chim Acta 504:149
30. Navarro-Villoslada F, Takeuchi T (2005) Bull Chem Soc Jpn 78:1354
31. Rossi C, Haupt K (2007) Anal Bioanal Chem 389:455
32. Patel A, Fouace S, Steinke JHG (2003) Chem Commun 1:88
33. Patel A, Fouace S, Steinke JHG (2004) Anal Chim Acta 504:53
34. Bompart M, Haupt K (2009) Aust J Chem 62:751
35. Chiefari J, Chong YK, Ercole F, Krstina J, Jeffery J, Le TPT, Mayadunne RTA, Meijs GF, Moad CL, Moad G, Rizzardo E, Thang SH (1998) Macromolecules 31:5559
36. Moad G, Rizzardo E, Thang SH (2006) Aust J Chem 59:669
37. Titirici MM, Sellergren B (2006) Chem Mater 18:1773
38. Du Zhenxia I, Zhifeng F (2006) Chin Chem Lett 17:549
39. Southard GE, Van Houten KA, Ott EW, Murray GM (2007) Anal Chim Acta 581:202
40. Southard GE, Van Houten KA, Murray GM (2007) Macromolecules 40:1395
41. Liu H, Zhuang X, Turson M, Zhang M, Dong X (2008) J Sep Sci 31:1694
42. Gonzato C, Courty M, Pasetto P, Haupt K (2011) Adv Funct Mater 21:3947

43. Otsu T (2000) J Polym Sci Pol Chem 38:2121
44. Otsu T, Yoshida M, Tazaki T (1982) Makromol Chem Rapid Commun 3:133
45. Wang HY, Kobayashi T, Fujii N (1997) J Chem Technol Biotechnol 70:355
46. Rückert B, Hall A, Sellergren B (2002) J Mater Chem 12:2275
47. Sellergren B, Rückert B, Hall A (2002) Adv Mater 14:1204
48. Rong F, Feng X, Li P, Yuan C, Fu D (2006) Chinese Sci Bull 51:2566
49. Pérez-Moral N, Mayes AG (2007) Macromol Rapid Commun 28:2170
50. Vaughan AD, Sizemore SP, Byrne ME (2007) Polymer 48:74
51. Su S, Zhang M, Li B, Zhang H, Dong X (2008) Talanta 76:1141
52. Kamigaito M, Ando T, Sawamoto M (2001) Chem Rev 101:3689
53. Kato M, Kamigaito M, Sawamoto M, Higashimura T (1995) Macromolecules 28:1721
54. Wang J-S, Matyjaszewski K (1995) J Am Chem Soc 117:5614
55. Matyjaszewski K, Xia J (2001) Chem Rev 101:2921
56. Salomon DH, Rizzardo E, Cacioli P (1985) US Patent US4581429
57. Hawker CJ, Bosman AW, Harth E (2001) Chem Rev 101:3661
58. Boonpangrak S, Whitcombe MJ, Prachayasittikul V, Mosbach K, Ye L (2006) Biosens Bioelectron 22:349
59. Teodorescu M, Matyjaszewski K (2000) Macromol Rapid Commun 21:190
60. Xia J, Zhang X, Matyjaszewski K (1999) Macromolecules 32:3531
61. Wei X, Li X, Husson SM (2005) Biomacromolecules 6:1113
62. Li X, Husson SM (2006) Biosens Bioelectron 22:336
63. Wei X, Husson SM (2007) Ind Eng Chem Res 46:2117
64. Zu BY, Pan GQ, Guo XZ, Zhang Y, Zhang HQ (2009) J Polym Sci A Polym Chem 47:3257
65. Sasaki S, Ooya T, Takeuchi T (2010) Polym Chem 1:1684
66. Zu BY, Zhang Y, Guo XZ, Zhang HQ (2010) J Polym Sci A Polym Chem 48:532
67. Sulitzky C, Ruckert B, Hall AJ, Lanza F, Unger K, Sellergren B (2002) Macromolecules 35:79
68. Vaughan AD, Zhang JB, Byrne ME (2010) AIChE J 56:268
69. Lu C, Zhou W, Han B, Yang H, Chen X, Wang X (2007) Anal Chem 79:5457
70. Pan GQ, Zu BY, Guo XZ, Zhang Y, Li CX, Zhang HQ (2009) Polymer 50:2819
71. Pan G, Ma Y, Zhang Y, Guo X, Li C, Zhang H (2011) Soft Matter 7:8428
72. Pichon V, Haupt K (2006) J Liq Chromatogr Relat Technol 29:989
73. Andersson LI (1996) Anal Chem 68:111
74. Bengtsson H, Roos U, Andersson LI (1997) Anal Comm 34:233
75. Haupt K, Dzgoev A, Mosbach K (1998) Anal Chem 70:628
76. Caro E, Marcé RM, Cormack PAG, Sherrington DC, Borrull F (2004) J Chromatogr A 1047:175
77. Haupt K, Mayes AG, Mosbach K (1998) Anal Chem 70:3936
78. Legido-Quigley C, Oxelbark J, De Lorenzi E, Zurutuza-Elorza A, Cormack PAG (2007) Anal Chim Acta 591:22
79. Díaz-Alvarez M, Turiel E, Martín-Esteban A (2009) Anal Bioanal Chem 393:899
80. Sun H-W, Qiao F-X (2008) J Chromatogr A 1212:1
81. Wulff G, Biffis A (2001) In: Sellergren B (ed) Molecularly imprinted polymers: man-made mimics of antibodies and their applications in analytical chemistry. Elsevier, Amsterdam, pp 71–111
82. Wulff G, Schönfeld R (1998) Adv Mater 10:957
83. Sellergren B (1994) Anal Chem 66:1578
84. Lübke C, Lübke M, Whitcombe MJ, Vulfson EN (2000) Macromolecules 33:5098
85. Urraca JL, Moreno-Bondi MC, Hall AJ, Sellergren B (2007) Anal Chem 79:695
86. Sadlej-Sosnowska N (1997) J Incl Phenom Mol Rec Chem 27:31
87. Piletsky SA, Andersson HS, Nicholls IA (1998) J Mol Recogn 11:94
88. Song SH, Shirasaka K, Hirokawa Y, Asanuma H, Wada T, Sumaoka J, Komiyama M (2010) Supramol Chem 22:149

89. Hishiya T, Shibata M, Kakazu M, Asanuma H, Komiyama M (1999) Macromolecules 32:2265
90. Hochuli E, Döbeli H, Schacher A (1987) J Chromatogr A 411:177
91. Hart BR, Shea KJ (2002) Macromolecules 35:6192
92. Hart BR, Shea KJ (2001) J Am Chem Soc 123:2072
93. Striegler S (2001) Tetrahedron 57:2349
94. Striegler S (2004) J Chromatogr B 804:183
95. Tse Sum Bui B, Haupt K (2010) Anal Bioanal Chem 398:2481
96. Andersson LI, Müller R, Mosbach K (1996) Macromol Res Comm 17:65
97. Fu GQ, Zhu J, Jiang YZ (2008) Anal Chem 80:2634
98. Verheyen E, Schillemans JP, van Wijk M, Demeniex MA, Hennink WE, van Nostrum CF (2011) Biomaterials 32:3008
99. Hjertén S, Liao JL, Nakazato K, Wang Y, Zamaratskaia G, Zhang HX (1997) Chromatographia 44:227
100. Janiak DS, Kofinas P (2007) Anal Bioanal Chem 389:399
101. Hansen DE (2007) Biomaterials 28:4178
102. Bossi A, Bonini F, Turner APF, Piletsky SA (2007) Biosens Bioelectron 22:1131
103. Shi HQ, Tsai WB, Garrison MD, Ferrari S, Ratner BD (1999) Nature 398:593
104. El Kirat K, Bartkowski M, Haupt K (2009) Biosens Bioelectron 24:2618
105. Linares AV, Vandevelde F, Pantigny J, Falcimaigne-Cordin A, Haupt K (2009) Adv Funct Mater 19:1
106. Uysal A, Demirel G, Turan E, Çaykara T (2008) Anal Chim Acta 625:110
107. Casey BJ, Kofinas P (2008) J Biomed Mater Res A 87A:359
108. Cutivet A, Schembri C, Kovensky J, Haupt K (2009) J Am Chem Soc 131:14699
109. Hoshino Y, Kodama T, Okahata Y, Shea KJ (2008) J Am Chem Soc 130:15242
110. Hoshino Y, Urakami T, Kodama T, Koide H, Oku N, Okahata Y, Shea KJ (2009) Small 5:1562
111. Hoshino Y, Koide H, Urakami T, Kanazawa H, Kodama T, Oku N, Shea KJ (2010) J Am Chem Soc 132:6644
112. Rachkov A, Minoura N (2000) J Chromatogr A 889:111
113. Rachkov A, Minoura N (2001) Biochim Biophys Acta 1544:255
114. Nishino H, Huang C-S, Shea KJ (2006) Angew Chem Int Ed 45:2392

Top Curr Chem (2012) 325: 29–82
DOI: 10.1007/128_2010_110
© Springer-Verlag Berlin Heidelberg 2011
Published online: 13 January 2011

Physical Forms of MIPs

Andrea Biffis, Gita Dvorakova, and Aude Falcimaigne-Cordin

Abstract The current state of the art in the development of methodologies for the preparation of MIPs in predetermined physical forms is critically reviewed, with particular attention being paid to the forms most widely employed in practical applications, such as spherical beads in the micro- to nanometer range, microgels, monoliths, membranes. Although applications of the various MIP physical forms are mentioned, the focus of the paper is mainly on the description of the various preparative methods. The aim is to provide the reader with an overview of the latest achievements in the field, as well as with a mean for critically evaluating the various proposed methodologies towards an envisaged application. The review covers the literature up to early 2010, with special emphasis on the developments of the last 10 years.

Keywords Membranes · Microgels · Microparticles · Microspheres · Molecular imprinting · Monoliths · Nanogels · Nanoparticles · Nanospheres

Contents

A. Biffis (✉) and G. Dvorakova
Dipartimento di Scienze Chimiche, Università di Padova, via Marzolo 1, 35131 Padova, Italy
e-mail: andrea.biffis@unipd.it

A. Falcimaigne-Cordin
Université de Technologie de Compiègne, UMR CNRS 6022, BP 20529, 60205 Compiègne, France
e-mail: aude.cordin@utc.fr

1   Introduction

As the number of technological applications of molecularly imprinted polymers (MIPs) continues to increase, there is a growing demand for MIPs tailored to specific physical forms. Traditionally, MIPs have been prepared as monoliths by bulk polymerisation of monomer mixtures containing the template molecule and a small amount of solvent acting as a porogen. The resulting polymer monolith had to be ground and sieved in order to produce sufficiently small particles with a reasonable fraction of accessible molecularly imprinted sites. This has been for very long time virtually the only method for MIP synthesis, but it is clearly a rough method which is not suitable for the large-scale production of MIPs. In fact, the grinding and sieving procedure is time-consuming and generally leads to a high amount of waste, which depends on the particle size and size distribution that is aimed at and can make up to 80% of the original bulk polymer. Furthermore, the resulting polymer particles are intrinsically irregular, characterised by a rather broad size and shape distribution, and therefore often unsuitable for more advanced applications. Finally, scale-up of this procedure could become problematic due to the intrinsic difficulty in dispersing the heat evolved by the polymerisation process leading to monolith production.

By the mid-nineties these limitations of the traditional preparation method had become apparent, so that various research groups started to investigate possible alternatives. For example, some researchers developed methods that allowed direct in situ synthesis of MIPs inside chromatographic columns or capillaries for electrophoresis, thereby avoiding the need for producing first a MIP powder (see Sect. 3.1.1 below). However, this methodology requires careful optimisation of the MIP preparation conditions for each kind of device to be filled, taking also into account the need for an efficient imprinting procedure. Therefore, it was soon recognised that a more general solution was the development of methods for the direct preparation of MIPs in a size- and shape controlled way. In the course of the last ten years numerous synthetic strategies have been proposed, which will form the subject of this chapter.

Up to now, most efforts have been directed towards the preparation of uniformly sized spherical MIP particles in the micrometre range. This is the obvious consequence of the need for this kind of materials as fillers for high-performance chromatographic columns, capillaries for electrophoresis, cartridges for solid-phase extractions and other applications requiring selective stationary phases. Additionally though, strategies for the preparation of other more sophisticated MIP forms, such as membranes, (nano)monoliths, films, micro- and nanostructured surfaces etc.

have been also implemented for more advanced applications. Finally, successful attempts to further downsizing MIP beads well into the nanometer range have been made: the preparation of imprinted "nanoobjects" such as nanoparticles, nanocapsules, nanofilaments, crosslinked polymer colloids (microgels) has been recently reported. In these materials, the surface-to-volume ratio of the MIP is clearly maximised, hence the accessibility of imprinted sites that would be otherwise buried inside the polymer matrix is improved. Furthermore, the availability of submicron-sized MIP beads enables novel applications such as their use as pseudostationary phases in capillary electrochromatography (CEC). Finally, a more fundamental reason of interest for imprinted nanoobjects lies is the production of MIPs comparable in size to natural systems capable of molecular recognition and catalysis like antibodies or enzymes. Successful transfer of the molecular imprinting approach to such a small scale represents a major step towards the development of "plastic analogues" of these biomolecules.

The aim of this chapter is to provide the reader with a comprehensive, although not necessarily exhaustive, overview of the methods proposed up to now for the preparation of predetermined physical forms of MIPs. The subject will be organised according to the type of form as well as according to the employed preparation strategy. In the course of the last years, other less extensive reviews on this research area have been published, and the interested reader is referred to them as well [1–5].

2 MIP Micro- and Nanobeads

As outlined in the introduction, the problem of the preparation of MIPs in beaded form was the first one to be addressed in view of the immediate applications of the resulting particles. Therefore, it is not surprising that the majority of reports on the preparation of MIPs in predetermined forms concerns spherical beads. As it will be apparent from the subsequent sections, research in this field has advanced considerably, moving from more traditional methods for the preparation of polymers in beaded form and arriving at developing original approaches specifically targeted to MIPs.

2.1 MIP Beads from Standard Polymerisation Techniques

The preparation of crosslinked polymers in beaded form is an issue that has been systematically addressed by polymer chemists over the years and a number of successful and well-established approaches are available. However, the specificity of the MIP preparation procedure renders a direct application of these methodologies not so obvious, so that virtually none of these approaches is automatically and generally applicable to the preparation of MIPs. Most notably, there are two issues that need consideration:

1. A good imprinting effect relies on the formation of a rather rigid macromolecular network, hence it requires highly crosslinked polymers; therefore, any preparation method for beaded MIPs must be compatible with the presence of a high amount of crosslinker.
2. A good imprinting effect also relies on the existence of stable assemblies between the template molecule and the functional monomers in the course of the polymerisation process; therefore, any preparation method for beaded MIPs must preserve such template-monomer assemblies.

As it will be apparent in the following, both these issues quite severely limit the possible application of standard polymerisation techniques to MIPs. On the other hand, they provide guidelines for critically evaluating the application of the various beaded polymer preparation methods to MIP synthesis.

2.1.1 Suspension Polymerisation

Suspension polymerisation is arguably the simplest and most widely employed preparation method for polymers in beaded form [6]. In this method, polymerisation occurs entirely within monomer droplets that are dispersed inside a monomer-immiscible phase; the stability of the droplet dispersion is often increased by addition of suitable stabilisers.

From the point of view of beaded MIP synthesis, this approach has advantages but also two important limitations. The positive features are that the methodology can be easily applied to the preparation of highly crosslinked polymers; furthermore, no diluents are needed for the monomer mixture. The limitations arise from the fact that the most commonly employed monomer-immiscible phase is water, which due to its polarity and high hydrogen-bonding capability is mostly unsuitable for molecular imprinting, since it severely affects the stability of the template-monomer assembly. Secondly, only rather large beads with size in the tens of micrometres rather than in the micro- or submicrometer range are accessible by suspension polymerisation. In spite of these limitations, several applications of this polymerisation strategy to MIP synthesis have been reported in recent years, and a significant fraction of these reports involves water as the continuous phase. For example, Lai et al. prepared in this way MIP microspheres for use as stationary phase in HPLC. Methacrylic acid (MAA) and ethylene dimethacrylate (EDMA, 86 mol%) were used as a functional monomer and crosslinker, respectively, with polyvinyl alcohol as stabiliser to assist polymerisation. Microspheres were imprinted with different templates like 4-aminopyridine [7], trimethoprim [8], antitumor drug piritrexim [9] or carotenoid pigment asthaxanthin [10]. The resulting MIP microspheres were 5–30 μm in size and exhibited specific affinity for the template molecule. However, no comparison with MIPs prepared by the conventional method was made. Similarly, other reports dealing with the development of MIPs for solid phase extraction (SPE) have appeared. The group of Takeuchi reported already in 1997 on the use of suspension polymerisation in water for the preparation of MIP microspheres (32–63 μm) imprinted with atrazine [11]. Hu et al. prepared MIP microspheres which were used as sorbents for selective SPE of phenobarbital from human urine and medicines [12]. In this case, microspheres were apparently obtained with a broad size distribution and sieving was required in order to isolate a 40–60 μm fraction. Finally, Shi et al. recently reported on the preparation of polymer microspheres (50–120 μm) imprinted with chloroamphenicol [13].

Although all these polymers presented enhanced affinity for the template molecule in comparison to nonimprinted controls, the presence of water as the continuous phase during polymerisation is expected to negatively affect the efficiency of the imprinting process, considering that in these cases imprinting is based on comparatively weak hydrogen bonding or dipolar interactions that are disrupted by water. A direct comparison with control bulk polymers prepared in the absence of water would be necessary to critically assess this point.

Better results can be expected using interactions between template and functional monomer that are not affected by water. For example, the group of Wulff reported in 2000 on the preparation of catalytic MIPs using *inter alia* suspension polymerisation in water [14]. Polymerisation was carried out from monomer mixtures containing methyl methacrylate (MMA, 11 mol%), EDMA as the crosslinker (80 mol%) and 9 mol% N,N'-diethyl-4-vinylbenzamidine, which forms in situ with diphenylphosphate a stable 1:1 adduct. This adduct, based on a stoichiometric noncovalent interaction in the form of a strong hydrogen bonding between the two components, can be regarded as a stable transition-state analogue of carbonate or carbamate hydrolysis, and it has been shown to be stable also in the presence of water. The resulting beads (8–375 μm in size, depending on the polymerisation conditions) were used as catalysts for the selective hydrolysis of diphenyl carbonate and diphenyl carbamate and were found to exhibit the same catalytic activity and even superior selectivity in comparison to bulk MIPs prepared under the same reaction conditions.

Apart from strong hydrogen bonds, hydrophobic and $\pi-\pi$ stacking interactions, which are unaffected or possibly even strengthened by the presence of water have been also very recently employed for the preparation of polymer beads imprinted with pyrene by suspension polymerisation in water [15], although it must be borne in mind that the inherent weakness of these interaction may limit their exploitation for imprinting purposes.

Another possible way of overcoming the limitations posed by the presence of water in the suspension polymerisation process is to substitute the continuous water phase with alternative solvents that could still act as dispersing medium for the monomer mixture but better preserve noncovalent interactions in the template-monomer assembly. For example, liquid fluorocarbons are chemically inert and do not affect interactions which are used in noncovalent imprinting. Use of such solvents for the preparation of MIP microbeads has been demonstrated already in 1996 by Mayes and Mosbach [16, 17]. A range of MIPs were prepared using Boc-L-phenylalanin as the template, MAA as the functional monomer and different kinds and amounts of crosslinkers and porogenic solvents. The resulting MIP microbeads were 5–50 μm in size, depending on the composition of the monomer mixture, and their performance as stationary phases in the HPLC separation of Boc-DL-phenylalanin was fully comparable to that of conventional bulk MIPs. The method was also employed for the synthesis of composite MIP microspheres containing superparamagnetic iron oxide particles [18]. The limitation of this method is that it needs expensive perfluorinated solvents and, most notably, special fluorinated polymer surfactants, which usually have to be custom-designed, synthesised and tested. Furthermore, although this method allows a reasonable control of particle size and delivers spherical particles in good yield, it is not able to generate monodisperse particles.

Liquid fluorocarbon was used as continuous phase by Pérez-Moral and Mayes [19] as well. They proposed a new method for rapid synthesis of MIP beads, in that they prepared 36 polymers imprinted for propranolol and morphine with different amounts of EDMA as a cross-linker and different functional monomers (MAA, acrylic acid, hydroxyethyl methacrylate, 4-vinylpyridine) directly in SPE cartridges. The properties of MIP microspheres prepared by this method were very similar in terms of size, morphology and extent of rebinding to microspheres prepared by "conventional" suspension polymerisation in perfluorocarbons as well as to bulk polymers prepared in the same solvent. The most notable advantages of this method are no waste production (no transfer of beads during washing steps) and possible direct use for a variety of screening, evaluation and optimisation experiments.

Another nonaqueous solvent which has been employed as continuous phase in suspension polymerisation is mineral oil, which was used by Kempe and Kempe for the synthesis of MIP libraries and especially for preparation of propranolol-imprinted polymer beads [20, 21]. The disadvantage of this solvent is again the wide range of resulting bead sizes (1–100 μm) which in turn requires extensive sieving in order to get a narrow size distribution. On the other hand, the advantage of mineral oil is that no addition of stabilisers is needed. In the study by Kempe and Kempe [21], droplets of pre-polymerisation solution (made out of MAA and trimethylolpropane trimethacrylate, TRIM, 50 mol% crosslinker) containing various template molecules were formed directly in mineral oil by vigorous agitation (8,000–24,000 rpm) followed by formation of solid spherical beads by photoinduced free-radical polymerization. The size, shape and yield of MIP beads was affected by the volume fraction of the pre-polymerisation solution in the suspension mixture as well as by the mixing rate and mixing time. The characteristics of the obtained beads (elemental composition, surface area, pore diameter and pore volume) were similar to the characteristics of particles prepared by conventional methods. However, the binding capacity of the beads were higher in most of the solvents and solvent mixtures tested.

Silicone oil was also recently employed by the group of Peng as an alternative nonaqueous solvent system for suspention polymerisation [22]. Silicone oil has the advantage of being cheap, inert and rather viscous, which enhances the stability of the monomer droplets in suspension. In this case, composite MIP beads were prepared including magnetic nanoparticles. Fe_3O_4 nanoparticles (average diameter

10 nm) were surface functionalised with methacryloxypropyltrimethoxysilane and subsequently copolymerised with EDMA and 4-vinylpyridine (crosslinker content 86 mol%) in the presence of the template molecule 2,4-dichlorophenoxyacetic acid. Redox initiation was employed at low temperature (10°C) in order to maximise the viscosity of the solvent and therefore the stability of the suspension. In this way, composite particles of about 20 µm diameter were obtained which exhibited enhanced absorption of the template molecule in comparison to nonimprinted controls.

As described above, MIP preparation via suspension polymerisation in mineral oil or in liquid perfluorocarbon yields beads with a wide distribution of particle sizes. In 2006, Zourob et al. [23] used a spiral-shaped microchannel as a micro-reactor for controlled suspension polymerization in liquid perfluorocarbon or mineral oil as the continuous phase. Propranolol-imprinted beads prepared using this micro-reactor were almost monodisperse as can be seen on Fig. 1. Because of the different crosslinkers used with different continuous phases, EDMA in liquid perfluorocarbon and TRIM in mineral oil, no direct comparison of the rebinding properties of beads obtained in the two solvents could be done, and no comparison with MIP polymers prepared by conventional bulk polymerisation was performed. In the same year, the group of Takeuchi independently published a related paper in

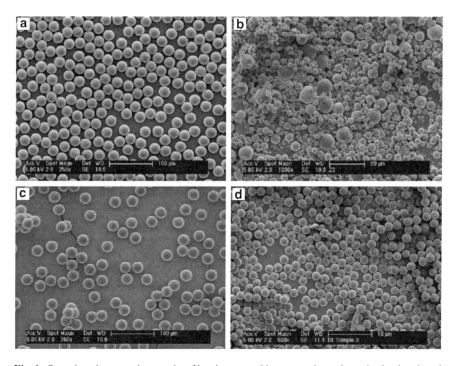

Fig. 1 Scanning electron micrographs of beads prepared by suspension polymerisation in mineral oil (*top*) or liquid fluorocarbon (*bottom*) using the microreactor (*left*) or the conventional approach (*right*). Reproduced with permission from [23]

which a microfluidic device was employed to carry out the synthesis of polymer beads (50 μm diameter) imprinted with atrazine by suspension polymerisation in water [24]. Although in this case no comparison between the outcome of the polymerisation with or without the microfluidic device was done, this testifies the growing interest in the application of microfluidics for the generation of highly uniform imprinted polymer beads.

Finally, in recent years an "inverse" suspension polymerisation approach using an organic solvent as the continuous phase and an aqueous solution as the discontinuous phase with ethyl cellulose as stabiliser has been occasionally adopted in the literature [25–28]. This approach is clearly limited to water-soluble monomers and templates but it proved indeed to work well with large template molecules (peptide sequences, proteins) that are expectedly capable of building up relatively stable interactions with polar monomers even in (concentrated) aqueous solutions. The resulting beads were in one case capable of recognising the template more efficiently than the corresponding bulk MIP, although it must be remarked that in the preparation of the latter the stabiliser (which could play an active role in the molecular recognition event) was not included [27]. More work is needed to assess the generality of this approach for MIP preparation against large biomolecules, a topic of steadily increasing interest among the scientific community.

2.1.2 Dispersion/Precipitation Polymerisation

Dispersion/precipitation polymerisation is another widely employed methodology for beaded polymer synthesis [29]. In this approach, the key point is the use of a solvent where the monomers are soluble but the resulting polymer is not. This implies that in the course of the polymerisation the growing polymer segregates from the solution while continuously capturing and consuming monomers and oligomers, eventually forming beads in the micro- or even submicrometer range under proper conditions. The simplicity of this methodology, the lack of need for stabilisers or other additives, its compatibility with high crosslinking degrees and the fact that the most commonly employed solvents are dipolar aprotic ones which preserve hydrogen bonds have made this approach very popular among MIP researchers. Consequently, it is not surprising that dispersion/precipitation has arguably become the most commonly employed preparation technique for beaded MIPs.

Application of this technique to MIPs was first proposed at the end of the nineties by the group of Ye and Mosbach [30]. They developed a general method for the synthesis of MIP beads in the submicrometer range, based on the use of acetonitrile as polymerisation medium, MAA as the functional monomer, EDMA or TRIM as crosslinker (80 mol%), which proved to work with different template molecules such as theophylline, caffeine, 17β-etradiol, (S)-propanolol and 2,4-dichlorophenoxyacetic acid [30–36]. A few years later, the group of Cormack and Sherrington extended the applicability of this method in that they found that, by tuning parameters such as the composition of the polymerisation medium (i.e. by using mixtures of acetonitrile and toluene in different relative amounts), the concentration of the monomer mixture and the amount of initiator, it was possible, using divinyl-benzene (DVB) as crosslinker, to prepare in a controllable way larger MIP particles of a few micrometres in size, suitable for direct application in HPLC or SPE [37, 38]. Building on these results, Ye and Mosbach more recently showed that using as crosslinker a mixture of DVB and TRIM in different ratios it was possible to tune the size of the resulting MIP particles in the range 130 nm–2.4 μm [36].

These pioneering investigations triggered an intensive use of this technique in the last years. The range of successfully employed templates was considerably extended to include inter alia fenuron [39], *trans*-acotinic acid [40], nicotine [41], di-(2-ethylhexyl)-phtalate [42], fluoroquinolones [43], and degradation products of chemical warfare agents [44]. Comparison with the corresponding bulk MIPs prepared by conventional means were only occasionally made [40, 42]: the MIP beads prepared by dispersion/precipitation generally proved superior in terms of specific affinity for the template.

The availability of MIP microparticles through this synthetic method has also stimulated the development of analytical techniques that make use of them as sensing elements. Apart from competitive radioassays [30] and immunoassays [32], which were already performed with ground bulk polymers, the small, regular size of the beads prepared by dispersion/precipitation polymerisation enables their use in CEC [45, 46], scintillation proximity assays [35], fluorescent polarisation assays [47], and chemiluminescence imaging [48].

There is just one significant drawback of this polymerisation technique from the point of view of MIP synthesis, which is the low monomer concentration (up to a few percents w/w) that is required in order to obtain beads and no continuous polymer phase. Under such high dilution conditions, the association equilibria that characterise the formation of the template-monomer assembly are shifted towards the free template and monomers, thereby negatively affecting the stability of the assembly and consequently the efficiency of the imprinting procedure. Neverthe-less, this problem may be overcome by a careful choice of sufficiently strong binding interactions between template and monomers [49]. Furthermore, in two recent independent reports it has been demonstrated that using highly viscous cosolvents in the polymerisation mixture much higher monomer concentrations can be employed, presumably as the consequence of the lower diffusional mobility of the growing polymer particles [50, 51]. Finally, it must be borne in mind that although the selectivity of the resulting imprinted beads may be lower than that of conventional imprinted polymers, their capacity is usually significantly higher due to the high surface to volume ratio of the beads. This in turn ensures accessibility of a higher fraction of the imprinted sites, and also overcomes the problem of cavity collapse after template removal, which notoriously plagues especially conventional MIPs prepared using weak noncovalent interactions. In this connection, special mention deserves the recent work of Shea et al., who demonstrated that it is possible to carry out MIP synthesis upon precipitation polymerisation in water, using cocktails of acrylamide-based monomers [52, 53]. The resulting beads are extremely small, in the tens of nanometers range, and although they are only lightly crosslinked their fully collapsed structure in water stabilises the imprinted cavities. Moreover, the employed amphiphilic template molecule (the peptide mellitin) locates itself during bead preparation at the interface between growing polymer particle and water, thereupon presumably acting as a surfactant and promoting particle stabilisation. As a consequence, the template can be simply and quantitatively removed by extensive dialysis, leaving selective cavities preferentially located at the surface of the beads. Most notably, the small size of the resulting beads makes them able to be transported by diffusion in e.g. blood capillaries, and it has been demonstrated that they are able to recognise and neutralise upon binding the toxic action of mellitin in vivo [53].

2.1.3 Emulsion Polymerisation

Emulsion polymerisation [54] is in principle related to the previously mentioned methodologies in that it is another way of exploiting the compartimentalisation of the monomer mixture or of the growing polymer chains for achieving the product in beaded form. Like in the case of suspension polymerisation, the monomer phase is dispersed inside a continuous phase made out of a monomer-immiscible solvent. In this case, however, the resulting dispersion is stabilised by a surfactant, which forms micelles around the monomer droplets. The presence of the surfactant allows to reach much smaller particle sizes, in the range of tens to hundreds of nanometers. Moreover, using the peculiar technique known as "miniemulsion polymerisation" [54, 55], the size of the resulting polymer particles exactly matches that of the droplets in the starting emulsion, which allows to predetermine the polymer particle size. Finally, unconventional polymer forms, like e.g. nanocapsules, are accessible by this technique [56, 57].

From the point of view of MIP synthesis, this procedure suffers from the same general drawback mentioned above for suspension polymerisation, namely the need for water as the continuous phase, which generally interferes with the imprinting procedure. Furthermore, the surfactant may also disturb the imprinting process, since it may become involved in interactions with the monomer and/or the template; the surfactant may also turn out to be difficult to remove from the resulting MIP. Finally, the general applicability of this methodology for the preparation of highly crosslinked polymer beads is to be demonstrated, since most investigations on this technique have been carried out with monomer mixtures having a crosslinker content not exceeding 20 mol%. Nevertheless, these drawbacks have not discouraged a few authors from attempting MIP synthesis via emulsion polymerisation, with some success.

Emulsions have been introduced in MIP synthesis by Japanese researchers in the early nineties. The aim was to create materials suitable for the recognition and binding of water-soluble substrates, most notably metal ions. Fatty acids, sometimes additionally functionalised with polymerisable groups, were employed both as surfactants and as functional monomers for ion binding. With the aid of these molecules, oil-in-water (O/W) emulsions of crosslinking and nonfunctional monomers (typically DVB-styrene) were prepared, which were subsequently polymerised in the presence of metal ions in the continuous water phase. In this way, it was possible to produce polymer particles in the submicron range functionalised at their surface with prearranged carboxylic acid groups (stemming from immobilised or copolymerised surfactant molecules) that showed some preference in binding the templating metal ion [58, 59]. In 2000, Yoshida et al. reported a variation of this approach for the preparation of MIP beads using multiple W/O/W emulsions. The resulting MIP particles were however quite large (15–40 μm), comparable in size to those accessible by suspension polymerisation [60].

In 2002, the group of Tovar reported for the first time on the preparation of MIP nanospheres by miniemulsion polymerisation [61]. The polymers were based on EDMA as the crosslinker and MAA as the functional monomer (crosslinking degree up to 80 mol%) and were imprinted against Boc-L-phenylalanin-anilide or its enantiomer. The resulting beads were 50–300 nm in size in the dry state and exhibited enhanced binding capacity for the template molecule in comparison to beads imprinted with the other enantiomer or to a nonimprinted control. Unfortunately, no competitive binding experiments were reported and, most notably, no comparison with a "traditional" MIP was made, which would have been useful to assess the efficiency of this procedure, in which water was used as continuous phase despite the fact that the formation of the template-monomer assembly was relying on comparatively weak hydrogen bonds and dipolar interactions.

In 2006, Ki and Chang reported on the preparation of MIP nanocapsules by the same technique [62]. They used however a much more stable template-monomer assembly, based on the formation of a reversible covalent bond upon reaction of an isocyanate group on a functional monomer with a hydroxyl function on the template, yielding a urethane linker between the two molecules. The template, in the present case oestrone, can be removed by thermal degradation of the urethane linker, and subsequent hydrolysis of the polymer-bound isocyanate group yields an amino group able to engage in hydrogen bonds with the hydroxyl group on the template (Fig. 2). In this way, Ki and Chang were able to prepare MIP nanocapsules

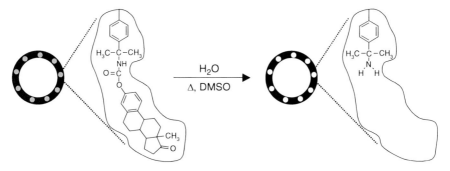

Fig. 2 Strategy for the generation of imprinted cavities inside nanocapsules employed by Ki and Chang. Reproduced with permission from [62]

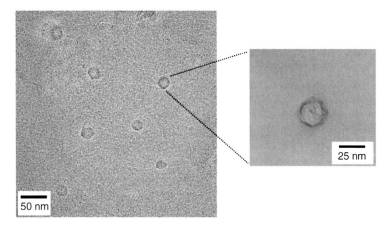

Fig. 3 TEM micrograph of the imprinted nanocapsules prepared by Ki and Chang. Reproduced with permission from [62]

(20–25 nm in diameter, with a wall thickness about 2.5 nm) imprinted with oestrone (Fig. 3).

The resulting nanocapsules exhibited an enhanced binding capacity for oestrone in comparison with nonimprinted controls; furthermore, the capacity of the nanocapsules to bind other structurally related steroids was significantly lower. Also in this case no competitive binding experiment was reported and, most notably, no comparison with a "traditional" MIP was made. On the other hand, it was demonstrated that template rebinding could affect the diffusion of smaller molecules through the nanocapsules. In particular, the rebinding of the template closed the "gates" which were opened in the nanocapsules by template removal, thereby impeding diffusion from the outside to the inside of the nanocapsules.

More recently, Priego-Capote et al. reported on the production of MIP nanoparticles with "monoclonal" behaviour by miniemulsion polymerisation [63]. In the synthetic method that they employed, they devised to use a polymerisable surfactant that was also able to act as a functional monomer by interacting with the template (Fig. 4). The crosslinker content was optimised at 81% mol/mol (higher or lower contents leading to unstable emulsions). In this way, the authors were able not only to produce rather small particles (80–120 nm in the dry state) but also to locate the imprinted sites on the outer particle surface. The resulting MIP nanobeads were very effective as pseudostationary phases in the analysis of (R,S)-propranolol by CEC.

Finally, very recently Shea et al. successfully employed inverse microemulsion polymerisation for the preparation of MIP beads in the tens of nanometers range using hydrophilic peptides as template molecules. In this case it was the template molecule which was prefunctionalised with a hydrophobic chain to orient it towards the surface of the growing bead during polymerisation. The rebinding efficiency of the resulting nanoparticles was however found to depend markedly on the nature of the employed template and to be lower than that recorded with beads of similar

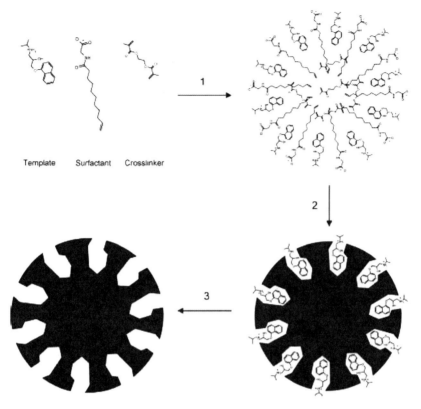

Fig. 4 Molecular imprinting at the nanoparticle surface by miniemulsion polymerisation. Reproduced with permission from [63]

size prepared by precipitation polymerisation using mellitin as template (see above) [64].

Emulsion polymerisation has been also extensively employed for the production of molecularly imprinted core-shell particles. This application is discussed in detail in Sect. 2.2.3.

2.1.4 Solution Polymerisation

Nanosized crosslinked polymers can also be prepared with a method not involving compartimentalisation of the monomer mixture nor phase separation (precipitation) of the growing polymer chains. Indeed, Staudinger reported already in the 1930s that upon copolymerisation of styrene and DVB in dilute solution no macroscopic gel phase was formed; instead, soluble crosslinked polymer particles ("microgels", according to a later definition by Baker [65]) were isolated, whose viscosity was exceptionally low given their high molecular weight [66]. What happens is that at high dilution intramolecular crosslinking becomes favoured compared to intermolecular crosslinking for entropic reasons (Ziegler's dilution law); moreover, the growing microgels become stabilised towards macrogelation by the osmotic repulsion forces generated by the interaction of solvated polymer chains and loops at the perifery of the microgel particles (steric stabilisation) [29]. Clearly, to achieve steric stabilisation the microgel chains must be efficiently solvated by the polymerisation solvent, so that the growing microgels can be considered to be in swelling equilibrium with the surrounding medium. The resulting crosslinked polymer particles are on the one hand more compact than linear macromolecules of the same molecular weight (which explains their low solution viscosity), on the other hand have generally smaller hydrodynamic radii (tens of nm) and are less rigid than particles prepared by compartimentalisation or phase separation: for example, microgels tend to collapse when dried, forming films, whereas particles prepared by the latter methods retain their spherical shape even in the dry state. For this reason, it appears appropriate to define "microgels" (or nanogels) crosslinked polymer particles prepared by solution polymerisation, as opposed to "microspheres" (or nanospheres) which should more properly indicate particles prepared by other methods involving compartimentalisation or phase separation (for an illustrative example of the morphological differences between microgels and microspheres as well as of the transition between the two kinds of particles with varying polymerisation conditions, see [67]).

The key advantage of solution polymerisation is that the onset of steric stabilisation makes it possible to obtain particles of very small sizes without having to add stabilisers to prevent macrogelation. Steric stabilisation is sufficient to stabilise the growing microgels against macrogelation if the monomer concentration is reduced below a critical value (critical monomer concentration, C_m). This value is dependent on many factors such as the polymerisation methodology and conditions, the crosslinking monomer content, and of course the nature of the solvent, which can be conveniently rationalised in terms of the solubility parameter δ [68]. In particular, a high content of crosslinking monomer has a detrimental effect on the stability of the growing microgels, since it reduces the length of solvated polymer chains and loops making out the stabilising shell at the perifery of the microgel particles. This represents a drawback in the preparation of highly crosslinked particles such as MIPs with this technique: while it is still possible to prepare highly crosslinked microgels by lowering the concentration of the monomer mixture [66], the C_m may be quite low (<5 mol%) and the resulting particles may be very polydisperse.

In spite of these limitations, radical solution polymerisation was extensively employed as a method for MIP synthesis by the group of Okhubo, who developed soluble MIPs as enzyme-analogue catalysts for the hydrolysis of esters [69–72]. However, MIPs prepared by this group were characterised by a surprisingly low crosslinking degree (9–35 mol%). Furthermore, no characterisation of the resulting soluble imprinted polymers was ever reported, and no comparison to nonimprinted analogues was ever made in order to critically assess the imprinting effect.

The first study proving the possibility to obtain an imprinting effect in microgels was reported by the group of Wulff in 2001 [73]. In this investigation, highly crosslinked microgels (70–90 mol%) based on methacrylate monomers and cross-linkers were imprinted by copolymerisation of the template monomer phenyl-2,3;4,6-bis-O-(4-vinylphenylboronyl)-α-D-mannopyranoside **1**.

1   **2**

After removal of template α-D-mannopyranoside **2** by hydrolysis, imprinted cavities containing two boronic acid units remained in the microgels. The resulting imprinted microgels were able to preferentially rebind the template in the presence of its enantiomer. The selectivity was significantly lower than with conventional MIPs, but nevertheless this was the first proof of concept for the applicability of the molecular imprinting approach to microgels. More recently, the group of Wulff published another paper dealing with imprinted microgels as artificial enzymes for the hydrolysis of carbonate diesters [74]. Highly crosslinked microgels (80–90 mol%) based on methacrylate monomers and crosslinkers were imprinted by copolymerisation of the template-monomer assembly which formed in situ between N,N'-diethyl-4-vinylbenzamidine and diphenylphosphate as a stable transition-state analogue of carbonate hydrolysis. In this paper, a strategy was developed in order to improve the imprinting effect in microgels. In particular, stepwise polymerisation at different monomer concentrations (starting from a more concentrated solution and diluting it before the gel point) as well as at different temperatures with successive initiator additions proved to be beneficial for the imprinting effect. The resulting microgel (10–20 nm size as determined by STEM) displayed a catalytic activity for carbonate hydrolysis about 300 times higher in comparison to solution and 20 times higher in comparison to a nonimprinted control.

Carbonate hydrolysis by molecularly imprinted microgels was also investigated by the group of Resmini [75, 76]. They prepared highly crosslinked microgels (70–90 mol%) based on N,N'-ethylene bisacrylamide as the crosslinker and acrylamide as the nonfunctional comonomer. N-Acryloyl arginine and N-acryloyl tyrosine were used as functional comonomers able to interact through hydrogen bonds with the template 4-nitro-4'-acetamido-diphenylphosphate. The resulting microgel displayed a catalytic activity for carbonate hydrolysis about one order of magnitude higher than solution.

The group of Zhao and Li reported on the preparation of imprinted microgels able to work as catalysts for the oxidation of homovanillic acid with hydrogen peroxide [77]. They prepared highly crosslinked microgels (85 mol%) based on EDMA as the crosslinker, acrylamide and 4-vinylpyridine as additional monomers and hemin as the catalytically active monomer. The substrate homovanillic acid was also used as the template. The resulting imprinted microgels were rather highly aggregated and therefore quite large (around 200 nm). Their catalytic activity under the best reported conditions was found to be three times higher than that of a nonimprinted control. Remarkably, even better results in terms of specific activity in pure water were more recently reported by the same group using lightly cross-linked, temperature- and pH-sensitive imprinted nanoparticles with size in the hundreds of nm, prepared by precipitation polymerisation using a protocol similar to that employed by Shea's group (see Sect. 2.1.2) [78].

Very recently the group of Resmini described the development of imprinted microgels mimicking a class I aldolase in the catalysis of the aldol condensation between 4-nitrobenzaldehyde and acetone [79]. They prepared highly crosslinked microgels (80 mol%) based on N,N'-methylenebisacrylamide as the crosslinker and containing as catalytically active monomer a proline derivative (Fig. 5).

Such monomer interacts with a suitable β-diketone forming an isolable enamide which effectively mimicks the transition state of the reaction. Copolymerisation of this adduct with the crosslinker and acrylamide followed by template removal yields small microgels (around 20 nm) that are able to catalyse the aldol condensation reaction about 19 times faster than nonimprinted controls.

Finally, very recently the group of Haupt employed solution polymerisation in a remarkable strategy for the synthesis of MIPs as enzyme inhibitors [80]. In their work, a benzamidine-containing monomer was employed, which functionality is known to be strongly bound by the active site of the enzyme trypsin. The enzyme-monomer complex was subsequently copolymerised with other monomers in water solution, thereby leading to molecularly imprinted microgels complementary to the active site of the enzyme. After separation of the microgels from the enzyme by electromigration, the polymers were indeed found to efficiently inhibit enzyme action, with an inhibition constant almost three orders of magnitude higher than that of the parent benzamidine monomer.

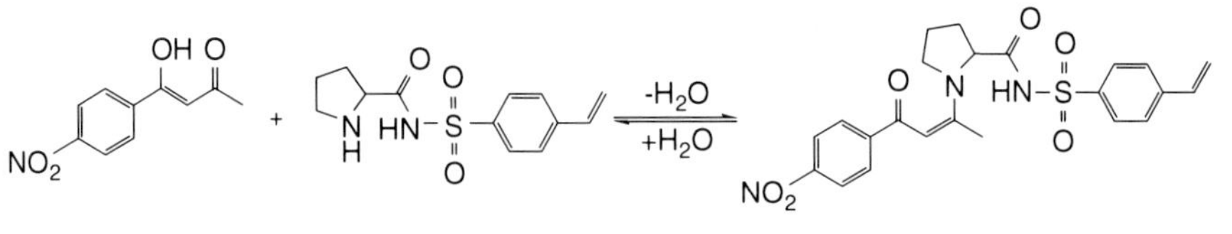

Fig. 5 Functional monomer and template-monomer assembly employed by the group of Resmini. Adapted from [79]

2.2 MIP Beads from Preformed Seeds

Having to prepare MIPs in beaded form, a possible alternative to their direct synthesis is to use preformed beaded "seeds" as scaffolds for the MIPs to be synthesised.

In this connection, different strategies are in principle possible. The first involves use of porous seeds that can be swollen or (in the case of a rigid porous seed) imbibed with the monomer mixture yielding the MIP. Subsequent polymerisation within the seeds yields a composite in beaded form containing the MIP component. The seed can be subsequently removed prior to MIP use or left in the composite.

The second strategy involves generation of a thin MIP layer on the surface of the seed. In this case, the MIP must be formed by polymerisation at the seed surface, ideally yielding particles covered with a MIP layer of controllable thickness. The biggest difficulty with this strategy is obviously in the achievement of selective MIP formation at the seed surface in the polymerisation process. Numerous elegant solutions for overcoming this problem have been proposed in the literature in the last 15 years. Indeed, methodologies have been implemented for forming thin MIP layers on widely different solid supports, ranging from beads to surfaces such as sensor interfaces, in a peculiar approach to MIPs known nowadays as "surface imprinting". As it will be apparent in the following, all these strategies have been extensively developed in order to cope with the requirements of the imprinting process.

2.2.1 Single- or Multistep Swelling and Polymerisation

This method involves a combination of stepwise swelling of preformed seed particles followed by polymerisation (Fig. 6).

Aqueous dispersions of organic polymer seeds, usually made out of polystyrene (PS) latexes of ca. 1 μm diameter prepared by emulsifier-free emulsion polymerisation [81] are generally employed for this purpose. The crucial step is the "activation" [82] of the preformed seed particles by addition of a suitable solvent during the first swelling step. The activation solvent, a water-insoluble, low molecular weight organic compound with plasticising properties such as dibutyl phthalate, increases the swelling capacity of the seeds up to more than 100 times when introduced before monomers are used to swell the seeds.

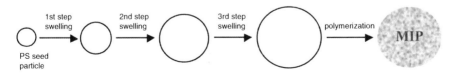

Fig. 6 Schematic description of multi-step swelling and polymerisation

Table 1 Two-step and multi-step swelling technique

	Two-step swelling	Multi-step swelling
1st step	• Water dispersion of PS particles Add microemulsion containing: • Activating solvent (dibutylphthalate) • Initiator • Surfactant (sodium dodecyl sulphate) • Distilled water Stir at room temperature, until disappearance of emulsion droplets	• Water dispersion of PS particles Add microemulsion containing: • Activating solvent (dibutyl phthalate) • Surfactant (sodium dodecyl sulphate) • Distilled water Stir at room temperature, until disappearance of emulsion droplets
2nd step	Add aqueous dispersion containing: • Cross-linker (usually EDMA) • Functional monomer • Porogenic solvent • Stabiliser (polyvinyl alcohol) • Template Stir at room temperature, 2–12 h	Add aqueous dispersion containing: • Initiator • Porogenic solvent • Stabiliser (polyvinyl alcohol) • Template Stir at room temperature, 2 h
3rd step		Add aqueous dispersion containing: • Cross-linker (usually EDMA) • Functional monomer • Stabiliser (polyvinyl alcohol) Stir at room temperature, 2 h

After the activation, PS seed particles are further swollen by contacting them with an aqueous dispersion containing monomers, template, initiator and optionally a porogenic solvent, in one or more steps (Table 1), and subsequently polymerised. Particle size, porosity and monodispersity can be controlled by the amount of the activating solvent, porogen, dispersion stabiliser and cross-linker/water ratio; if necessary, a desired particle size can be reached also by repeating several times the swelling and polymerisation steps. In this way monodisperse polymer particles with sizes larger than those obtainable via direct dispersion/precipitation polymerisation and in the range suitable for direct application as stationary phases can be prepared.

The first published MIP prepared by stepwise swelling and polymerisation technique was reported by Hosoya et al. [83]. In this work, uniformly sized macroporous polymer-based stationary phases for HPLC were prepared via a typical two-step swelling and polymerisation method in an aqueous media using EDMA as cross-linker, acrylic acid as a functional monomer and 1,5-diaminonaphthalene or 1,8-diaminonaphthalene as a template molecule. After polymerisation, the linear polystyrene seed was removed by extensive washings, leaving MIP beads with a highly porous structure. Even if this preparation method was not optimised, chemical yields were nearly quantitative, size monodispersity was quite good and separation factors between the two diaminonaphthalenes used as templates were quite similar to those previously reported for rod-type columns prepared in a non-aqueous medium [84].

In the following years, many imprinted polymers were prepared by two- or multi-step swelling and polymerisation method using mostly EDMA as a cross-linker and various functional monomers interacting with the chosen template through noncovalent interactions. The employed templates included inter alia β-blockers (S)-naproxen [85] and (S)-ibuprofen [86], (S)-nilvadipine [87], green tea catechins [88], D-chloro- and D-bromopheniramin [89], triazine herbicides [90], and clenbuterol [91].

Imprinted polymers prepared by stepwise swelling and polymerisation have been mostly prepared by thermal or photoinitiated polymerization using conventional azo or peroxidic initiators; however, redox initiator systems like benzoylperoxide/N,N'-dimethylaniline [92, 93] have been also employed. In this case, redox polymerization can be performed at low temperature (0°C instead of 50–80°C commonly employed with the other initiator systems) by adding after completion of the swelling steps an emulsion prepared from the second component of the initiator system (N,N'-dimethylaniline), a part of the cross-linker (EDMA), sodium dodecylsulfate and distilled water. Haginaka et al. in 1998 evaluated the effect of the type of initiation on MIPs imprinted with (S)-naproxen [93]. In this work, EDMA was used as a cross-linker, 4-vinylpyridine as a functional monomer and toluene as a porogenic solvent. The resulting MIP beads were packed into a stainless-steel column and their chromatographic properties were evaluated in aqueous-rich mobile phases. Acidic ((S)-naproxen), neutral (benzoin) and basic (propranolol) analytes were all retained less on the material prepared by thermal initiation, and enantioseparation factors were also lower. On the other hand, beads prepared by thermal initiation provided higher resolution and also a higher number of theoretical plates.

The stepwise swelling and polymerisation technique for the synthesis of MIP beads can be easily adapted to the preparation of restricted access media (RAM) for chromatographic applications. RAMs are used for example in the enrichment and pretreatment of proteinaceous analytes in HPLC. In principle RAMs enable small molecules (a drug, its metabolites) to reach hydrophobic, ion exchange or affinity sites within the RAM while large molecules (e.g. proteins) are excluded from such sites and elute with the void volume without decomposition. Haginaka et al. prepared RAM-MIPs, MIPs imprinted with (S)-naproxen [94, 95], (S)-ibuprofen [95] and racemic propranolol [96] and selectively covered with a hydrophilic external layer. In all these cases after the third swelling step the polymerization was started at 50°C. After 4 h, a mixture of glycerol monomethacrylate (GMMA) and glycerol dimethacrylate (GDMA) (GMMA:GDMA 1:1) containing potassium peroxodisulfate were added. The temperature was increased to 70°C and the polymerisation was allowed to proceed for 20 h (Fig. 7).

Following the hydrophilic surface modification of the MIP beads, the retention factors of the various employed analytes were decreased, whereas similar or (in the case of (S)-naproxen) even higher enantioselectivities were observed. Furthermore, the recovery of bovine serum albumin was complete on modified MIPs while it was only 10–40% on unmodified MIPs. Although the column efficiency of modified beads was lower, possibly due to mass transfer limitation by hydrophilic external layers in an aqueous-rich eluent, these preliminary results demonstrated the applicability of RAM-MIPs in direct serum injection assays [95].

Since the multi-step swelling and polymerisation method is time consuming, in 2005 Chen et al. [97] prepared uniformly sized MIP beads for separation of

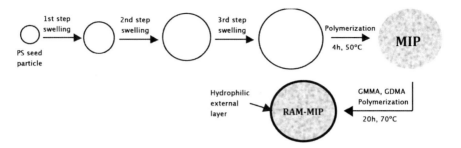

Fig. 7 Schematic representation of the preparation of RAM-MIP

sulfamethazine (SMZ) by one-step swelling and polymerisation. In this case, polystyrene seed particles prepared by dispersion polymerization with diameter of 3 μm were used as a shape template. In the sole swelling step, a water dispersion of PS seed particles was admixed with a dispersion stabiliser (PVA) and surfactant (SDS) and then an emulsion prepared from a functional monomer (MAA), cross-linker (EDMA), template (SMZ), dispersion stabiliser (PVA), surfactant (SDS), initiator (AIBN), porogenic solvents (acetonitrile, chloroform) and distilled water was added dropwise over 2 h with slow stirring at room temperature. After complete absorption of the emulsified mixture by PS seed particles, polymerisation yielded particles of 6,2 μm in diameter which were successfully used as a stationary phase for SMZ in organic and especially in aqueous mobile phases, where SMZ-imprinted conventional MIPs perform with poor efficiency [98, 99]. A similar strategy was more recently employed by the same group for the preparation of MIP beads imprinted with metsulfuron-methyl (MSM), one of the sulfonylureas herbicides [100]. The resulting beads were successfully integrated into a SPE device for MSM enriching in real drinking water samples.

2.2.2 Beaded Porous Silica Supports

A porous, rigid support in beaded form (generally an inorganic support such as silica) can be in principle utilised as scaffold for the preparation of MIPs like the swellable, organic polymer seeds reviewed in the preceding section. Beaded silica with a small bead size (starting from about 5–10 μm) and narrow size distribution is widely available commercially due to its extensive use as stationary phase in HPLC. Furthermore, the preparation of MIPs within such beads can be very straightforward: in its simplest form, it is sufficient to imbibe the support with the monomer mixture containing the template, the initiator and optionally some solvent, paying only attention to remain below the point of "incipient wetness" of the support (i.e. to fill the support pores with the monomer mixture without leaving a significant quantity of it outside). Subsequent polymerisation yields composite particles as a free-flowing powder, which can be directly further processed. A better adhesion of the MIP to the silica support can be obtained through surface derivatisation of the support with polymerisable groups prior to use, e.g. by reaction with methacryloxypropyltrimethoxysilane.

Given the simplicity of this approach, it is not surprising that first examples of its application date back to the mid-eighties [101–105]. However, such early examples had almost no follow-ups, possibly due to some disadvantages that were soon recognised in using these composite materials, namely their comparatively low capacity and bad accessibility. At the turn of the century, however a few novel contributions revitalised the research in this area. First of all, it was shown that the silica scaffold could be removed by e.g. etching with HF without disrupting the structure of the bead; furthermore, it was pointed out that linking the template molecule to the pore surface of the scaffold prior to polymerisation ensures that, after removal of the scaffold, the imprinted sites are located at the surface of the macropores within the MIP beads (Fig. 8).

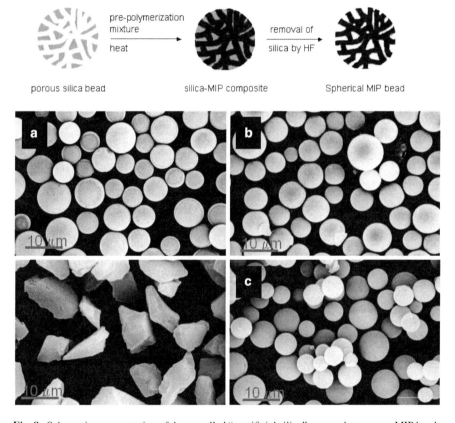

Fig. 8 Schematic representation of the so-called "sacrificial silica" approach to porous MIP beads and examples of SEM micrographs of the resulting beads: (**a**) silica beads; (**b**) composite beads (silica + MIP); (**c**) MIP beads after silica etching; (**d**) ground bulk MIP employed as control. Reproduced with permission from [107]

This methodology, sometimes termed the "sacrificial silica" approach, was proposed for the first time by the group of Mosbach in 2000 using theophylline as a model template [106, 107] as well as by the group of Sellergren [108–111], who investigated it more thoroughly varying several details of the experimental procedure as well as using different template monomers, in particular short peptide sequences which generated imprinted cavities useful for binding larger polypeptides including the same sequence (epitope imprinting).

An alternative way of using silica beads as supports for MIPs is to suspend silica beads bearing polymerisable groups grafted at their surface in a solution containing the monomers making out the MIP as well as the template; subsequent polymerisation should produce a MIP layer grafted to the surface of the macropores of the support, without filling them completely, as in the previous case. This approach, first proposed by Mosbach in the mid-nineties and subsequently employed especially by Japanese researchers [112–115], should produce composite particles with a better accessibility of the imprinted sites, but its main limitation is clearly the difficulty in controlling the polymerisation process: more often than not, the resulting composite beads coagulate to the point that grinding and sieving is required in order to roughly part them. Ideally, polymerisation should take place at the silica surface exclusively, without coagulation nor independent formation of polymer particles. A general solution to this problem is to favour this by generating the radicals giving rise to the polymerisation process at the silica surface. The simplest way of attempting to achieve this goal is to premix a suspension of the silica beads bearing grafted polymerisable groups with a radical initiator and add the monomers only in a subsequent step [116]. However, a much more effective way of ensuring radical generation at the silica surface exclusively is to heterogenise a radical initiator, or an iniferter, on the surface itself. This approach was pioneered by the group of Sellergren [117–120] and has been extensively optimised using standard templates and monomer mixtures, though it still presents the drawback of having to prederivatize the beads with the initiator/iniferter before polymerisation. This technique has been termed "grafting from" as opposed to "grafting to" defining the preparation of MIP layers on silica particles by the more conventional polymerisation with the initiator in solution. Remarkably, use of a heterogenised iniferter makes it possible to conduct the polymerisation in a "living" fashion, which allows to grow successive layers of polymer of different nature on the same support particles [119]. More recently, the same group has also utilised surface-initiated RAFT (reversible addition-fragmentation chain transfer) polymerisation for the same purpose [120].

The synthetic strategies presented in this section appear to be quite successful in yielding well-defined, efficiently imprinted spherical beads in the size range of tens of micrometres. Nevertheless, up to now such methodologies has not yet set themself through as a general method of beaded MIP synthesis. Possibly the reason is that other approaches currently appear more promising and flexible, particularly in terms of the obtainable particle sizes.

2.2.3   Core-Shell Methodologies

The preparation of MIP beads with a core-shell structure, with the imprinted layer making out the particle shells, is perhaps the less straightforward among the synthetic strategies presented up to now. On the other hand, it is expected that the resulting beads exhibit the best performance in technological applications, since the resulting particles (1) are small but regular and mechanically stable due to the presence of the core; (2) have the imprinted cavities located exclusively in the outer shell of the particle, which facilitates their accessibility; (3) possess a core which can be tailored in size and properties independently from the imprinted shell, so that additional functionalities can be imparted to the resulting composite (e.g. magnetic, fluorescent and so on).

In principle, core-shell particles of this kind can be prepared from preformed seeds, such as amorphous silica nanoparticles, bearing polymerisable groups grafted at their surface. Such particles can be suspended in a solution containing the monomers making out the MIP as well as the template, subsequent polymerisation leading as in the case of porous silica beads to the formation of a MIP layer around the seed. However, in this case it is even more difficult to control the polymerisation process in order to avoid coagulation or independent formation of polymer particles. Nevertheless, with some peculiar templates and monomers this appears to be possible without special precautions. For example, Gao et al. reported recently on the preparation of trinitrotoluene (TNT)-imprinted core-shell nanoparticles made from silica nanoparticle cores [121]. In this case, the capability of the template to strongly interact via charge-transfer complex formation with both the silica-bound alkylacrylamide groups and the functional monomer acrylamide, thereby helping in keeping it close to the surface, makes it apparently possible to obtain exclusive polymerisation at the silica nanoparticles surface (Fig. 9).

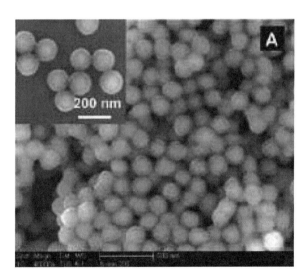

Fig. 9 SEM micrograph of core-shell MIPs obtained by Gao et al. through MIP grafting onto silica nanoparticles. Reproduced with permission from [121]

The thickness of the imprinted polymer shell can be also tuned in the range 10–40 nm by changing the relative amounts of functionalised silica nanoparticles and polymer shell precursors. The resulting core-shell particles exhibit enhanced capacity of rebinding the TNT template over 2,4-dinitrotoluene in comparison to particles prepared by precipitation polymerisation. Nevertheless, this strategy, although leading to impressive results, cannot be easily applied to other templates and monomers.

Formation of core-shell particles can be favoured, as in the case of porous silica seed, by radical generation at the seed surface. Perez-Moral and Mayes [122] explored this approach by using seeds made out of an organic lightly crosslinked polymer and functionalised with a dithiocarbamate iniferter. In this case, it cannot be excluded that the lightly crosslinked core becomes at least partially swollen by the monomer solution, thereupon causing a certain degree of interpenetration between core and shell polymer. Nevertheless, the approach was successful, yielding particles about 90 nm in size which were capable of specific rebinding of standard templates such as morphine, propranolol and naproxen. Furthermore, the living nature of the employed polymerisation system allowed to grow several superimposed shells of different nature on the same core particle.

The most effective way to obtain polymerisation at the seed surface exclusively is arguably to compartimentalise the seeds as well as the monomer mixture, confining each seed together with some monomer mixture in a small microenvironment prior to polymerisation. Use of (mini)emulsions appears to be currently the best method for achieving this goal. It must be borne in mind however that the shortcomings of this thecnique when applied to the synthesis of simple MIP beads (Sect. 2.1.3) still hold when the preparation of core-shell particles is considered, particularly for what it concerns the detrimental presence of water.

The first examples of molecularly imprinted core-shell particles prepared by emulsion polymerisation were reported by the group of Whitcombe [123, 124]. They employed as seeds organic polymer latexes prepared by emulsion polymerisation in water, as well as in an aqueous ferrofluid (Fe_3O_4 nanoparticles dispersed in water) [123]. After addition of surfactant and of a monomer mixture containing as template a cholesterol molecule linked via a cleavable carbonate ester bond to a polymerisable group, the emulsion was polymerised resulting in the formation of imprinted core-shell particles (50–100 nm). Removal of the cholesterol template by carbonate ester hydrolysis left behind a phenol group in the imprinted cavities able to engage in hydrogen bonds with the hydroxyl group of the template, which imparted to the cavities the capacity to rebind cholesterol selectively. In another investigation by the same group [124], the cholesterol template was functionalised through the same carbonate ester linker with a surfactant chain. The resulting molecule was incorporated together with other polymerisable surfactant molecules (the surfactant chain being connected to the polymerisable group by the same carbonate linker) into the stabilising shell of the emulsion, with the hydrophobic cholesterol moiety placed on the inner side, in contact with the monomer mixture surrounding the seed (Fig. 10). Subsequent polymerisation resulted in the preparation of core-shell particles (ca. 60–290 nm in size, depending on the composition of

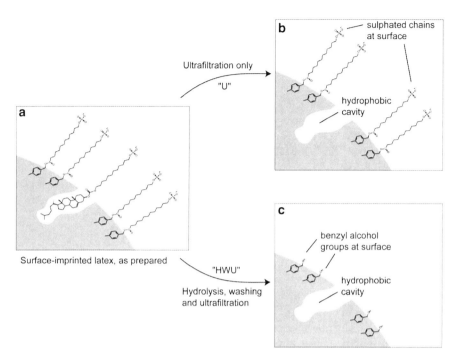

Fig. 10 Schematic description of the strategy followed by Whitcombe et al. for the preparation of core-shell MIP nanoparticles with imprinted sites located on the particle surface. Reproduced with permission from [124]

the stabilising surfactant layer) with the imprinted sites located at the outer surface of the shell. Removal of the template as well as of the surfactant chains by carbonate hydrolysis left hydrophobic cavities embedded in a hydrophilic particle surface. The imprinting effect is in this case solely based on hydrophobic interactions between the template and the cavities, but nevertheless it was demonstratedly present.

Core-shell imprinted particles from emulsion polymerisation in water using organic polymer latexes as seeds were also prepared by Carter and Rimmer [125–127]. In the buildup of the imprinted polymer shell, they used an approach borrowed from earlier studies by Japanese groups on ionic imprinting [58]. In particular, they used as functional "monomer" oleyl phenyl hydrogen phosphate, which bears no polymerisable group (the internal double bound of the oleyl chain is known to copolymerise poorly by radical pathway) but is expected to become immobilised in the imprinted polymer shell by absorption of the hydrophobic tail as well as possibly by covalent binding to the growing polymer chains following chain transfer reactions. As standard template molecules theophylline, caffeine, propranolol, atenolol as well as a few tripeptides were employed. The caffeine-imprinted MIP was found to bind preferentially caffeine over theophylline.

However, the theophylline-imprinted polymer beads also bound more caffeine than theophylline, pointing out the fact that nonspecific hydrophobic interactions may play a role in the absorption phenomenon.

A similar approach using a polymerisable functional monomer (*N*-acryloyl-L-phenylalanine) was employed by the group of Jeong for the synthesis of core-shell MIP beads imprinted with glucose [128]. Interestingly, they included in the shell composition an oligo-*N*-isopropylacrylamide macromer which imparted thermo-sensitive properties to the resulting particles. The particles exhibited enhanced glucose rebinding in comparison to nonimprinted controls, and the rebinding extent was comparable to that of a conventional MIP of similar composition. However, the extent of rebinding apparently exceeded by far the amount of functional monomer present, which incidentally was employed in very low amount (0.06 equivalents) with respect to the template. Therefore, nonspecific binding and/or specific binding through interactions with the macromer should contribute to the overall performance of the MIP.

Additional examples of the synthesis of core-shell MIP beads by aqueous emulsion polymerisation were provided by Perez-Moral and Mayes [129], who investigated beads imprinted with propranolol prepared from organic polymer seeds in the presence or in the absence of toluene as diluent for the monomer mixture. The presence of toluene was found to enhance specific binding of the template, although the degree of rebinding was lower than that observed for imprinting of propranolol with the same monomers and very similar monomer ratios using other established polymerisation methods (see Sect. 2.4).

Very recently several research group started to take advantage of the stepwise nature of the synthesis of core-shell MIPs by building additional functionality into the resulting nanoparticles. For example, Chinese researchers employed Fe_3O_4 nanospheres as cores by first directly covalently binding the template bovine haemoglobin to them via an imine bond and subsequently copolymering a shell of γ-aminopropyltrimethoxysilane and tetraethyl orthosilicate at the nanoparticle surface: the resulting core-shell nanoparticles (100–150 nm size) exhibited magnetic properties as well as enhanced binding and selectivity towards the template protein [130]. A related approach was used by the group of Li [131], who started from preformed silica nanoparticles containing CdSe quantum dots. In this case, the silica shell was formed by inverse microemulsion polymerisation of γ-aminopropyltriethoxysilane and tetraethyl orthosilicate solutions containing lambda-cyhalothrin as the template molecule. The resulting nanocomposite particles (90 nm size) exhibited a remarkable quenching of the fluorescence of the embedded quantum dots when the template molecule was rebound compared to other related pyrethroids. Finally, the group of Haupt decorated EDMA/MAA cores with Au colloids before synthetising a thin MIP shell onto them by seeded emulsion polymerisation using (*S*)-propranolol as the template [132]. This construction results in signal enhancement when the composite particles are employed as optical nanosensors due to the possibility to exploit to this scope the Surface Enhanced Raman Scattering (SERS) by the embedded Au nanoparticles.

2.3   MIP Beads from Preformed (Co)Polymers

The preparation of MIP beads from preformed (co)polymers, as opposed to preparation from monomer mixtures, has significant precedent in the literature. Indeed, this method was already described in 1971 by Takagishi and Klotz, who crosslinked linear polyethyleneimine in dilute solution in the presence of methyl orange obtaining soluble polymer nanoparticles (microgels) which exhibited enhanced rebinding for the methyl orange molecule [133]. Subsequently though, this methodology has been only occasionally adopted, the main reason being presumably the difficulty in generating precursor polymers that on the one hand have the necessary variety of functional groups to interact with different potential templates, on the other hand are easily and efficiently crosslinked to yield the final MIP matrix. These limitations also help explaining why for many years polyethylene imine was nearly exclusively employed as starting polymer. Indeed, other examples of the strategy first described by Takagishi and Klotz were afterwards reported by the groups of Williams [134] and Okhubo [135]. The latter employed also poly (N-vinylimidazole) as an alternative starting polymer [136] and prepared in this way MIP microgels as enzyme-analogue catalysts for the hydrolysis of esters (see also Sect. 2.1.4).

In the nineties, the group of Maeda and Takagi reported on an approach for the development of ion-imprinted polymer particles starting from a preformed copolymer of styrene, butyl acrylate and MAA prepared by emulsifier-free emulsion polymerisation [137, 138]. The resulting polymer latexes were swollen with DVB and contacted with an aqueous solution of a metal ion template, which allegedly interacted with the carboxylic acid moieties present within the latex orienting them at the latex surface. Subsequent polymerisation of DVB "froze" the orientation of the carboxylic acid groups, yielding submicron-sized particles which showed some preference in binding the templating metal ion.

More recently, alternative strategies for the preparation of MIP beads from preformed (co)polymers have been proposed. One possibility is to use the so-called "phase inversion" approach, which is generally employed for the preparation of MIP membranes (see also Sect. 4.1.2), for the preparation of MIP beads. In this methodology, a solution containing the polymer and the template is added dropwise to a large amount of a solvent which is compatible with the solvent of the first solution but not with the polymer dissolved therein. In this way, the solution droplets instantaneously coagulate to form beads, from which the template molecule can be subsequently liberated thus forming the imprinted cavities. Using this methodology, polyethersulfone beads imprinted with Bisphenol A have been successfully prepared [139]. Remarkably, in this approach no chemical crosslinking of the polymer chains is necessary, which allows to overcome at least one of the two above mentioned limitations to the use of preformed polymers for the preparation of MIP beads. On the other hand, beads prepared by this technique were very large (in the mm range) and therefore unsuitable for many applications. Much better results using the phase inversion methodology have been recently obtained by the group of

Lin [140, 141]. They employed commercial *poly*-(ethylene-co-vinyl alcohol) and standard proteins as templates, which resulted in much smaller particles sizes (100–200 nm). The authors also introduced quantum dots in the solution, which yielded nanocomposite beads where the binding event could be monitored by measuring the quenching in the quantum dot fluorescence.

A more sophisticated approach makes use of diblock copolymers which self-assemble to core-corona micelles in block-selective solvents [142]. If the block making out the core bears functional groups able to bind template molecules as well as polymerisable groups, it is possible to introduce the template and subsequently to crosslink the core thereby producing imprinted cavities in the core. Such an approach has been demonstrated using nucleobases as templates. Crosslinked micelles with a 50 nm diameter could be prepared, which exhibited a degree of template rebinding much higher than nonimprinted controls and also slightly higher than that of a corresponding traditional MIP control. The main limitation of this strategy is clearly the synthetic effort required to prepare the functionalised diblock copolymer.

Preformed polymers can also be employed to prepare imprinted core-shell particles [143]. The group of Chang recently prepared a poly(amic acid) bearing oestrone as a template molecule covalently bound to the polymer through a urethane linker (see Fig. 2). A layer of this polymer was subsequently deposited on silica particles (10 μm diameter) prefunctionalised with amino groups at their surface. Thermal imidisation of the polymer yielded finally a polyimide shell (thickness about 100 nm) on the silica particles. Subsequent template removal yielded the imprinted cavities, which exhibited selective rebinding of oestrone in HPLC experiments.

2.4 Comparison of the Various Techniques

As it was shown in the previous sections, there is at present a wide variety of available methods for the preparation of molecularly imprinted micro-and nano-particles. Little has been done however to compare directly two or more of these various synthetic strategies in order to better ascertain their advantages and limitations, and in particular their capability to yield efficient MIPs.

The first study of this kind was carried out by Chinese researchers in 2003 [144]. They prepared MIP beads for the SPE of tyrosine by simple suspension in water as well as by two-step swelling and suspension polymerisation. They found no substantial difference in the rebinding capacity of the beads prepared by the two methods. A more thorough analysis of various synthetic approaches to MIP beads was conducted a year later by Perez-Moral and Mayes [145]. They took a standard monomer mixture with propranolol as the template molecule and polymerised it by bulk polymerisation, suspension polymerisation, precipitation polymerisation, two-step-swelling polymerisation and emulsion core-shell poly-merisation (see also Sect. 2.2.3). Care was taken to keep the polymerisation conditions as similar as possible. The imprinting efficiency was evaluated by radioligand binding assay in two different solvents, namely toluene and water. Unfortunately, no clear-cut conclusion could be drawn about the polymer beads with the highest efficiency of imprinted sites, since the behaviour of the various polymers turned out to be completely different in the two solvents, the MIP prepared by precipitation polymerisation possessing e.g. the highest efficiency in toluene and the worst in water.

In the same year, the group of Mayes published another contribution [146], in which they compared beaded MIPs prepared by suspension polymerisation in perfluorocarbon with beads obtained by MIP grafting onto a porous silica support and with a conventional ground MIP. The beads were imprinted with propranolol and their performance was tested in HPLC separations of many different β-blockers as well as in Turbulent Flow Chromatography (TFC) applications. The conventional MIP turned out to be still the best performer averaging over all the analytes tested, although the silica-grafted MIP provided improved peak shape and complete enantiomeric resolution of (R,S)-propranolol in significantly shorter times. In turn, MIP beads prepared by suspension polymerisation were the fastest, easiest and most efficient to synthetise, and they also provided an improvement in peak shape over the ground MIP as well as some degree of enantioseparation of all the β-blockers tested.

The group of Mizaikoff published in 2006 a study on MIPs for β-estradiol in which they indirectly compared MIP beads prepared by them via precipitation polymerisation with beads prepared by two other groups via multistep swelling and polymerisation [147]. The performance of the beads prepared by precipitation polymerisation was claimed to be superior in terms of higher separation and selectivity factors against β-estradiol and oestrone.

In a contribution published in 2005 the group of Sellergren compared the performance of composite MIPs imprinted with propazine and prepared from the same porous silica support using the pore-filling technique (with and without final removal of the silica) as well as the layer grafting approach with an immobilised iniferter [148]. The MIP composite prepared by layer grafting was found to exhibit the best performance when employed as stationary phase in HPLC, although extensive optimisation of the imprinted layer thickness was required.

Finally, in 2007 a joint publication by several research groups working on molecular imprinting described the preparation and direct comparison of different stationary phases for HPLC imprinted with the local anaesthetic bupivacaine [149]. The screened phases included beaded MIPs prepared by precipitation polymerisation and by surface grafting, as well as a conventional ground MIP and a capillary monolith (see the subsequent Section). In this case, no imprinting effect was obtained in particles prepared by precipitation polymerisation. Once more, the conventional MIP gave the best response in terms of imprinting efficiency, whereas use of the other MIP forms resulted in faster analysis due to reduced nonspecific binding, at the price of a reduced capacity of the stationary phase.

3 Molecularly Imprinted Monoliths

The first monolithic materials initially emerged in the 1960s, but it is during the last 20 years that monoliths have been intensively developed in a variety of fields and particularly in analytical chemistry for separation techniques. Nowadays, these macroporous materials are widely used and have found numerous applications in different chromatographic modes such as liquid chromatography (LC) or CEC, as indicated by several reviews [150, 151]. Less commonly, monolithic materials can also be applied, for example, to solid-phase extraction, combinatorial synthesis and for enzyme immobilisation.

The success of monoliths as chromatographic stationary phases stems from the numerous advantages of these materials compared to traditional packed columns in terms of chromatographic performance. Indeed, a monolith can be defined as a single continuous bed with a hierarchical meso- and macroporous structure that does not contain the interparticular voids of packed columns. These structural features make it possible to achieve a high and controlled porosity and therefore a better permeability of the monolithic material, which in turn improves linear flow velocity and provides faster separation with lower backpressure. In addition, monoliths have a lower resistance to mass transfer providing a higher separation efficiency even at high flow rates. Consequently, the hydrodynamic flow and mass transport in monolithic materials differ considerably from those of packed columns. Moreover, monoliths are versatile supports since they can be prepared by polymerisation from diverse inorganic and organic precursors such as silica gel, acrylic gel or polystyrene-DVB gel, and tuned with different functional groups for specific applications.

Due to their specific molecular recognition properties, MIPs have found their main application in analytical chemistry. As outlined in the introduction, the common preparation method of MIPs as bulk polymers, which are subsequently crushed, ground and sieved to obtain particles, is not well adapted to achieve a high separation performance. Thus, the preparation of monolithic MIPs seemed particularly attractive for separation science due to the permeability properties, the easy in situ preparation and the absence of retaining frits. On the other hand, the use of the monolith format is still limited and the strategy of MIP monolith preparation has been little developed in recent years.

3.1 Strategies for the Preparation of Molecularly Imprinted Monolithic Phases

This section is devoted to the strategies for the syntheses of monolithic MIP and to the influence of the pre-polymerisation mixture and preparation protocol on the monolith properties. Monolithic MIPs can be prepared by in situ polymerisation of functional monomers and crosslinkers in the presence of the template. This most commonly employed method has been extensively studied. Today, monolithic MIP can also be obtained by post-grafting of MIPs on monolithic matrices or by non-hydrolytic sol-gel processes for silica-based MIP monoliths.

3.1.1 In Situ Polymerisation

The first reported technique for the preparation of monolithic MIPs used an in situ polymerisation protocol, which is a simple and easy procedure to perform.

The in situ polymerisation consists of filling a capillary or a column with the pre-polymerisation mixture containing the template, the functional monomer, the cross-linker, the initiator and the porogenic solvent (Fig. 11). Then the column is heated or submitted to UV radiation for polymerisation. In the in situ thermally initiated polymerisation process, the tube with the pre-polymerisation mixture is submerged in a controlled-temperature water bath, whereas for in situ photoinitiated polymerisation, a UV-transparent capillary or column is needed. The resulting continuous rod of polymer is washed with an appropriate solvent to remove the template and the excess of monomer.

In 1993, the first preparation of continuous rods of macroporous imprinted polymer was reported by Matsui et al. for the chromatographic separation of stereoisomers and enantiomers [84]. In this case, monomer mixtures containing MAA and EDMA dissolved in cyclohexanol and dodecanol were used to imprint phenylalanine anilide and diaminonaphthalene by thermal in situ polymerisation. The resulting monolith showed higher capacity factors for the corresponding template isomer than for its antipode. Since then, the main developments on imprinted monoliths have been focused on optimising the conditions of polymerisation and chromatographic separation for improved molecular recognition and flow-though properties and on understanding the phenomena involved in molecular recognition with preformed monolithic MIPs.

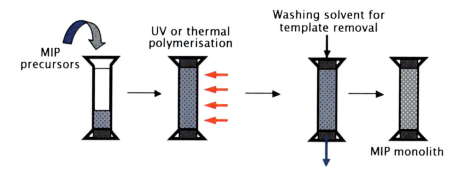

Fig. 11 In situ polymerisation procedure for MIP monolith preparation

Preparation of "Superporous" Monoliths

Although the preparation of monolithic MIPs is quite simple, the application of monolithic materials in separation technology requires the presence of large pores to provide a good flow through the phase for efficient passage of sample and solvent. In the case of MIPs, the major challenge is to find the appropriate conditions enabling the simultaneous generation of meso- and macropores in the system and of the binding site at molecular scale.

Schweitz et al. have reported two different processes for the preparation by in situ polymerisation of superporous imprinted monolith applicable in CEC [152–155]. In the first approach the macroporosity of the imprinted monolith was generated by prematurely stopping the polymerisation after a specific time by simply removing the unreacted monomers and the radical source [153, 156]. This technique could be used when the polymers were synthesised in solvents yielding dense polymers. Thus, monolith imprinting with propanolol was accomplished by UV in situ polymerisation at $-20\,°C$ using MAA as the functional monomer, TRIM as cross-linker and toluene as porogen in a silica capillary silanised with methacryloyl silane for polymer adhesion [152]. The SEM images of the resulting polymer are presented in Fig. 12. The morphology of the monolith consists of particle aggregates (1–20 μm wide) interconnected by large pores (7 μm wide), ensuring good flow-through properties. The main disadvantage of this technique is the lack of reproducibility because of the need for a precise control of the polymerisation kinetics, which was achieved with difficulty. In the second approach, which is now more frequently used, the macropores were created by adding a porogenic solvent for decreasing the polymer density [153, 154, 157–165]. The porogenic solvent should be a thermodynamically poor solvent for the polymer.

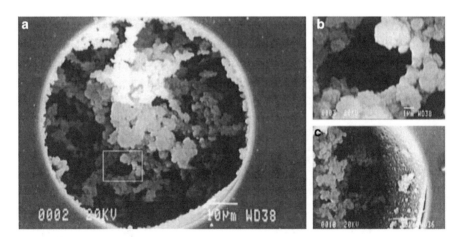

Fig. 12 Scanning electron micrographs of a polymer-filled capillary column. (**a**) Micrometre-sized globular units of macroporous MIP surrounded by 1–20 μm wide interconnected superpores. (**b**) A superpore, about 7 μm wide, magnified from the square present in (**a**). (**c**) The covalent attachments of the polymer to the capillary wall. Reproduced with permission from [152]

For the enantiomeric separation of propanolol, MIP monoliths have been rendered porous by the addition of isooctane in toluene at 2%. The poor solvent content is a crucial parameter for controlling the porosity of the MIP monolith, a higher concentration of poor solvent leading to a more porous but also more fragile material. Actually, a combination of these two techniques, where the selection of the poor solvent and the timing of polymerisation is optimised, can also be employed for the preparation of preformed imprinted monoliths [166, 167].

The preparation of MIP capillary columns by in situ polymerisation involves the initial activation of capillary walls by silanisation with MPS before filling the capillary with the prepolymerisation mixture according to the procedure described by Hjerten [168], which allows the covalent anchoring of the polymer, making the use of frits unnecessary. Moreover, the hydrolysis of MPS following the copolymerisation with 4-VP and EDMA in the preparation of (S)-ibuprofen-imprinted monoliths provided negative surface charges, which were responsible for the strong electroosmotic flux observed [169].

In order to accelerate the separation of molecules on a monolithic phase, its length can be restricted without necessarily causing a decrease in the performance of the monolithic phase. Indeed, the complete separation of R and S enantiomers of propanolol was later achieved using a shorter column (8.5 cm) in less than 30 s [155]. In addition, Huang et al. have observed that the column performance in the separation of enantiomers and diastereoisomers was unaffected by decreasing the length of a molecularly imprinted monolithic column while maintaining its volume at the same level, whereas the column length in chromatography usually affects the separation of analytes [170]. Shorter columns allow for the use of higher flow rates with lower backpressure. Thus, the diastereoisomers of cinchonidine could be separated 30 times faster by increasing the flow rate from 0.5 mL/min to 7 mL/min with good resolution.

Operating Conditions Affecting the Imprinted Monolith Performance

The performance of the molecularly imprinted monolith in terms of molecular recognition and flow-through properties depends on several factors, especially the density and the porosity of the polymer. In order to obtain a monolith with high selectivity and high permeability, some preparation conditions must be optimised, in particular the composition of the prepolymerisation mixture including the amount of template, the type and amount of functional monomer, crosslinker, porogenic solvent and the initiator, and the polymerisation conditions such as initiation process and polymerisation time.

The choice of the porogenic solvent is obviously one of the crucial parameters for controlling the porosity of the molecularly imprinted monolithic polymers. Therefore, the selected porogen should produce large pores to allow good flow-though properties of the resulting polymer, and at the same time have good solvency for the template, the monomer and the crosslinker. On the other hand, porogens that are too polar should be avoided since they negatively affect the stability of the template-monomer assembly and consequently the selectivity of the resulting MIP. In this sense, several porogen mixtures have been tested for the synthesis of MIP monoliths. The use of toluene alone or with a co-porogen including isooctane or dodecanol has been demonstrated to be advantageous for the generation of a good column permeability and high MIP selectivity [154, 157, 163–167]. However, these mixtures are not always suitable for the dissolution of template molecules, especially in the case of more polar compounds. To overcome this difficulty, an increase in the monomers-to-porogen ratio in the prepolymerisation mixture has been proposed [158]. Furthermore, the imprinting efficiency of the column was very sensitive to the amount of coporogen, which must be optimised. Thus, the ratio of isooctane to toluene could provide either a decrease [161] or an increase [166] of the imprinting factor according to the template. Indeed, several studies indicate that the morphology of the monolith is altered by varying the porogen composition or the concentration of the poor solvent. This modification of polymer density and porosity affecting both the flow-though properties and the position of recognition sites is responsible for the improvement of the separation performance of MIP monoliths compared to conventional MIPs [157].

The nature of the functional monomer can sometimes be of importance. For example, if the MIP monolith is to be used as stationary phase for electro-chromatography, the presence of charged groups may be essential for the generation of an electroosmotic flow. These can be simply generated by the use of MAA as the functional monomer [171], but in some cases a combination with other monomers was necessary to improve the imprinting effect [172, 173].

Relatively few investigations have been carried out on the nature of the cross-linking monomer, and EDMA is almost invariably employed. However, TRIM has been found to be a better crosslinker for the production of a monolith imprinted with ropivacaine [152].

Variation in monomer and crosslinker content in terms of type and amount not only leads to MIPs with different composition but also produces different porous structures [154]. For this reason, the effect of monomer/crosslinker ratio has also been well studied, since only a proper crosslinker content can afford high selectivity [157, 162, 164, 171]. An increase in the amount of the crosslinker results in an improved selectivity of the resulting MIP, as the increased rigidity of the polymer enhances the stability of the recognition sites. However, at higher concentration of either monomer or crosslinker, too dense polymer is produced, preventing good flow through properties [158]. For monolith preparation, the optimum crosslinking degrees have been found to lie between 80% and 93%. Although this parameter strongly affects the monolith characteristics, a factorial experimental design showed that its influence in monolith performance is less important compared with that of initiator and template concentration [167].

Besides the choice of monomers and solvent, the ratio of template molecule to functional monomer not only affects the imprinting effect [174] but also the morphology of MIP monoliths. Several authors have observed differences in the monolith structure (polymer morphology, pore size distribution, flow characteristics) between the non-imprinted control polymer and the MIP, derived from the presence of the template [158, 175]. For example, an MIP imprinted with ceramide III was compared to the corresponding non-imprinted polymer. The latter presented larger pores and a skeleton composed of clusters formed by microspheres, while this structural feature did not appear in the imprinted monolith [175]. This phenomenon has been attributed to the amphiphatic nature of the ceramide III template which could enhance the solubilisation of the co-polymer in the porogenic solvent and retard phase separation. Similarly, a 4-hydroxybenzoic acid-imprinted monolith had an extremely different morphology from the non-imprinted polymer, which was characterised by a smoother surface, presumably stemming from differences in polymerisation kinetics [158]. In addition, an alteration of the MIP morphology related to the amount of the template molecule has also been observed [153, 161]. These results suggest that the presence or not, and the type of template have an influence on the polymerisation process. In this respect, the multiple templates imprinting technique has been developed for the preparation of nilvadipine-imprinted monolith using cbz-L-tryptophan as co-template [176]. Adding a co-template affected the polymerisation process, allowing the formation of macro-through pores needed for the subsequently performed enantiose-paration of nilvadipine.

MIP monoliths can be prepared by either photopolymerisation or thermally initiated polymerisation. For the former, the effect of polymerisation time has been evaluated mainly with respect to the flow-through properties of the resulting monolith [153, 167]. When a longer time was used, the flushing of the imprinted monolithic column was impossible due to a higher density of the polymer, whereas a too low polymerisation time yielded a very low amount of polymer in the column, resulting in a poor separation efficiency of the monolith. Thermal initiation is by far the most widely used method for the preparation of monolithic columns due to a better control of polymerisation conditions and the possibility of avoiding UV-transparent columns. In the case of thermal polymerisation, the morphology and the selectivity of the imprinted monolith depend both on the time and temperature of the reaction [157, 164].

Recently, controlled/living radical polymerisation has been proposed for the synthesis of imprinted monolithic materials [177, 178]. RAFT has been employed in the preparation of clenbuterol and enrofloxacin imprinted monolith using diben-zyl trithiocarbonate (DBTTC) as chain transfer agent. Compared to conventional free radical polymerisation, a higher selectivity and a better column efficiency were obtained. The addition of the chain transfer agent in the polymerisation mixture strongly affected the structure of the monolithic polymer in terms of pore size, surface area, dimension and size distribution of the globular domains, with a more homogenous structure for the monolith prepared by RAFT [177].

3.1.2   Post-Grafting of MIPs on Monolithic Matrices

Recently, a method for preparing MIP monolithic columns for electrophoresis, chromatography and solid-phase extraction has been developed, which uses a preformed polymeric monolith, onto which an MIP with specific recognition sites is subsequently grafted [149, 179–181].

This technique starts with the preparation of a porous monolithic core, which can be either an organic or inorganic matrix. For example, in situ UV or thermally initiated polymerisation of a solution of 40% TRIM in a mixture of 2,2,4-trimethyl-pentane and toluene (70/30, w/w) as porogen [149, 179] yielded a porous organic monolith, onto which an MIP could then be grafted. Alternatively, an inorganic monolith was prepared by a sol-gel process starting with tetramethoxysilane (TMOS) or methyltrimethoxysilane (MTMS), using in some cases polyethylene glycol as porogenic compound. Subsequent chemical modification of the surface of the silica monolith by silanisation with 3-(trimethoxysilyl)propylmethacrylate allowed covalent attachment of the MIP layer. The composite silica-based monolith appeared to have higher chromatographic efficiency and longer lifetime than the organic-based monolith [182], an MIP monolith or an MIP-beads based column [181]. The core monolith was then filled with a prepolymerisation mixture composed of the template, functional monomer, crosslinker and initiator in a suitable solvent, and an MIP layer was grafted by copolymerisation onto the surface of the monolith. An image of the MIP layer obtained at the inner surface of the preformed monolith is shown in Fig. 13.

This material has been successfully applied for the enantioseparation of Tröger's base and tetrahydropalmatine, and for the separation of anaesthetic drugs, where the

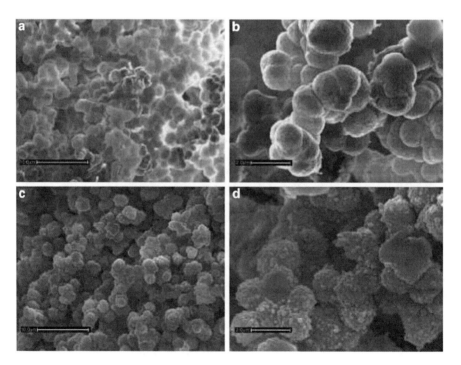

Fig. 13 Scanning electron microscopy images of non-grafted core monolith at (**a**) 3,000× and (**b**) 10,000× magnification, and of grafted BV-mMIP monolith at (**c**) 3,000× and (**d**) 10,000× magnification. Reproduced with permission from [179]

monolithic stationary phase yielded faster separation than a conventional column filled with a ground bulk polymer [149]. Recently, the selective separation of the proteins lysozyme and BSA has been performed on a hybrid monolithic column composed of an MIP coated onto the surface of a silica monolithic core [181]. The imprinted polymer coating was synthesised by copolymerisation of acrylamide as functional monomer and bisacrylamide as crosslinker in water. This hybrid monolithic column exhibited higher chromatographic resolution and better specificity and selectivity than the corresponding organic-based monolithic column.

Compared to the in situ polymerisation of a monolith, the grafting approach does not need re-optimisation of the protocol in order to obtain an appropriate porosity and flow properties for the monolith when monomer or template is changed. Moreover, the properties of the core materials are generally preserved and the imprint generated on the surface of the materials only requires a minimum amount of template and provides well-accessible recognition sites.

3.1.3 Silica-Based MIP Monolith

The use of silica as imprinting matrix has been studied for many years (see also Sect. 2.2.2). Silica matrices have interesting mechanical properties, high permeability, good solvent resistance and monolithic formats are commercially available [183]. However, the common hydrolytic sol-gel process normally requires curing and ageing steps, resulting in the cracking and shrinkage of the monolith during the post-treatment of materials. For this reason, monolithic phases of molecularly imprinted silica are rarely described in the literature.

Recently, a novel methodology has been proposed by Wang et al. based on a room temperature ionic liquid-mediated nonhydrolytic sol-gel process [182, 184]. The synthesis of imprinted monoliths was carried out with an acidic functional monomer, such as MAA, MPS, which acted as the crosslinker, and a hydrophobic room temperature ionic liquid, such as $BMIM^+PF_6^-$, which was used as a porogenic solvent and to reduce gel shrinkage. The silica matrix was formed by two simultaneous processes. One reaction is a condensation of MPS catalysed by a carboxylic acid, the functional monomer MAA, to form a silica sol-gel network. The other reaction is a thermally initiated polymerisation of the functional monomer and the siloxane using the radical initiator AIBN. The morphology of this imprinted silica-based monolith is shown by the SEM images presented Fig. 14. Following this approach, imprinted monoliths have been developed for chiral separation of acidic (naproxen) and basic (zolmitriptan) templates by CEC.

3.2 Application of MIPs in Monolithic Format

Although potentially useful for sample preparation by solid-phase extraction, monolithic MIP phases had initially only been applied as stationary phase for LC

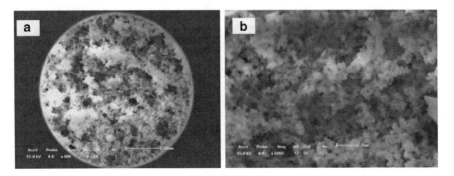

Fig. 14 Scanning electron microscopy images of the silica-based MIP monolith: cross-section of the formed monolith magnified (**a**) 800× and (**b**) 5,000×. Reproduced with permission from [184]

and capillary electrophoresis [183, 185–190]. During the last few years, however, a small number of authors have investigated the potential of MIP monoliths for solid-phase microextraction [191].

3.2.1 Liquid Chromatography and Capillary Electrochromatography

The molecular imprinting strategy can be applied for the recognition of different kinds of templates from small organic molecules to biomacromolecules as proteins. Some examples of separations investigated with MIP monoliths in CEC and LC are shown in Table 2. The influence of the imprinted monolithic phase preparation procedure and of the separation conditions on the selectivity and chromatographic efficiency have been widely studied [154, 157, 161, 166, 167, 192]. The performance of imprinted monoliths as chromatographic stationary phase has also been compared to that of the traditional bulk polymer packed column [149, 160]. It was shown that the monolithic phases yielded faster analyses and improved chiral separations.

3.2.2 Solid-Phase Extraction

The fabrication of imprinted monolithic solid-phase microextraction fibres has been developed for the selective extraction and preconcentration of diacetylmorphine and its structural analogues, triazines, bisphenol A, anaesthetics, and antibiotics followed by GC or HPLC analysis [156, 163, 179, 196, 197]. In addition, the on-line coupling of the imprinted monolith as a preconcentration column with a conventional analytical column has been proposed for the enrichment and cleanup of environmental and food samples [163]. However, at present, the capacity of the imprinted fibres and thus the degree of recovery of analytes are very variable and obviously need some improvement. For example, the recoveries of triazines after SPME with an imprinted monolith prepared by in situ polymerisation of MAA as

Table 2 Summary of MIP monolith format application for CEC and LC

Application	Polymerisation	Template	Monomers + solvent (S)	Sample separation	Ref.
CEC	In situ photo-polymerisation	(S)-Propranolol	MAA, MMA, BMA, EPMA /TRIM S: toluene/ isooctane	rac-Propranolol and mixture of aromatic compounds	[154, 155]
		Thiabendazole	MAA/EDMA S: toluene/ isooctane	Thiabendazole in citrus samples	[167]
	In situ thermal polymerisation	4-Aminopyridine	MAA/EDMA S: acetonitrile	4-AP and 2-AP isomers	[192]
		4-Hydroxy-benzoic acid	MAA/EDMA S: toluene/ isooctane	4-HBA and 2-HBA isomers and thiourea	[158]
		(R)-Binaphthol	MAA/EDMA S: toluene/ isooctane	rac-Binaphthol	[166]
		(S)-Naproxen	MAA/EDMA S: toluene/ isooctane	rac-Naproxen	[193]
		Felodipine	MAA, 4-VP/ EDMA S: toluene/ isooctane	Dihydropyridine calcium antagonists	[161]
		(S)-Ibuprofen	4-VP/EDMA S: toluene/ isooctane	rac-Ibuprofen	[169]
	In situ thermal-polymerisation	L-Phe	OPA/AM/MAA/ 2-VP/EDMA. S: propanol	Oligopeptides	[194]
	Post-grafting on silica monolith	(L)-THP	MAA/EDMA S: dodecanol/ toluene	rac-THP, rac-Tröger's base	[180]
	RTIL-mediated nonhydrolytic sol-gel	Zolmitriptan	MAA/MPS S: BMIM$^+$PF$_6^-$	rac-Zolmitriptan	[184]
HPLC	In situ thermal polymerisation	(+)-Nilvadipine	4-VP/EDMA S: toluene/ dodecanol	rac-Nilvadipine	[176]
		N-CBZ-L-phe	AM/EDMA S: toluene/ dodecanol	rac-Phenylalanine	[164]
		Theophylline	AM/EDMA S: toluene/ dodecanol	Xanthine derivatives	[172]
		Sulfamethoxazole	MAA, 4-VP, AM/ EDMA S: DMF	Sulphonamide analogues	[162]
		Caffeic acid	MAA/EDMA S: THF/isooctane	Chlorogenic acid separation from phenols	[159]
		(S)-Nateglinide	MAA, AM/ EDMA S:cyclohexanol/ dodecanol	rac-Nateglinide	[195]

(*continued*)

Table 2 (continued)

Application	Polymerisation	Template	Monomers + solvent (S)	Sample separation	Ref.
		Cinchonine, cbz-L-Trp, Fmoc-L-Trp	MAA, 4-VP/ EDMA S: toluene/ dodecanol	Cinchonine/ cinchonidine, cbz-rac-Trp, Fmoc-rac-Trp	[157]

functional monomer and EDMA as cross-linker dissolved in toluene reached only 50% because of the low capacity of the fibre [156]. In contrast, a bisphenol A imprinted precolumn, prepared by in situ copolymerisation of 4-VP and EDMA in toluene/dodecanol, presented a complete recovery of hydroquinone and resorcin; however, the selectivity seemed to be low since other phenolic compounds could be extracted with a recovery of 54.2 % [163].

Recently, water-compatible imprinted monoliths were applied as specific sorbents for the analysis of fluoroquinolones in milk samples [198]. A pefloxacin-MIP monolith was prepared in capillaries by using MAA as functional monomer, di(ethylene glycol) dimethacrylate as a cross-linker and methanol-water as the porogenic solvent. The MIP monolith showed recognition properties in an aqueous environment for several fluoroquinolones with recoveries above 90%, with a relatively high non-specific adsorption on the NIP. Nevertheless, HPLC analysis of spiked milk after MIP-microextraction presents no interfering peak compared to a traditional C18-SPE.

4 MIP Membranes

Similarly to MIP monoliths, MIP membranes are of interest for their highly porous morphology inducing a high permselectivity. Indeed, porous imprinted membranes could overcome the problems associated with the limited accessibility of the recognition sites of the traditional bulk imprinted polymers as well as with the lack of selectivity of usual commercial membranes.

The transport properties across an MIP membrane are controlled by both a sieving effect due to the membrane pore structure and a selective absorption effect due to the imprinted cavities [199, 200]. Therefore, different selective transport mechanisms across MIP membranes could be distinguished according to the porous structure of the polymeric material. Meso- and microporous imprinted membranes facilitate template transport through the membrane, in that preferential absorption of the template promotes its diffusion, whereas macroporous membranes act rather as membrane absorbers, in which selective template binding causes a diffusion delay. As a consequence, the separation performance depends not only on the efficiency of molecular recognition but also on the membrane morphology, especially on the barrier pore size and the thickness of the membrane.

Two types of MIP membranes have been proposed, which either consist exclusively of an MIP, the specific membrane morphology being formed simultaneously with the imprinting process, or are prepared in or on a membrane support with suitable morphology. Imprinted membranes can be obtained by a number of methods including in situ polymerisation, phase inversion, surface imprinting or grafting polymerisation of MIP on a preformed membrane surface. In particular, composite membranes have attracted great attention because of the combination of mechanical properties of the starting membrane and the selectivity of the MIP.

4.1 Self-Supported MIP Membranes

Self-supported MIP membranes can be seen as an alternative format to the traditional MIP particles for applications in separation and sensor technology, avoiding the limitations of mass transfer across conventional MIP materials. Two main approaches have been used for the preparation of membranes composed of an MIP: in situ polymerisation and polymer solution phase inversion.

4.1.1 In Situ Polymerisation

Early papers in the field describe MIP membranes imprinted with an adenine derivative, synthesised by bulk in situ polymerisation of the monomer mixture (MAA and EDMA) containing the template. The study of the transport properties of an imprinted membrane showed a selective permeability for adenine compared to another nucleotide [201]. However, low fluxes (around 10^{-8} mol/cm^2/h) and membrane brittleness were observed, probably due to the high cross-linking degree and the compact membrane structure, which was found to be composed of 50–100 nm sized polymer nodules. A significant improvement of membrane flexibility and stability has been observed by addition of an oligourethane-acrylate oligomer to the polymerisation mixture [202–204]. Concerning the membrane permeability, high water fluxes through the imprinted membrane could be achieved when a porogenic solvent or linear polymers, e.g. poly(ethylene glycol), were added to the prepolymerisation mixture [205, 206]. Addition of linear polymers as co-porogens that were extracted after membrane synthesis resulted in the formation of a semi-interpenetrating network generating a high macroporosity.

Self-supported photoresponsive MIP membranes synthesised using a photosensitive monomer, p-phenylazo-acrylanilide, and dansylamine as template, have been applied as recognition element in sensors [207, 208]. The binding capacity of the membrane for dansylamide could be reversibly modified by changing the illumination of membrane. Other examples are NADH and NADPH imprinted membranes synthesised on piezoelectric crystals and ISFET devices [209], and a potentiometric sensor for atrazine [210].

A hybrid inorganic-organic imprinted membrane has been developed for the enantioseparation of D,L-phenylalanine [211]. The membrane was composed of sodium alginate and poly(APTES), respectively, as organic and silica phases. Imprinted recognition sites for D-phenylalanine were formed in the alginate matrix upon the sol-gel process of APTES occurring in the presence of the template. The selectivity of the imprinted membrane was strongly affected by the APTES content. Indeed, this parameter controlled the flexibility of the membrane and the stability of the imprinted cavities. The mechanism of D-phenylalanine transport across the imprinted membrane was shown to be a facilitated permeation process.

4.1.2 Phase Inversion Polymerisation

In an alternative approach, MIP membranes can be obtained by generating molecularly imprinted sites in a non-specific matrix of a synthetic or natural polymer material during polymer solidification. The recognition cavities are formed by the fixation of a polymer conformation adopted upon interaction with the template molecule. Phase inversion methods have used either the evaporation of polymer solvent (dry phase separation) or the precipitation of the pre-synthesised polymer (wet phase inversion process). The major difficulties of this method lay both in the appropriate process conditions allowing the formation of porous materials and recognition sites and in the stability of these sites after template removal due to the lack of chemical cross-linking.

Dry Phase Separation

A dry phase inversion process has been employed for the preparation of chiral membranes recognising boc-L-Trp [212, 213]. Polystyrene modified with a peptide recognition group was employed as matrix for amino acid imprinting. After blending of the template and the matrix polymer containing the specific recognition site, molecular imprinting was achieved by polymer solidification through solvent evaporation. Enantioselectivity and high fluxes have been obtained for the imprinted membrane compared to a non-imprinted membrane. Highly ordered, imprinted porous membranes could be produced by evaporating polymer solutions containing a volatile solvent under high humidity conditions [214]. Scanning electron micrographs (Fig. 15) show the morphology of the MIP and control membranes prepared with poly(styrene-acrylonitrile) as matrix polymer and benzylhydantoin as template. The membrane is composed of an upper layer with ordered pores and a sponge-like inner structure.

Recently a metal ion-imprinted membrane was developed for selective separation of silver ions [215]. Chitosan and polyvinyl alcohol were blended in the presence of

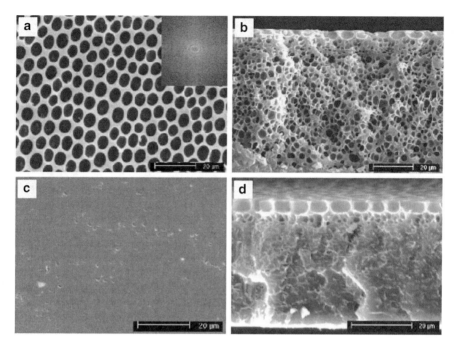

Fig. 15 Scanning electron microscopy images of (**a**) the top surface and (**b**) the cross-section of an MIP membrane, (**c**) the bottom surface of the MIP membrane (the surface contacting the glass substrate during solidification), (**d**) the cross-section of the control membrane. The *inset* in (**a**) is a Fourier transform of the top surface SEM image. Reproduced with permission from [214]

silver ions as template. A thermal drying step resulted in solvent evaporation and thermal cross-linking of the membrane.

Wet Phase Inversion Process

In the wet phase inversion process, the solidification step promoting the entrapment of the template involves the immersion of the polymer blend matrix interacting with the template in a non-solvent for precipitation or coagulation. For example, the imprinting of theophylline was achieved in a presynthesised acrylonitrile-acrylic acid copolymer by precipitation of this cast solution in water [216]. The imprinted membranes obtained had an asymmetric structure with about 70 μm thickness composed of a large porous layer with a finger-like structure supporting a dense top layer. This membrane morphology, of which an example is shown in Fig. 16, is typically induced by the wet phase inversion process because of the instantaneous demixing of the cast solvent and water, which is the most commonly employed coagulant. The presence of the template, its concentration and the nature of the coagulation medium seriously affect the pore structure of the membrane and

Fig. 16 SEM image of the
cross section of a
theophylline-imprinted
membrane: asymmetric
structure of imprinted
membrane prepared by the
wet phase inversion process.
Reproduced with permission
from [217]

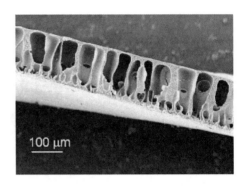

100 μm

consequently its permeability and binding capacity [216–218]. An enhancement of permeability was achieved by adding a pore inducing agent such as polyethylene glycol [219]. Recently, supercritical CO_2 has been used as non-solvent in the phase separation process instead of water for the synthesis of uracil-imprinted membranes [220]. The morphology of the imprinted membrane prepared with supercritical CO_2 is totally different from that obtained with the conventional phase inversion process based on water. A denser, regular cellular structure and a homogeneous porosity were observed without the typical fingers-like morphology, resulting in a higher adsorption capacity.

A wide range of polymers has been used as membrane materials including polyacrylonitrile [216, 221–224], polysulfone [219, 225], nylon [218] and dextran [226]. Mixtures of polymers have also been employed, one polymer providing the functional groups for template binding and the other acting as a matrix polymer in order to tune the polarity of the membrane environment to the template [219, 225].

MIP membranes were prepared for the recognition of water soluble L-glutamine [218], tetracycline [221], uric acid [223], theophylline [216, 222] and naringin [227]. More recently, imprinted membranes for the selective recognition of proteins were also reported [217, 226].

A bisphenol A imprinted PES hollow fibre membrane obtained by the dry-wet spinning method was shown to be an interesting alternative to a flat sheet membrane because of its higher surface area [228].

The main limitation of this wet phase inversion technique for membrane imprinting is that the use of the resulting imprinted membranes is limited to an aqueous medium, since the membrane generally swells in organic solvents, resulting in a loss of recognition sites.

4.2 Composite Imprinted Membranes

In spite of the straightforward fabrication of free-standing MIP membranes, their generally rather low permeability and their extremely diverse morphologies

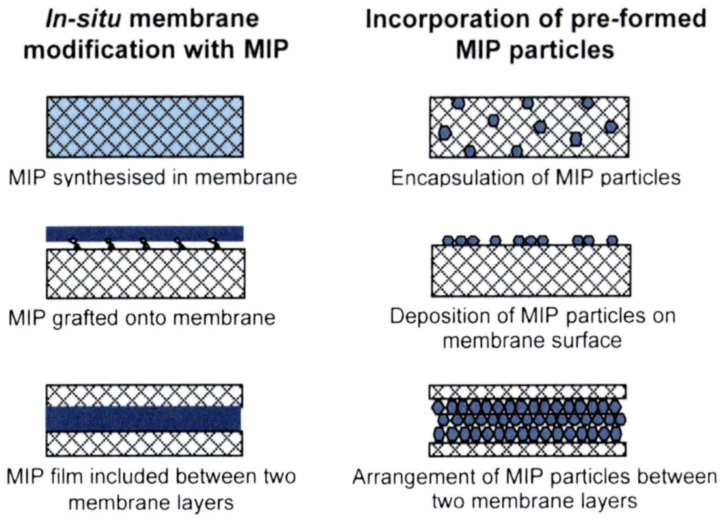

In-situ membrane modification with MIP

MIP synthesised in membrane

MIP grafted onto membrane

MIP film included between two membrane layers

Incorporation of pre-formed MIP particles

Encapsulation of MIP particles

Deposition of MIP particles on membrane surface

Arrangement of MIP particles between two membrane layers

Fig. 17 Summary of composite imprinted membrane preparation methods

depending on numerous factors significantly limit their applicability. One solution to these problems is the use of composite membranes. These are composed of an appropriate porous membrane support which is subsequently functionalised with a thin layer or with particles of an MIP. The selection of the base membrane can take into account parameters like pore size, surface area, hydrophobicity, binding capacity and permeability. The incorporation of the MIP can be obtained through two approaches, including for each several techniques summarised in Fig. 17.

4.2.1 Membrane Modification with In Situ Synthesised MIP

Molecularly imprinted composite membranes have been developed based on the functionalisation of a commercial membrane with an MIP in order to improve the mechanical stability of the imprinted polymer phase, similarly to the preparation of MIP composite beads, discussed in Sect. 2.2.2.

MIP Synthesis in a Support Membrane

The support membrane is soaked in the prepolymerisation mixture consisting of functional monomer, crosslinker and template. An MIP layer is formed on or in the porous support by in situ photo or thermal polymerisation of the monomer mixture. This was first demonstrated by Haupt and coworkers on an amino acid specific MIP synthesised in the pores of a polypropylene membrane. Composite membranes were applied for separation [229–234] for example, in the form of a hollow fibre microextraction system for the pre-treatment of milk samples [234], and for the slow release of antibiotics [235].

Several commercial flat membrane supports have been used including polyethylene [235], polyamide [230, 231], cellulose [236, 237], PVDF [229, 238], PTFE [222, 239], polyurethane [235] and porous alumina [240]. Fibres and hollow fibres of glass [241], polypropylene [233, 234] or PVDF [242] were also employed as support.

MIP Grafting onto a Support Membrane

Grafting a thin MIP film onto a support membrane is another alternative to the low permeability of self-standing MIP membranes. Here, a polymerisation initiator is grafted onto the support membrane, allowing the surface-initiated photopolymerisation of an MIP layer with controlled thickness. For example, the functionalisation of a commercial porous polypropylene membrane with an MIP specific for desmetryn was achieved by photopolymerisation initiated with benzophenone using 2-acrylamido-2-methylpropanesulphonic acid as functional monomer and methylenebisacrylamide as the cross-linker in aqueous media [243]. Composite membranes imprinted with atrazine synthesised by photopolymerisation on a polypropylene support have been found to display a better imprinting effect and selectivity for the template than the corresponding bulk MIP, arguably because of better accessibility of the recognition sites and favourable interactions of the template molecule with the support [244].

MIP composite membranes were developed for a wide range of template molecules including herbicides [245, 246], nucleotides [247, 248] and drugs [249–252].

MIP Embedding Between Two Membranes

A few studies have reported the embedding of an MIP film between two membranes as a strategy for the construction of composite membranes. For example, a metal ion-selective membrane composed of a Zn(II)-imprinted film between two layers of a porous support material was reported [253]. The imprinted membrane was prepared by surface water-in-oil emulsion polymerisation of divinylbenzene as polymer matrix with 1,12-dodecanediol-O,O'-diphenylphosphonic acid as functional host molecule for Zn(II) binding in the presence of acrylonitrile–butadiene rubber as reinforcing material and L-glutamic acid dioleylester ribitol as emulsion stabiliser. By using the acrylonitrile–butadiene rubber in the polymer matrix and the porous support PTFE, an improvement of the flexibility and the mechanical strength has been obtained for this membrane.

4.2.2   Incorporation of Pre-Synthesised MIP Particles into Membranes

As an alternative to in situ MIP synthesis, preformed MIP particles can be attached onto, or incorporated into, a support membrane. This strategy presents the advantage of decoupling the MIP synthesis from membrane production, thus avoiding the need of compromises for both steps.

Supported MIP Particles on Membrane

Only a few attempts were reported concerning the arrangement of MIP particles between two porous membranes, or their deposition on a single membrane. For example, Lehmann et al. used MIP nanoparticles with diameters between 50 nm and 300 nm imprinted with boc-L-phenylalanin-anilide obtained by miniemulsion polymerisation; the selective rebinding properties as well as the hydrodynamic properties of the nanoparticles stacked between two polyamide membranes were studied [254].

The group of Ciardelli prepared MIP nanoparticles imprinted with theophylline or caffeine by precipitation polymerisation and coated them onto a poly-MMA-*co*-AA membrane (Fig. 18) [217, 255, 256].

Encapsulation of MIP Particles into Membrane

Composite MIP membranes can also be obtained by incorporation of MIP particles into the membrane polymer matrix by mixing the particles in an appropriate solvent with the membrane-forming polymer that is then solidified by the phase inversion process. Membranes thus prepared were used for separation, with targets such as tetracycline [257], theophylline [255], methylphosphonic acid [258], bisphenol [259], indole derivatives [260], propanolol [261], luteolin [262] and norfloxacin

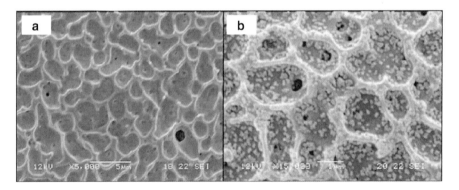

Fig. 18 Scanning electron micrographs of the surface of P(MMA-*co*-AA) membrane (**a**) before and (**b**) after MIP particle deposition. Reproduced with permission from [217]

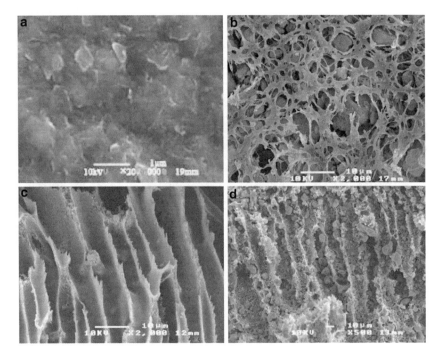

Fig. 19 Scanning electron micrographs of the top surface and cross-section of polysulfone membranes (**a,c**) without and (**b,d**) with embedded luteolin-MIP particles. Reproduced with permission from [262]

[263]. The mechanical stability and morphology of the resulting membrane depended on the polymeric matrix and the presence of additives such as plasticisers or pore forming agents, e.g. PEG, as well as on the MIP particle size and concentration (Fig. 19) [258, 261, 262].

This technique has also been employed for the preparation of a catalytic imprinted membrane by coating a cellulose membrane with a polymer incorporating particles imprinted with the transition-state analogue of a dehydrofluorination reaction [264]. The application of such an MIP composite membrane as the recognition element in an optical sensor has been reported for digitoxin analysis in serum samples by embedding digitoxin-MIP particles in polyvinyl chloride film in presence of plasticizer by the dry inversion process [265].

References

1. Flavin K, Resmini M (2009) Anal Bioanal Chem 393:437
2. Schillemans JP, van Nostrum CF (2006) Nanomedicine 1:437
3. Pérez-Moral N, Mayes AG (2004) In: Piletsky S, Turner A (eds) Molecular imprinting of polymers. Landes Bioscience, Austin

4. Tovar GEM, Kräuter I, Gruber C (2003) Top Curr Chem 227:125
5. Mayes AG (2001) In: Sellergren B (ed) Molecularly imprinted polymers. Elsevier, Amsterdam
6. Lovell PA, El-Aasser MS (eds) (1997) Emulsion polymerization and emulsion polymers. Wiley, Chichester
7. Lai J-P, Lu X-Y, Lu C-Y, Ju H-F, He X-W (2001) Anal Chim Acta 442:105
8. Lai J-P, Cao X-F, Wang X-L (2002) Anal Bioanal Chem 372:391
9. Lai J-P, He X-W, Chen F (2003) Anal Bioanal Chem 377:208
10. Lai J-P, Jiang Y, He X-W, Huang J-C, Chen F (2004) J Chromatogr B 804:25
11. Matsui J, Okada M, Tsuruoka M, Takeuchi T (1997) Anal Commun 34:85
12. Hu S-G, Wang S-W, He X-W (2003) Analyst 128:1485
13. Shi X, Wu A, Zheng S, Liu R, Zhang D (2007) J Chromatogr B 850:24
14. Strikovsky AG, Kasper D, Grün M, Green BS, Hradil J, Wulff G (2000) J Am Chem Soc 122:6295
15. Baggiani C, Anfossi L, Barravalle P, Giovannoli C, Giraudi G (2007) Anal Bioanal Chem 389:413
16. Mayes A, Mosbach K (1996) Anal Chem 68:3769
17. Ansell RJ, Mosbach K (1997) J Chromatogr A 787:55
18. Ansell RJ, Mosbach K (1998) Analyst 123:1611
19. Perez-Moral N, Mayes AG (2006) Biosens Bioelectron 21:1798
20. Kempe H, Kempe M (2004) Macromol Rapid Commun 25:315
21. Kempe H, Kempe M (2006) Anal Chem 78:3659
22. Wang X, Ding X, Zheng Z, Hu X, Cheng X, Peng Y (2006) Macromol Rapid Commun 27:1180
23. Zourob M, Mohr S, Mayes AG, Macaskill A, Pérez-Moral N, Fielden PR, Goddard NJ (2006) Lab Chip 6:296
24. Kubo A, Shinmori H, Takeuchi T (2006) Chem Lett 35:588
25. Pang XS, Cheng GX, Li RS, Lu SL (2005) Zhang YH Anal Chim Acta 550:13
26. Pang XS, Cheng GX, Lu SL, Tang EJ (2006) Anal Bioanal Chem 384:225
27. Pan J, Xue X, Wang J, Xie H, Wu Z (2009) Polymer 50:2365
28. Abbate V, Frascione N, Bansal SS (2010) J Polym Sci A Polym Chem 48:1721
29. Barrett KEJ, Thomas HR (1975) In: Barrett KEJ (ed) Dispersion polymerization in organic media. Wiley, London
30. Ye L, Cormack AG, Mosbach K (1999) Anal Commun 36:35
31. Ye L, Weiss R, Mosbach K (2000) Macromolecules 33:8239
32. Surugiu I, Ye L, Dzgoev A, Danielsson B, Mosbach K, Haupt K (2000) Analyst 125:13
33. Ye L, Mosbach K (2001) J Am Chem Soc 123:2901
34. Ye L, Cormack AG, Mosbach K (2001) Anal Chim Acta 435:187
35. Ye L, Surugiu I, Haupt K (2002) Anal Chem 74:959
36. Yoshimatsu K, Reimhult K, Krozer A, Mosbach K, Sode K, Ye L (2007) Anal Chim Acta 584:112
37. Wang J, Cormack PAG, Sherrington DC, Khoshdel E (2003) Angew Chem Int Ed 42:5336
38. Wang J, Cormack PAG, Sherrington DC, Khoshdel E (2003) Pure Appl Chem 79:1505
39. Tamayo GF, Casillas JL, Martin-Esteban A (2003) Anal Chim Acta 482:165
40. Jiang Y, Tong A-J (2004) J Appl Polym Sci 94:542
41. Sambe H, Hoshina K, Moaddel R, Wainer IW, Haginaka J (2006) J Chromatogr A 1134:88
42. Lai J-P, Yang M-L, Niessner R, Knopp D (2007) Anal Bioanal Chem 389:405
43. Turiel E, Martin-Esteban A, Tadeo JL (2007) J Chromatogr A 1172:97
44. Malosse L, Buvat P, Ades D, Siove A (2008) Analyst 133:588
45. Schweits L, Spegel P, Nilsson S (2000) Analyst 125:1899
46. de Boer T, Mol R, de Zeeuw RA, de Jong GJ, Sherrington DC, Cormack PAG, Ensing K (2002) Electrophoresis 23:1296
47. Hunt CE, Pasetto P, Ansell RJ, Haupt K (2006) Chem Commun 1754

48. Wang L, Zhang Z, Huang L (2008) Anal Bioanal Chem 390:1431
49. Boonpagrak S, Prachayasittikul V, Bülow L, Ye L (2005) J Appl Polym Sci 99:1390
50. Jin Y, Jiang M, Shi Y, Lin Y, Peng Y, Dai K, Lu B (2008) Anal Chim Acta 612:105
51. Horvath V, Lorantfy B, Toth B, Bognar J, Laszlo K, Horvai G (2009) J Sep Sci 32:3347
52. Hoshino Y, Kodama T, Okahata Y, Shea KJ (2008) J Am Chem Soc 130:15242
53. Hoshino Y, Koide H, Urakami T, Kanazawa H, Kodama T, Oku N, Shea KJ (2010) J Am Chem Soc 132:6644
54. Chern CS (2008) Principles and applications of emulsion polymerization. Wiley, Chichester
55. Landfester K, Bechthold N, Tiarks F, Antonietti M (1999) Macromolecules 32:5222
56. Landfester K (2001) Adv Mater 13:765
57. Tiarks F, Landfester K, Antonietti M (2001) Langmuir 17:908
58. Kido H, Miyajima T, Tsukagoshi K, Maeda M, Takagi M (1992) Anal Sci 8:749
59. Koide Y, Senba H, Shosenji H, Maeda M, Takagi M (1996) Bull Chem Soc Jpn 69:125
60. Yoshida M, Hatate Y, Uezu K, Goto M, Furusaki S (2000) J Polym Sci A Polym Chem 38:689
61. Vaihinger D, Landfester K, Kräuter I, Brunner H, Tovar GEM (2002) Macromol Chem Phys 203:1965
62. Ki CD, Chang JY (2006) Macromolecules 39:3415
63. Priego-Capote F, Ye L, Shakil S, Shamsi SA, Nilsson S (2008) Anal Chem 80:2881
64. Zheng Z, Hoshino Y, Rodriguez A, Yoo H, Shea KJ (2010) ACS Nano 4:199
65. Baker WO (1949) Ind Eng Chem 41:511
66. Staudinger H, Husemann E (1934) Chem Ber 68:1620
67. Frank RS, Downey JS, Yu K, Stöver HDH (2002) Macromolecules 35:2728
68. Brandrup J, Immergut EH (eds) (1989) Polymer handbook, 3rd edn. Wiley, New York
69. Ohkubo K, Urata Y, Hirota S, Honda Y, Fujishita Y, Sagawa T (1994) J Mol Catal 93:189
70. Ohkubo K, Funakoshi Y, Urata Y, Hirota S, Usui S, Sagawa T (1995) J Chem Soc, Chem Commun 2143
71. Ohkubo K, Funakoshi Y, Sagawa T (1996) Polymer 37:3993
72. Ohkubo K, Sawakuma K, Sagawa T (2001) Polymer 42:2263
73. Biffis A, Graham NB, Siedlaczek G, Stalberg S, Wulff G (2001) Macromol Chem Phys 202:163
74. Wulff G, Chong B-O, Kolb U (2006) Angew Chem Int Ed 45:2955
75. Maddock SC, Pasetto P, Resmini M (2004) Chem Commun 536
76. Pasetto P, Maddock SC, Resmini M (2005) Anal Chim Acta 542:66
77. Chen Z, Hua Z, Wang J, Guan Y, Zhao M, Li Y (2007) Appl Catal A Gen 328:252
78. Chen Z, Xu L, Liang Y, Zhao M (2010) Adv Mater 22:1488
79. Carboni D, Flavin K, Servant A, Gouverneur V, Resmini M (2008) Chem Eur J 14:7059
80. Cutivet A, Schembri C, Kovenski J, Haupt K (2008) J Am Chem Soc 131:14699
81. Smigol V, Svec F, Hosoya K, Wang Q, Frechet JMJ (1992) Angew Makromol Chem 195:151
82. Ugelstad J, Kaggerud KH, Hansen FK, Berger A (1979) Makromol Chem 180:737
83. Hosoya K, Yoshizako K, Tanaka N, Kimata K, Araki T, Haginaka J (1994) Chem Lett 1437
84. Matsui J, Kato T, Takeuchi T, Suzuki M, Yokoyama K, Tamiya E, Karube I (1993) Anal Chem 65:2223
85. Haginaka J, Sanbe H (2001) J Chromatogr A 913:141
86. Haginaka J, Sanbe H, Takehira H (1999) J Chromatogr A 857:117
87. Fu Q, Sanbe H, Kagawa C, Kunimoto K-K, Haginaka J (2003) Anal Chem 75:191
88. Haginaka J, Tabo H, Ichitani M, Takihara T, Sugimoto A, Sambe H (2007) J Chromatogr A 1156:45
89. Haginaka J, Kagawa C (2004) J Chromatogr B 804:19
90. Sambe H, Hoshina K, Haginaka J (2007) J Chromatogr A 1152:130
91. Masci G, Aulenta F, Crescenzi V (2002) J Appl Polym Sci 83:2660
92. Haginaka J, Takehira H, Hosoya K, Tanaka N (1997) Chem Lett 555

93. Haginaka J, Takehira H, Hosoya K, Tanaka N (1998) J Chromatogr A 816:113
94. Haginaka J, Takehira H, Hosoya K, Tanaka N (1999) J Chromatogr A 849:331
95. Haginaka J, Sanbe H (2000) Anal Chem 72:5206
96. Sanbe H, Haginaka J (2003) Analyst 128:593
97. Chen Z, Zhao R, Shangguan D, Guoquan L (2005) Biomed Chromatogr 19:533
98. Zheng N, Fu Q, Li YZ, Chang WB, Wang ZM, Li TJ (2001) Microchem J 69:153
99. Zheng N, Li YZ, Chang WB, Wang ZM, Li TJ (2002) Anal Chim Acta 452:277
100. Liu X, Chen Z, Zhao R, Shangguan D, Guoquan L, Chen Y (2007) Talanta 71:1205
101. Norrlöw O, Glad M, Mosbach K (1984) J Chromatogr 299:29
102. Glad M, Norrlöw O, Sellergren B, Siegbahn N, Mosbach K (1985) J Chromatogr 347:11
103. Wulff G, Oberkobusch D, Minarik M (1985) React Polym 3:261
104. Glad M, Reinholdsson P, Mosbach K (1995) React Polym 25:47
105. Plunkett D, Arnold FH (1995) J Chromatogr A 708:19
106. Ylmaz E, Haupt K, Mosbach K (2000) Angew Chem Int Ed 39:2115
107. Ylmaz E, Ramstrom O, Moller P, Sanchez D, Mosbach K (2002) J Mater Chem 12:1577
108. Sellergren B, Büchel G (1999) WO01/32760 A1
109. Titirici MM, Hall AJ, Sellergren B (2002) Chem Mater 14:21
110. Titirici MM, Hall AJ, Sellergren B (2003) Chem Mater 15:422
111. Titirici MM, Sellergren B (2004) Anal Bioanal Chem 378:1913
112. Kempe M, Glad M, Mosbach K (1995) J Mol Recognit 8:35
113. Burow M, Minoura N (1996) Biochem Biophys Res Commun 227:419
114. Hirayama K, Burow M, Morikawa Y, Minoura N (1998) Chem Lett 731
115. Hirayama K, Sakai Y, Kameoka K (2001) J Appl Polym Sci 81:3378
116. Matsui T, Osawa T, Shirasaka K, Katayama M, Hishiya T, Asanuma H, Komiyama M (2006) J Incl Phen Macroc Chem 56:39
117. Sulitzky C, Rückert B, Hall AJ, Lanza F, Unger K, Sellergren B (2002) Macromolecules 35:79
118. Rückert B, Hall AJ, Sellergren B (2002) J Mater Chem 12:2275
119. Sellergren B, Rückert B, Hall AJ (2002) Adv Mater 14:1204
120. Titirici MM, Sellergren B (2006) Chem Mater 18:1773
121. Gao D, Zhang Z, Wu M, Xie C, Guan G, Wang D (2007) J Am Chem Soc 129:7859
122. Perez-Moral N, Mayes AG (2007) Macromol Rapid Commun 28:2170
123. Perez N, Whitcombe MJ, Vulfson EN (2000) J Appl Polym Sci 77:1851
124. Perez N, Whitcombe MJ, Vulfson EN (2001) Macromolecules 34:830
125. Carter SR, Rimmer S (2002) Adv Mater 14:667
126. Carter S, Lu S-Y, Rimmer S (2003) Supramol Chem 15:213
127. Carter SR, Rimmer S (2004) Adv Funct Mater 14:553
128. Ko DY, Lee HJ, Jeong B (2006) Macromol Rapid Commun 27:1367
129. Perez-Moral N, Mayes AG (2004) Langmuir 20:3775
130. Kan X, Zhao Q, Shao D, Geng Z, Wang Z, Zhu J-J (2010) J Phys Chem B 114:3999
131. Li H, Li Y, Cheng J (2010) Chem Mater 22:2451
132. Bompart M, De Wilde Y, Haupt K (2010) Adv Mater 22:1
133. Takagishi T, Klotz IM (1972) Biopolymers 11:483
134. Weatherhead RH, Williams A, Stacey KA (1993) J Mol Catal 85:33
135. Ohkubo K, Urata Y, Honda Y, Nakashima Y, Yoshinaga K (1994) Polymer 35:5372
136. Ohkubo K, Urata Y, Hirota S, Honda Y, Sagawa T (1994) J Mol Catal 87:L21
137. Yu KY, Tsukagoshi K, Maeda M, Takagi M (1992) Anal Sci 8:701
138. Tsukagoshi K, Yu KY, Maeda M, Takagi M (1993) Bull Chem Soc Jpn 66:114
139. Yang K, Liu Z, Mao M, Zhang X, Zhao C, Nishi N (2005) Anal Chim Acta 546:30
140. Lin H-Y, Ho M-H, Lee M-H (2009) Biosens Bioelectron 25:579
141. Lee M-H, Chen Y-C, Ho M-H, Lin H-Y (2010) Anal Bioanal Chem. doi:10.1007/s00216-010-3631-x
142. Li Z, Ding J, Day M, Tao Y (2006) Macromolecules 39:2629

143. Kim TH, Ki CD, Cho H, Chang T, Chang JY (2005) Macromolecules 38:6423
144. Zhang L, Cheng G, Fu C (2003) React Funct Polym 56:167
145. Perez-Moral N, Mayes AG (2004) Anal Chim Acta 504:15
146. Fairhurst RE, Chassaing C, Venn RF, Mayes AG (2004) Biosens Bioelectron 20:1098
147. Wei S, Molinelli A, Mizaikoff B (2006) Biosens Bioelectron 21:1943
148. Tamayo FG, Titirici MM, Martin-Esteban A, Sellergren B (2005) Anal Chim Acta 542:38
149. Oxelbark J, Legido-Quigley C, Aureliano CSA, Titirici MM, Schillinger E, Sellergren B, Courtois J, Irgum K, Dambies L, Cormack PAG, Sherrington DC, De Lorenzi E (2007) J Chromatogr A 1160:215
150. Guiochon G (2007) J Chromatogr A 1168:101
151. Hilder EF, Svec F, Frechet JMJ (2004) J Chromatogr A 1044:3
152. Schweitz L, Andersson LI, Nilsson S (1997) Anal Chem 69:1179
153. Schweitz L, Andersson LI, Nilsson S (1997) J Chromatogr A 762:401
154. Schweitz L, Andersson LI, Nilsson S (2002) Analyst 127:22
155. Schweitz L, Andersson LI, Nilsson S (2001) Anal Chim Acta 435:43
156. Turiel E, Tadeo JL, Martin-Esteban A (2007) Anal Chem 79:3099
157. Huang X, Zou H, Chen X, Luo Q, Kong L (2003) J Chromatogr A 984:273
158. Liu ZS, Xu YL, Yan C, Gao RY (2004) Anal Chim Acta 523:243
159. Li H, Liu Y, Zhang Z, Liao H, Nie L, Yao S (2005) J Chromatogr A 1098:66
160. Kim H, Guiochon G (2005) Anal Chem 77:93
161. Deng Q, Lun Z, Shao H, Yan C, Gao R (2005) Anal Bioanal Chem 382:51
162. Liu X, Ouyang C, Zhao R, Shangguan D, Chen Y, Liu G (2006) Anal Chim Acta 571:235
163. Ou J, Hu L, Hu L, Li X, Zou H (2006) Talanta 69:1001
164. Yan H, Row KH (2006) Biotechnol Bioprocess Eng 11:357
165. Ou J, Dong J, Tian T, Hu J, Ye M, Zou H (2007) J Biochem Biophys Methods 70:71
166. Liu ZS, Xu YL, Wang H, Yan C, Gao RY (2004) Anal Sci 20:673
167. Cacho C, Schweitz L, Turiel E, Perez-Conde C (2008) J Chromatogr A 1179:216
168. Hjerten S (1985) J Chromatogr 347:191
169. Deng QL, Lun ZH, Gao RY, Zhang LH, Zhang WB, Zhang YK (2006) Electrophoresis 27:4351
170. Huang X, Qin F, Chen X, Liu Y, Zou H (2004) J Chromatogr B 804:13
171. Liu ZS, Xu YL, Yan C, Gao RY (2005) J Chromatogr A 1087:20
172. Sun H, Qiao FX, Liu GY (2006) J Chromatogr A 1134:194
173. Seebach A, Seidel-Morgenstern A (2007) Anal Chim Acta 591:57
174. Ou J, Tang S, Zou H (2005) J Sep Sci 28:2282
175. Zhang M, Xie J, Zhou Q, Chen G, Liu Z (2003) J Chromatogr A 984:173
176. Haginaka J, Futagami A (2008) J Chromatogr A 1185:258
177. Bompart M, Haupt K (2009) Aust J Chem 62:751
178. Liu H, Zhuang X, Turson M, Zhang M, Dong X (2008) J Sep Sci 31:1694
179. Courtois J, Fischer G, Sellergren B, Irgum K (2006) J Chromatogr A 1109:92
180. Ou J, Li X, Feng S, Dong J, Dong X, Kong L, Ye M, Zou H (2007) Anal Chem 79:639
181. Lin Z, Yang F, He X, Zhao X, Zhang Y (2009) J Chromatogr A 1216:8612
182. Wang HF, Zhu YZ, Yan XP, Gao RY, Zheng JY (2006) Adv Mater 18:3266
183. Wu R, Hu L, Wang F, Ye M, Zou H (2008) J Chromatogr A 1184:369
184. Wang HF, Zhu YZ, Lin JP, Yan XP (2008) Electrophoresis 29:952
185. Svec F, Kurganov AA (2008) J Chromatogr A 1184:281
186. Quaglia M, Sellergren B, De Lorenzi E (2004) J Chromatogr A 1044:53
187. Nilsson J, Spégel P, Nilsson S (2004) J Chromatogr B 804:3
188. Turiel E, Martin-Esteban A (2005) J Sep Sci 28:719
189. Liu H, Ho Row K, Yang G (2005) Chromatographia 61:429
190. Huang YP, Liu ZS, Zheng C, Gao RY (2009) Electrophoresis 30:155
191. Svec F (2006) J Chromatogr B 841:52
192. Yan W, Gao R, Zhang Z, Wang Q, Jiang C, Yan C (2003) J Sep Sci 26:555

193. Xu YL, Liu ZS, Wang HF, Yan C, Gao RY (2005) Electrophoresis 26:804
194. Lin CC, Wang GR, Liu CY (2006) Anal Chim Acta 572:197
195. Yin J, Yang G, Chen Y (2005) J Chromatogr A 1090:68
196. Djozan D, Baheri T (2007) J Chromatogr A 1166:16
197. Sun X, He X, Zhang Y, Chen L (2009) Talanta 79:926
198. Zheng MM, Gong R, Zhao X, Feng YQ (2010) J Chromatogr A 1217:2075
199. Piletsky SA, Panasyuk TL, Piletskaya EV, Nicholls IA, Ulbricht M (1999) J Memb Sci 157:263
200. Ulbricht M (2004) J Chromatogr B 804:113
201. Mathew-Krotz J, Shea KJ (1996) J Am Chem Soc 118:8154
202. Sergeyeva TA, Piletsky SA, Brovko AA, Slinchenko EA, Sergeeva LM, El'skaya AV (1999) Anal Chim Acta 392:105
203. Sergeyeva TA, Piletska OV, Piletsky SA, Sergeeva LM, Brovko OO, El'ska GV (2008) Mater Sci Eng C 28:1472
204. Sergeyeva TA, Gorbach LA, Slinchenko OA, Goncharova LA, Piletska OV, Brovko OO, Sergeeva LM, El'ska GV (2010) Mater Sci Eng C 30:431
205. Sergeyeva TA, Piletsky SA, Piletska EV, Brovko OO, Karabanova LV, Sergeeva LM, El'skaya AV, Turner APF (2003) Macromolecules 36:7352
206. Sergeyeva TA, Brovko OO, Piletska EV, Piletsky SA, Goncharova LA, Karabanova LV, Sergeeva LM, El'skaya AV (2007) Anal Chim Acta 582:311
207. Minoura N, Idei K, Rachkov A, Uzawa H, Matsuda K (2003) Chem Mater 15:4703
208. Minoura N, Idei K, Rachkov A, Choi YW, Ogiso M, Matsuda K (2004) Macromolecules 37:9571
209. Pogorelova SP, Zayats M, Bourenko T, Kharitonov AB, Lioubashevski O, Katz E, Willner I (2003) Anal Chem 75:509
210. D'Agostino G, Alberti G, Biesuz R, Pesavento M (2006) Biosens Bioelectron 22:145
211. Wu H, Zhao Y, Nie M, Jiang Z (2009) Sep Purif Technol 68:97
212. Yoshikawa M, Ooi T, Izumi JI (2001) Eur Polym J 37:335
213. Itou Y, Nakano M, Yoshikawa M (2008) J Memb Sci 325:371
214. Lu Y, Zhao B, Ren Y, Xiao G, Wang X, Li C (2007) Polymer 48:6205
215. Wang X, Zhang L, Ma C, Song R, Hou H, Li D (2009) Hydrometallurgy 100:82
216. Wang HY, Kobayashi T, Fujii N (1996) Langmuir 12:4850
217. Silvestri D, Barbani N, Cristallini C, Giusti P, Ciardelli G (2006) J Memb Sci 282:284
218. Richter A, Gruner M, Bel Bruno JJ, Gibson UJ, Nowicki M (2006) Colloids Surf A 284–285:401
219. Ramamoorthy M, Ulbricht M (2003) J Memb Sci 217:207
220. Kobayashi T, Leong SS, Zhang Q (2008) J Appl Polym Sci 108:757
221. Trotta F, Baggiani C, Luda MP, Drioli E, Massari T (2005) J Memb Sci 254:13
222. Kobayashi T, Wang HY, Fujii N (1998) Anal Chim Acta 365:81
223. Cristallini C, Ciardelli G, Barbani N, Giusti P (2004) Macromol Biosci 4:31
224. Wang HY, Xia SL, Sun H, Liu YK, Cao SK, Kobayashi T (2004) J Chromatogr B 804:127
225. Ramamoorthy M, Ulbricht M (2004) Sep Purif Technol 39:211
226. Silvestri D, Cristallini C, Ciardelli G, Giusti P, Bardani N (2005) J Biomater Sci Polym Ed 16:397
227. Trotta F, Drioli E, Baggiani C, Lacopo D (2002) J Membr Sci 201:77
228. Zhao C, Yu B, Qian B, Wei Q, Yang K, Zhang A (2008) J Membr Sci 310:38
229. Wang XJ, Xu ZL, Feng JL, Bing NC, Yang ZG (2008) J Membr Sci 313:97
230. Zhu X, Su Q, Cai J, Yang J, Gao Y (2006) J Appl Polym Sci 101:4468
231. Koter I, Ceynowa J (2003) J Mol Catal B Enzym 24–25:17
232. El-Toufaili FA, Visnjevski A, Brüggemann O (2004) J Chromatogr B 804:135
233. Dzgoev A, Haupt K (1999) Chirality 11:465
234. Liu M, Li M, Qiu B, Chen X, Chen G (2010) Anal Chim Acta 663:33
235. Sreenivasan K (2005) Macromol Biosci 5:187
236. Chen C, Chen Y, Zhou J, Wu C (2006) Anal Chim Acta 569:58

237. Wang P, Hu W, Su W (2008) Anal Chim Acta 615:54
238. Bing NC, Xu ZL, Wang XJ, Yang ZG, Yang H (2007) J Appl Polym Sci 106:71
239. Petcu M, Schaare PN, Cook CJ (2004) Anal Chim Acta 504:73
240. Zhang Y, Gao X, Xiang L, Zhang Y, Diniz da Costa JC (2010) J Memb Sci 346:318
241. Ceolin G, Navarro-Villoslada F, Moreno-Bondi MC, Horvai G, Horvath V (2009) J Comb Chem 11:645
242. Wang XJ, Xu ZL, Bing NC, Yang ZG (2008) J Appl Polym Sci 109:64
243. Piletsky SA, Matuschewski H, Schedler U, Wilpert A, Piletska EV, Thiele TA, Ulbricht M (2000) Macromolecules 33:3092
244. Schneider F, Piletsky S, Piletska E, Guerreiro A, Ulbricht M (2005) J Appl Polym Sci 98:362
245. Sergeyeva TA, Matuschewski H, Piletskya SA, Bendig J, Schedler U, Ulbricht M (2001) J Chromatogr A 907:89
246. Panasyuk-Delaneya T, Mirskya VM, Ulbricht M, Wolfbeis OS (2001) Anal Chim Acta 435:157
247. Kochkoban V, Hilal N, Windsor PJ, Lester E (2003) Chem Eng Technol 26:463
248. Hilal N, Kochkodan V (2003) J Memb Sci 213:97
249. Bodhibukkana C, Srichana T, Kaewnopparat S, Tangthong N, Bouking P, Martin GP, Suedee R (2006) J Control Release 113:43
250. Hattori K, Hiwatari M, Iiyama C, Yoshimi Y, Kohori F, Sakai K, Piletsky SA (2004) J Memb Sci 233:169–173
251. Donato L, Figoli A, Drioli E (2005) J Pharm Biomed Anal 37:1003
252. Suedee R, Bodhibukkana C, Tangthong N, Amnuaikit C, Kaewnopparat S, Srichana T (2008) J Control Release 129:170
253. Araki K, Maruyama T, Kamiya N, Goto M (2005) J Chromatogr B 818:141
254. Lehmann M, Brunner H, Tovar GEM (2002) Desalination 149:315
255. Silvestri D, Borrelli C, Giusti P, Cristallini C, Ciardelli G (2005) Anal Chim Acta 542:3
256. Ciardelli G, Borrelli C, Silvestri D, Cristallini C, Barbani N, Giusti P (2006) Biosens Bioelectron 21:2329
257. Suedee R, Srichana T, Chuchome T, Kongmark U (2004) J Chromatogr B 811:191
258. Prathish KP, Prasad K, Prasada Rao T, Suryanarayana MVS (2007) Talanta 71:1976
259. Takeda K, Kobayashi T (2006) J Memb Sci 275:61
260. Takeda K, Uemura K, Kobayashi T (2007) Anal Chim Acta 591:40
261. Jantarat C, Tangthong N, Songkro S, Martin GP, Suedee R (2008) Int J Pharm 349:212
262. Zhang Y, Gao X, Wang Y, Zhang Y, Lu GQ (2009) J Memb Sci 339:100
263. Liu W, Wang B (2009) J Appl Polym Sci 113:1125
264. Kalim R, Schomäcker R, Yüce S, Brüggemann O (2005) Polym Bull 55:287
265. Gonzales GP, Hernando PF, Durand Alegria JS (2009) Anal Chim Acta 638:209

Top Curr Chem (2012) 325: 83–110
DOI: 10.1007/128_2011_308
© Springer-Verlag Berlin Heidelberg 2011
Published online: 20 December 2011

Micro and Nanofabrication of Molecularly Imprinted Polymers

Marc Bompart, Karsten Haupt, and Cédric Ayela

Abstract Molecularly imprinted polymers (MIPs) are tailor-made receptors that possess the most important feature of biological antibodies and receptors – specific molecular recognition. They can thus be used in applications where selective binding events are of importance, such as chemical sensors, biosensors and biochips. For the development of microsensors, sensor arrays and microchips based on molecularly imprinted polymers, micro and nanofabrication methods are of great importance since they allow the patterning and structuring of MIPs on transducer surfaces. It has been shown that because of their stability, MIPs can be easily integrated in a number of standard microfabrication processes. Thereby, the possibility of photopolymerizing MIPs is a particular advantage. In addition to specific molecular recognition properties, nanostructured MIPs and MIP nanocomposites allow for additional interesting properties in such sensing materials, for example, amplification of electromagnetic waves by metal nanoparticles, magnetic susceptibility, structural colors in photonic crystals, or others. These materials will therefore find applications in particular for chemical and biochemical detection, monitoring and screening.

Keywords Biosensor · Lithography · Microbiochip · Microfabrication · Molecularly imprinted polymer · Nanocomposite · Nanofabrication · Nanomaterial · Synthetic receptor

M. Bompart and K. Haupt (✉)
Compiègne University of Technology, UMR CNRS, 6022, BP 20529, 60205-Compiègne, France
e-mail: m_bompart@yahoo.fr; karsten.haupt@utc.fr

C. Ayela
Laboratoire de l'Intégration du Matériau au Système, 351 Cours de la Libération, 33405
TALENCE, Cedex, France
e-mail: cedric.ayela@ims-bordeaux.fr

Contents

1　Molecularly Imprinted Polymers

Synthetic receptors capable of binding a target molecule with similar affinity and specificity to antibodies or enzymes have been a long-term goal of bioorganic chemistry. One technique that is being increasingly adopted for the generation of such biomimetic receptors is molecular imprinting of synthetic polymers [1–5]. This is a process where functional and cross-linking monomers are copolymerized in the presence of an imprint molecule (the target molecule or a derivative thereof) that acts as a molecular template. The functional monomers initially form a complex with the imprint molecule and, following polymerization, their functional groups are held in position by the highly cross-linked polymeric structure. Subsequent removal of the imprint molecule reveals binding sites that are complementary in size and shape to the template. In that way, a molecular memory is introduced into the polymer, which is now capable of selectively rebinding the target (Fig. 1).

Molecularly imprinted polymers (MIPs) have been used in many different applications, such as affinity separation matrices [6, 7], antibody mimics in immunoassays [8–11], recognition elements in biosensors [12–16], selective

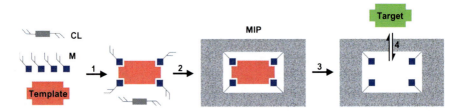

Fig. 1 General principle of molecular imprinting. The molecular template is mixed with functional monomers (M) and cross-linking monomers (CL), and a complex is formed by autoassembly (1). The system is polymerized (2) resulting in a polymer with molecularly imprinted sites (MIP). The molecular template is removed (3), liberating cavities that can specifically recognize and bind the target molecule (4)

solid-phase extraction (SPE) media [17–19], for directed synthesis and catalysis [20–22], slow drug release [23–26], and others. Compared with their natural counterparts (antibodies or enzymes), MIPs are considered advantageous for many of these applications. They are tailor-made materials that can be obtained for a large number of different targets. Moreover, they typically have a superior chemical and physical stability, which facilitates their storage and handling as well as their integration in standard industrial fabrication procedures. They can be regenerated and are potentially reusable, and may thus be less costly than biomacromolecules.

For some of the above-mentioned applications, in particular for sensors and biochips, but also for drug delivery materials, MIPs in the form of micro or nanostructures are required. For example, microbiochip production requires micropatterning of the recognition element on the substrate surface, while a hierarchically nanostructured morphology may be an advantage for drug delivery materials. There have been considerable advances during the past decade concerning micro and nanostructuring of MIPs. In the following, the main techniques and approaches are reviewed.

2 MIP Microfabrication Methods

Micro and nanobiochips are currently attracting significant interest since large scale arrays as the core element of high-throughput parallel-processing systems are capable of handling a large number of samples at a time [27, 28]. Miniaturization also enhances the portability of detection devices, while fabrication costs are reduced. Due to the easier integration of MIPs, as compared to biomacromolecules, in standard microfabrication processes, the use of tools inspired from micro and nanotechnologies permits the elaboration of micro and nanopatterned functional films of MIPs. Patterning tools can be referenced as electrical, optical, or mechanical ones, the choice of the particular tool being guided by the type of polymer, device, or the application. Among others, popular techniques include photolithography [29, 30], mechanical microspotting [31, 32], or soft lithography [33, 34], offering precise deposition and structuring of MIP films at the micro and nanoscales. Multiplexing, shape control, resolution and processing time are among the main issues in this context.

2.1 Electrical Deposition

Theoretically, electrical patterning is one the simplest method to structure materials since they can be patterned directly on the surface of an electrode. Creating conducting microelectrodes is, nowadays, fast and simple using micro and nanotechnology tools. Deposition and etching, or deposition followed by lift-off, are the conventional methods [35]. Other solutions based on electrodeposition of metals were also proposed [36]. Patterning methods based on electropolymerization present the advantage of being fast techniques where microstructures are created at low cost with a relative simplicity. To pattern polymers electrically, they must be electropolymerizable. Electropolymerization of MIPs is an interesting alternative to radical polymerization to create MIP micropatterns as it allows the monitoring and regulation of the density and the thickness of the polymeric film [37, 38]. For this, several parameters can be adjusted such as the number of cyclic scans or the voltage. Different types of monomers have been studied such as polypyrrole [39], overoxidized polypyrrole [40], polyphenol [41], polythiophene [38], and poly(o-phenylenediamine) [42] that can be easily oxidized from the aqueous solutions of their monomers. The growth of these types of monomers depends on the nature of the synthesized polymer. If the employed polymer is conducting, the polymeric growth is practically unlimited; on the other hand, if the polymeric chain is nonconducting, growth would be self-limited resulting in very thin polymeric layers (10–100 nm) [43]. Recent examples of electropolymerization of MIPs include electrochemical polymerization of dimethoate as template in a poly(o-phenylenediamine) as monomer where silver nanoparticles were added [44]. Analyte binding and extraction cycles were characterized by cyclic voltammetry. Other examples show that different types of monomers can be used as much as different materials for the microelectrodes. Most frequently used electrodes are gold and platinum [45], glassy carbon [44], or carbon fibers [46]. Also, functionalization of integrated transducers is now done for the development of biomimetic biosensors, and mass-sensitive transduction elements such as Quartz Crystal Microbalance (QCM) are of great interest in this context [38]. A recent example of a nanostructured electrochemical sensor based on an MIP was published by Cai et al. They showed that arrays of carbon nanotube tips embedded in a polydimethylsiloxane layer (Fig. 2) with a molecularly imprinted nonconducting polymer coating could recognize proteins with subpicogram per liter sensitivity using electrochemical impedance spectroscopy. The sensors were specific for human ferritin, human papillomavirus derived E7 protein, and calmodulin [47].

2.2   Mechanical Spotting

Mechanical spotting allows one to deposit precisely a solution with volumes from microliters down to picoliters. The tools available are generally composed of dispensing objects mounted on an automated stage, allowing a precise localization of the resulting patterns. Initially, mechanical spotting was introduced in industrial processes for the fabrication of biochips. Commercially available systems are composed of arrays of needles containing the bioreceptors that are deposited on the substrate by capillarity [31, 32]. With these systems, volumes in the microliter range are commonly deposited. However, due to the high cost of biological samples, and for the sake of miniaturization to increase portability of the devices, tools where the quantity of biological material is largely reduced are necessary [48].

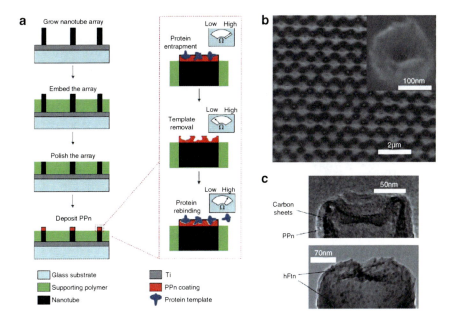

Fig. 2 Fabrication of an MIP-based protein nanosensor. (**a**) Schematic of nanosensor fabrication and template protein detection. The supporting polymer (SU8-2002) is spin-coated on a glass substrate containing nanotube arrays. Template proteins trapped in the polyphenol (PPn) coating are removed to reveal the surface imprints. *Inset*: hypothetical sensor impedance responses at critical stages of fabrication and detection. (**b**) Scanning electron microscopy image of a polished nanotube array after PPn coating. *Inset*: cross-section of nanotube tip after polishing. (**c**) TEM images of PPn-coated nanotube tip without (*top*) and with (*bottom*) human ferritin. Reproduced with permission from [47]. Copyright (2010) Nature Publishing Group: Nature Nanotechnology

In this context, micro and nanomachining processes show great potential for the fabrication of micro/nano cantilever-based dispensers and sensors [49]. One domain that is currently drawing considerable interest is the field of biosensors based on Micro Electro Mechanical Systems (MEMS), where free-standing structures such as cantilevers or micromembranes are used as a transducing platform together with sensitive layers [50, 51]. Although they were initially proposed as transducers for biosensing applications, cantilevers can also be used as liquid dispensing microtool for dip-pen lithography [52]. An ink containing the functional material is transferred through a channel or an aperture etched in the cantilever onto a directed position of the substrate. Figure 3 shows an example of arrays of silicon microcantilevers of a design comparable to quill pens, fabricated through silicon micromachining techniques. They are composed of a reservoir for successive depositions without reloading, connected to a fluidic channel of 5 μm in width for the patterning of dots in the nanoliter range. Each chip contains 12 cantilevers, where 10 of them are used for deposition, and 2 for positioning. Using this device, proteins, but also molecularly imprinted polymers [53], were spotted onto glass slides. In fact, arrays of dots containing MIP precursors were spotted that were then

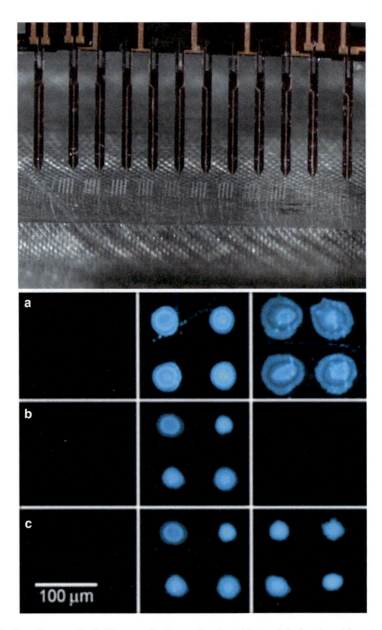

Fig. 3 *Top*: Photograph of silicon cantilevers used as depositing tool during deposition on a glass slide [53]. *Bottom*: Fluorescence microscope images of microdot matrices of the nonimprinted control polymer (*left*), positive control with the copolymerized template (*middle*), and imprinted polymer (*right*) synthesized with DMSO-diglyme-1% PVAc as the porogen (**a**) directly after deposition, (**b**) after washing and elution of the template, and (**c**) after reincubation in a solution of 30 μM fluorescein. Reprinted with permission from [53]. Copyright (2007) American Chemical Society

Fig. 4 (a) SEM image of a cantilevered nanopipette used for nanofountain pen (NFP). *Inset*: a zoom on aperture created at cantilever's tip for deposition. Reprinted with permission from [61]. Copyright (2005] American Chemical Society. (b) Fluorescence image of dots and lines of a fluorescein-imprinted polymer written with NFP. (c) AFM scan of MIP microdots [60]

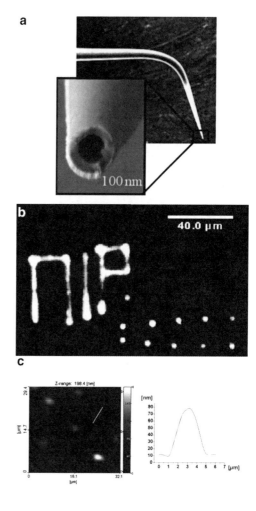

in-situ polymerized. This resulted in arrays of MIPs dots in the higher micrometer range templated with fluorescein and 2,4-D (Fig. 3, bottom). The selectivity of the dots was demonstrated by competitive binding assays, and by comparison to nonimprinted control dots.

To obtain smaller dots, dip-pen nanolithography (DPN) or nanofountain pen (NFP) methods have been developed. Both methods were inspired from atomic force microscopy (AFM). Dip-pen nanolithography consists in dipping an AFM probe in an ink [54]. The ink is then transferred to the substrate by capillary forces. With nanofountain pens, AFM tips are replaced by cantilevered nanopipettes containing an aperture [55]. The pipettes are filled with the solution and the liquid flows to the substrate when the opened tip is brought into contact with the surface. Using these methods, proteins [56], enzymes [57], DNA [58], and polymers [59] have been successfully deposited. In collaboration with Gheber's group, we have been able to deposit MIPs with NFP, resulting in features in the low micrometer range [60]. Cantilevered nanopipettes 500–600 μm in length with an aperture of 100–600 nm diameter (Fig. 4) were filled with an MIP precursor solution and mounted on an AFM system, composed of a flat scanner and two coaxial optical microscopes, for examination of the sample simultaneously from above and below during deposition. However, to be able to write dots and lines, the MIP recipe had to be optimized, since many parameters, in particular the wettability of the surface by the solution, the viscosity, and the surface tension influence the writing and the feature size, in addition to contact time. After thorough optimization, features of a few micrometers in diameter could be written. In order to demonstrate molecular imprinting, a fluorescent template was used to allow for fluorescence microscopy-based binding assays.

Ink-jet printing and drop-on-demand technology as another microspotting technique has become familiar as an industry compatible tool with a deposited volume ranging from microliters to nanoliters [62, 63]. Ink-jet is perfectly compatible with polymers and widely used for organic electronic devices [64], organic actuators [65], or lenses. Surprisingly, this promising method has so far not been used for patterning MIPs.

2.3 Optical Patterning

The possibility of photopolymerizing MIPs allows the application of optical patterning methods. Often, to functionalize a surface with an MIP, a precursor mixture is deposited on the surface by a standard technique like spin-coating before polymerization with UV under an oxygen-free atmosphere. Spin-coating of thin MIP films in this way was reported by Haupt and coworkers [66, 67]. More recent examples include the functionalization of a quartz crystal microbalance device for mass-sensor applications [68, 69]. Optical patterning of films can be achieved by local photopolymerization. For this, two main strategies can be applied: using a photo-mask to illuminate only parts of interest or using a localized light beam. Microstereolithography is a method for manufacturing tridimensional shapes by means of localized photopolymerization using a sharply focused laser beam [70]. Shea and coworkers have applied this technique to MIPs. A UV laser beam focused to 1–2 μm was directed onto a substrate coated with a thin layer of MIP precursors mixture that was repeatedly moved in the x–y plane to design a pattern, resulting in the localized layer-by-layer photopolymerization of MIP microstructures imprinted with 9-dansyl adenine (Fig. 5) [71]. To achieve 3D microstructuring, the effective area of photopolymerization needs to be effectively controlled. Important are the concentration and molar extinction coefficients of the initiator. An additional UV absorber was used in this example.

The use of a focused infrared laser beam on a surface results in local heating, which has been applied to synthesize MIPs locally by thermal polymerization [72]. In this case, a common thermo-initiator, azobisisobutyronitrile, was used together with a standard MIP recipe, to generate MIP microdots in a microfluidic device.

Fig. 5 MIP microstructure
(600 × 600 μm^2) fabricated
by microstereolithography.
Reproduced with permission
from [71]. Copyright 2003
WILEY-VCH Verlag GmbH
& Co

The 116-μm focused beam raised the temperature locally up to 70 °C. An $X–Y–Z$ moving stage permitted the creation of arrays of dots of 350 μm diameter.

Optical patterning through photo-masking should allow for the large-scale fabrication of MIP chips. In contrast to the above serial methods based on localized beams, photo-masking is a parallel microfabrication method [29, 30]. A consequence is the short time necessary to create microarrays, making these methods suitable for industrial processes. UV photolithography is intensively used in micro and nanotechnologies to pattern photoresists that are used as masks for metal deposition or materials etching [30]. A photoresist is uniformly deposited on a substrate by spin-coating, baked, and placed under a mask aligner. The subsequent photolithography step consists in defining locally illuminated areas in the resist by using a chromium/glass mask located at or close to the photoresist. Consequently, illuminated areas will remain on the substrate or will be dissolved after development, depending on polarity of the photopolymer. Photolithography was introduced in 2004 for patterning MIPs at the microscale [73]. A voltammetric biosensor was functionalized by an MIP against albuterol where patterns of 50 μm were achieved. However, conventional photoresist was used as the MIP matrix rather than the common methacrylate or vinyl-based copolymers, which limits the applicability of the technique. More recently, the same group published another example of combination between MIPs and photolithographic printing [45], describing the integration of acrylic MIP patterns into a miniaturized three-electrode cell, using a photolithographic approach. By prepolymerizing an MIP made of benzyl methacrylate, methacrylic acid, and 2-hydroxyethyl methacrylate, they obtained a sensing layer that could discriminate the target albuterol from the interfering analogs such as clenbuterol and terbutalanine. Compared to their previous work, they enhanced the pattern resolution to 20 μm, and on-chip MIP parallelization seemed to be within reach. The group of Byrne [74] demonstrated photopatterning by photolithography being potentially applicable to polymer gel matrices. A novel copolymer containing poly(ethylene glycol) dimethacrylate and acrylamide was developed and used as a matrix to be imprinted with D-glucose. The polymer gel patterns exhibited thicknesses of about 13 μm and a number of geometrical configurations. Incubation of these patterns with a fluorescent glucose derivative indicated that binding was not only on the surface of the structures, but homogeneously distributed throughout the matrix.

In 2009, Ayela and coworkers proposed the large scale fabrication of MIP microbiochips using an optimized photo-masking process [75]. A resolution of

MIP features down to 1.5 µm on 4-in silicon wafers was reached (Fig. 6). A wide range of micrometric patterns with different geometries can be obtained, such as lines, spirals, circle matrices, and circular, squared, or hexagonal patterns (Fig. 6 bottom). Multiplexed chips containing three different polymers were also fabricated, paving the road to mass production of biomimetic chips. Fluorescence microscopy was used to test for the binding of fluorescent model analyte to the micropatterns.

Projection photolithography is a related method that was developed in the early 1980s. The photoresist and the mask are kept at some distance from each other and the image from the mask is focused onto the photoresist by an optical system. It is thereby possible to achieve patterns with appreciable resolution, while avoiding problems that can occur due to contact between the mask and the photoresist. Pattern resolution is limited by diffraction of the light at the edges of the opaque zone of the mask. The only example so far of MIP patterning using projection printing was published by Linares et al. [76]. A fluorescence microscope was used for projection, the photomask replacing the field diaphragm of the microscope. Illumination is thus possible using the excitation light source of the microscope, and the pattern is reduced in size by the magnification factor of the objective. The MIP

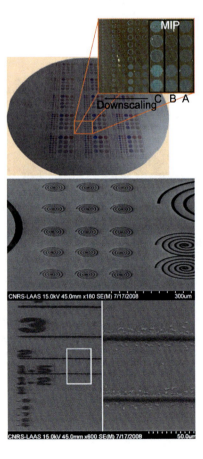

Fig. 6 *Top*: A 4-in. silicon substrate with multiplexed MIPs. Features defining one chip are repeated on the entire wafer. *Inset*: A, MIP templated with dansyl-L-Phe; B, MIP templated with boc-L-Phe; C, nonimprinted control polymer [75]. *Bottom*: Examples of MIP patterns obtained by photolithography. With this method, a wide variety of shapes is achieved with resolution down to 1.5 µm [75] Reproduced by permission of the Royal Society of Chemistry

photoresist was a standard mixture of vinyl-based monomers. Using a 10× micro-scope objective and a 0.5 mm feature size on the mask, dot sizes of 70 μm were obtained, although much smaller structures are possible. Moreover, if the pattern was projected on a nanoporous alumina oxide substrate coated with the MIP precursors, nanostructured dots were obtained composed of parallel nanofilaments, 150 nm in diameter and 4 μm long. In that way, the surface-to-volume aspect ratio of the MIP dots is considerably increased, and the binding site accessibility improved compared to flat dots. Indeed, the technique allowed for the imprinting of a protein (myoglobin) in the filament surface.

2.4   Nanopatterning by Soft Lithography

Soft lithography has raised considerable interest in recent years as a low-cost alternative to photolithography or electron beam lithography [34, 77]. Soft lithog-raphy was introduced by G.M Whitesides in the late 1990s as an injection molding process [78], before developing further to the equally powerful technique known as micro transfer molding (μTm) [79]. Approaches based on soft lithography are based on a common principle. A stamp, generally made of polydimethylsiloxane (PDMS), is fabricated using a patterned mold as master. The resulting stamp carries the negative 3D image of the master. It can then be used for micro molding in capillaries (MIMIC) [80], or as a deposition tool in the case of micro contact printing (μCP) and micro transfer molding (μTm) [81]. With these methods, numerous materials have been deposited structured at micro and nanoscales, including metals, polymers [81], and biomolecules [82].

The first successful example of soft lithography to generate MIP microstructures was reported by Yan and Kapua using the MIMIC technique [83]. The MIP precursors solution with the herbicide 2,4-D as the molecular template was introduced at one end of the PDMS stamp that was placed in conformal contact with the glass substrate, and capillary forces drew the solution into the micro-channels. After polymerization, MIP micropatterns with a cross-sectional dimen-sion of 20×20 μm^2 were obtained. However, for a soft lithography approach, the achieved resolution was quite poor. In addition, large areas cannot be covered using the MIMIC method since viscosity and surface tension limit the transfer of the polymer precursors into the capillaries. Other solutions based on microcontact printing were found to enhance resolution and surface coverage. A first example is the two-dimensional surface imprinting of polymers to obtain MIP micropatterns (Fig. 7) [84]. A template molecule (theophylline) was covalently anchored to the PDMS stamp surface, and the stamp was then brought into contact with an Si/SiO$_2$ surface where an MIP precursors solution had been applied. The polymer was then cross-linked thermally in presence of the stamp. After polymerization and removal of the stamp, target-specific recognition sites at the surface of the polymer micropatterns were revealed. Arrays of lines of 25 μm in width were produced

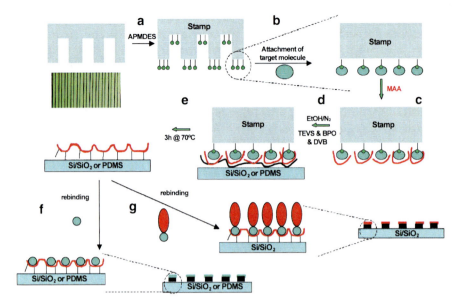

Fig. 7 Schematic illustration of the principle of 2D molecular imprinting. (**a**) Amination of the PDMS stamp. (**b**) Attachment of the carboxylic derivative of theophylline on the aminated stamp. (**c**) Inking with the recognition monomer (MAA). (**d**) Rinsing and inking with a mixture of anchoring monomer (TEVS), radical initiator (BPO) and cross-linking agent (DVB). (**e**) Stamping on Si/SiO$_2$ or a PDMS substrate and polymerization. (**f**). Removal of the stamp and rebinding of theophylline. (**g**) Rebinding of fluorophore-labeled theophylline. Reprinted with permission from [84]. copyright (2007) American Chemical Society

using this method and the molecularly imprinted polymer exhibited selectivity for theophylline, as revealed by binding assays.

Still, the templating of MIPs combined with three-dimensional structuring at the nanoscale remains challenging. A solution was proposed recently using a μTm process to create arrays of nanopatterns at large scale [85]. After fabrication of the PDMS stamp, its surface was made hydrophilic by oxygen plasma treatment to enhance its wettability before inking with the precursors solution consisting of trimethylolpropane trimethacrylate (TRIM) as the cross-linker, 2-carboxyethyl acrylate (CEA) as the functional monomer, diglyme as the porogenic solvent containing 1% poly(vinyl acetate) as co-porogen, and the molecular template boc-L-Phe (Fig. 8). The inked stamp was then dried under nitrogen and scraped with a flat piece of PDMS to remove excess of solution. A conformal contact between the stamp and a silanized glass substrate was then established, before UV photopolymerization at 365 nm. As a result, arrays of MIP nanolines characterized by a width of 660 nm, a height of 140 nm, and a pitch of 1 μm were obtained, and a relatively large area of 400 × 400 μm^2 could be covered (Fig. 8b). Fluorescence microscopy was used after incubation with the fluorescent analog dansyl-L-Phe, demonstrating the generation of specific cavities by molecular imprinting.

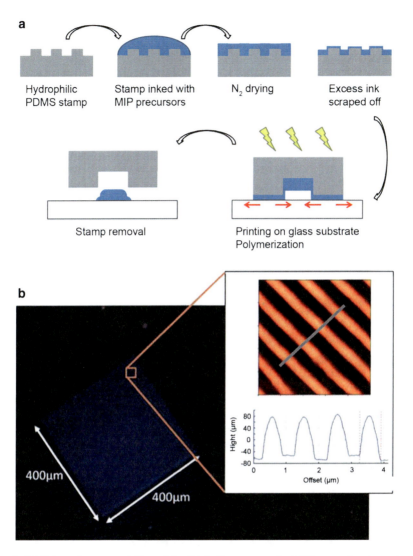

Fig. 8 (**a**) Schematic representation of the fabrication process for MIP nanopatterning using a hydrophilic PDMS stamp. (**b**) Dark field microscopy image of an MIP nanopatterned by soft lithography. *Inset*: AFM topography scan of the MIP nanopattern [85]

To enlarge the panel of soft lithography approaches to create MIP nanopatterns, nanoimprint lithography (NIL) was introduced very recently. As for microcontact printing, patterns resulted from combination of a bottom-up approach based on molecular imprinting and a top-down nanostructuring method by using NIL. Nanoimprint lithography is now being increasingly used to fabricate nano-structured polymer resists [86, 87]. One of the NIL patterning processes, UV-NIL, involves the in situ polymerization of monomer precursors confined with a small reaction volume defined by the structures of the NIL stamp. The final resolution of material patterns generated by NIL can be as low as a few nanometers, limited only by the physical resolution of the pattern on the stamp.

Combining a methacrylate MIP precursor solution templated with *S*-propranolol and NIL according to a simple protocol, arrays of MIPs were successfully fabricated with a resolution down to 100 nm (Fig. 9) [88]. Both thermo and photopolymerization of the precursor solution were tested, but UV initiated nanoimprint gave better results in a shorter time (8 min for global processing). This approach is thus in agreement with industrial processes where high aspect ratio is necessary to fabricate sensitive biochips, while short processing time is suitable for future mass production.

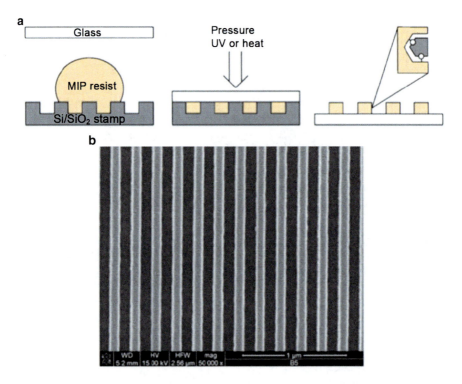

Fig. 9 (**a**) Patterning MIP nanostructures using reactive NIL. A monomer mixture containing the molecular template is deposited on a patterned stamp. A substrate is pressed against the stamp, causing the reaction mixture to fill up the empty space defined by the stamp pattern. The monomers are polymerized by UV or thermal polymerization. Releasing the stamp (demolding) furnishes an MIP pattern containing molecularly imprinted binding sites. (**b**) SEM images of a patterned MIP fabricated by UV-NIL. Reprinted with permission from [88]. The Royal Society of Chemistry

2.5 Near-Field-Assisted Optical Lithography

Basic research on near-field optics started from several independent groups in Europe, US and Japan, in the early 1980s. Rapid progress was made in the area as from the 1990s [89]. Near-field optics concerns phenomena involving nonpropagating inhomogeneous fields (evanescent electromagnetic waves) and their interactions with matter [90]. Optical near-fields are located at the source region of optical radiation, or at the surface of materials interacting with free radiation, and in many situations they are explored for their ability to localize optical energy to lenghthen scales smaller than the diffraction limit of roughly $\lambda/2n$, with λ being the wavelength of light and n the refractive index. Lougnot's group was first to report the fabrication of very thin polymer parts by the use of evanescent wave photopolymerization (PEW) [91–93]. The concept is based on total reflection of a laser beam on a substrate in order to create an evanescent field in this substrate (Fig. 10). When a photosensitive monomer mixture was placed on top of the substrate, the thickness of the monomer layer in which actinic light is confined was similar to the penetration depth of the evanescent field. Thus, polymerization results in polymeric parts with a thickness of only a few tens of nanometers. In 2007, Soppera et al. illustrated the potential of evanescent fields to fabricate 1D and 2D periodically patterned thin polymer films, lines with sub-100 nm width being easily accessible [95]. Similarly, the evanescent field generated by surface plasmon resonance at a metal surface can be used for localizing photopolymerization of thin films on this surface [96]. Fuchs et al. have recently reported the first proof of concept of MIP photopatterning using near-field assisted optical lithography [94]. The MIP precursors mixture had to be specifically optimized to be compatible with this technique. It uses photopolymerization of an MIP under visible light (405 nm). Compromises had to be found with respect to a high viscosity of the

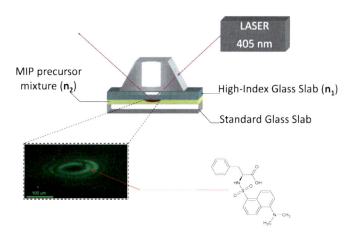

Fig. 10 Principle of polymerization by evanescent wave, and fluorescence microscopic image of an ultrathin MIP dot imprinted with dansyl-L-Phe. Adapted with permission from [94]. Copyright (2011) American Chemical Society

precursor solution being required for near-field polymerization by evanescent waves, and the necessity to use solvents in the mixture to render the MIP porous. The MIP microdots exhibited sub-100 nm thicknesses and a mesoporous morphology. They were imprinted with the amino acid derivative Z-(L)-Phe, and incubation tests with the fluorescent derivative dansyl-(L)-Phe revealed specific target binding and a certain degree of enantioselectivity.

3 MIP Nanomaterials

A recent trend goes towards the synthesis of MIPs in the form of nanomaterials or nanostructured materials, or the combination of MIPs with other materials in nanocomposites. In that way, materials with enhanced properties can be obtained. Because of the small dimensions and large numbers of features per volume, very high surface-to-volume ratios are possible. This shortens diffusion distances and improves access to the binding sites by the target molecules, since all binding sites are situated at or close to the material's surface. In addition, nanostructures can be obtained with very regular tunable shapes and sizes. Combining the imprinting process with neat supramolecular chemistry, researchers have even been able to synthesize well-defined monomolecular templated materials. Ishi-i et al. created "imprints" at the surface of fullerenes. They introduced two boronic acid groups into fullerene using saccharides as template molecules, and observed stereoselective rebinding of the saccharide [97]. Later, Zimmerman's group developed a very interesting approach for the "imprinting" of a molecular memory inside a dendrimer [98]. Their method involved the covalent attachment of dendrons to a porphyrin core (the template), cross-linking of the end-groups of the dendrons, and removal of the porphyrin template by hydrolysis. This technique yields homogeneous binding sites, allows quantitative template removal, produces only one binding site per polymer molecule, and the materials are soluble in common organic solvents.

The most basic form of MIP nanomaterials is the spherical nanoparticle, obtained by a number of techniques such as microemulsion polymerization [99–101], and polymerization in diluted solutions resulting in nanospheres and microgels [102–106]. Microgels (also sometimes referred to as nanogels) are particularly interesting, since they represent soluble, though cross-linked, MIPs with a size in the low nanometer range, close to that of proteins.

3.1 Nanocomposite MIPs

Composite MIP nanomaterials are of increasing interest, since the inclusion in particular of inorganic additives into the MIP matrix provides additional useful properties, such as magnetic susceptibility, fluorescence, plasmonic enhancement, or simply an increased stability of the material.

3.1.1 Magnetic Nanoparticles

Nanocomposite particles have been reported bearing inorganic cores of superparamagnetic iron oxide, which allows their easier handling and separation from a suspension with a magnet [107]. The use of controlled/living radical polymerization methods such as iniferters, RAFT, or ATRP is particularly indicated for the synthesis of such core–shell composites [108]. For example, Lu et al. prepared a bisphenol A imprinted magnetic composite by funtionalizing Fe_3O_4 nanoparticles with the ATRP initiator 2-bromo-2-methyl-N-(3-(triethoxysilyl)propyl) propanamide [109]. Gai et al. imprinted the enzyme lysozyme using 2-bromo-2-methyl propanamide-modified Fe_3O_4 via ATRP as well [110]. Others used RAFT polymerization for the controlled synthesis of magnetic MIP particles [111, 112]. In all examples of controlled polymerization, very thin MIP layers only a few nanometers thick were synthesized around the magnetic core particle. It has also been shown that, owing to the living character of RAFT polymerization, an additional outer layer can be added to the particles, by attaching additional functional groups thus enabling fine-tuning of their surface properties, [112].

3.1.2 Quantum Dots

Quantum dot nanocrystals are another example of an inorganic nanomaterial that can be incorporated in an MIP. Quantum dots (QDs) are semiconductor colloidal crystals of cadmium selenide, cadmium sulfide, indium arsenide, or indium phosphide. An immediately visible optical feature of colloidal quantum dots is their color. Illuminated with UV light, quantum dots of the same material, but with different sizes, can emit light of different colors. The larger the dot, the redder (lower energy) its fluorescence spectrum becomes. Conversely, smaller dots emit bluer (higher energy) light. Detection of analyte molecules by utilizing the fluorescent/luminescent chromophoric nature of semiconductor nanoparticles is a well-established method in biological science. They are more effective than organic dyes as fluorescent labels, due to their broader excitation spectra and better stability against photodegradation. The hydrophobicity and the solvent-dependent quantum yields of organic dyes are other issues that can be circumvented by replacing organic dyes by quantum dots as reporter molecules in a variety of bioassays [113, 114]. There have been a few reports on the combination of quantum dots with MIPs. Lin et al. [115] incorporated QDs in MIPs, resulting in a sensor material where the binding of the target analyte quenched the photoluminescence (PL) emission of the quantum dots. MIPs against several templates were synthesized with methacrylic acid and ethylene glycol dimethacrylate as functional monomer and cross-linker. The CdSe/ZnS semiconductor nanocrystals used were derivatized with 4-vinylpyridine and mixed with the other MIP precursors before polymerization. Diltemiz et al. [116] proposed a thiol ligand-capping method with polymerizable methacryloylamido-cysteine attached to QDs. Methacryloylamidohistidineplatinum (MAH-Pt(II)) was used as a metal-chelating monomer to bind to guanosine templates of DNA via metal coordination–chelation interactions. The binding affinity of the guanosine imprinted nanocrystals was determined by a Langmuir isotherm, yielding micromolar dissociation constants with guanosine and guanine. More recently, Li et al. reported molecularly imprinted silica nanospheres with embedded CdSe quantum dots for the optosensing of pyrethroids [117].

3.1.3   Carbon Nanotubes

Other nanomaterials that were incorporated into MIPs are carbon nanotubes (CNT). They can be useful for their fluorescence, and for their semiconductor properties. Lee et al. [118] developed a biosensor system based on a CNT field effect transistor using MIPs as recognition elements. A 10-nm MIP film templated with theophylline was polymerized onto CNT via an acrylated Tween 20 surfactant. Binding of theophylline to the MIP composite was better than to the nonimprinted control composite, and theophylline bound selectively over the structural analog caffeine. Kan et al. [119] reported an MIP composite containing multiwalled CNTs (MWNTs) for use as the selective layer in an electrochemical sensor. The MIP was prepared with dopamine as the template and methacrylic acid and trimethylolpropane trimethacrylate as monomers. Vinyl groups introduced on MWNTs surface were found to be the key for the formation of the MWNTs-MIP composite, the thickness of the MIP layer being 15–20 nm (Fig. 11). Attenuated total reflection Fourier transform infrared spectroscopy, transmission electron microscopy, scanning electron microscopy, and thermogravimetric analysis were used to characterize the composite. The authors demonstrated that MWNTs-MIPs not only possessed a rapid dynamic adsorption but also exhibited a higher selectivity toward dopamine compared to epinephrine. The electrochemical sensor fabricated by modifying MWNTs-MIPs on a glassy carbon electrode was used to detect dopamine at concentrations between 0.5 and 200 μM. Others grafted an acrylic MIP onto CNT by surface-initiated polymerization with iniferters [120]. A more recent example of a CNT-MIP composite was published by Cai et al. [47]. They reported arrays of carbon nanotube tips embedded in a polydimethylsiloxane layer (Fig. 2) where a molecularly imprinted nonconducting polymer specific for proteins was generated on the tips of the CNTs by electropolymerization [47]. A free-standing carbon nanotube array was used as a 3D platform for surface-molecularly imprinted polypyrrole for the sensing of caffeine [121].

Metal nanoparticles have also been included into MIPs. Such particles can be used, for example, as nanoantennae for the enhancement of electromagnetic waves (plasmonic enhancement). It has been shown by He et al. [122] that a thin layer (20–120 nm) of testosterone-imprinted silica could be synthesized around 350 nm silver particles in a controlled way. The composite material showed specific binding of the testosterone target. Matsui et al. [123] reported a molecularly imprinted polymer with immobilized Au nanoparticles as a sensing material for spectrometry. The sensing mechanism is based on the variable proximity of the Au nanoparticles

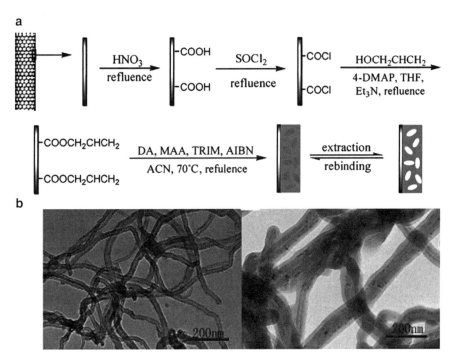

Fig. 11 (**a**) Schematic representation of the synthesis procedure of MIP-coated MWNTs. (**b**) TEM image of MWNT (*left*) and MWNT-MIP (*right*). Adapted with permission from [119]. Copyright (2008) American Chemical Society

immobilized in the imprinted polymer, which swells upon selective binding, causing a blue-shift in the plasmon absorption band of the immobilized Au nanoparticles (Fig. 12a). The same group also reported a dopamine-imprinted MIP film with embedded gold nanoparticles that was deposited on a continuous gold layer for surface plasmon resonance (SPR) measurements [124]. The sensing is based on swelling of the imprinted polymer gel that is triggered by analyte binding (Fig. 12b). The swelling causes an increase in distance between the gold nanoparticles and the gold layer, shifting the dip of the SPR curve to a higher SPR angle. The modified sensor chip showed an SPR response as a function of dopamine concentration. The gold nanoparticles were shown to be effective for enhancing the signal intensity compared to a sensor chip without gold nanoparticles included in the MIP.

Bompart et al. [125] reported the synthesis of composite nanoparticles of approximately 500 nm diameter consisting of a polymer core, a layer of gold nanoparticles attached to the core, and a few nanometers thick MIP outer layer (Fig. 13). These particles were used as individually addressable nanosensors, where surface-enhanced Raman spectroscopy was used to detect the binding of the target analyte, the beta-antagonist propranolol. A 1,000× improved detection limit was

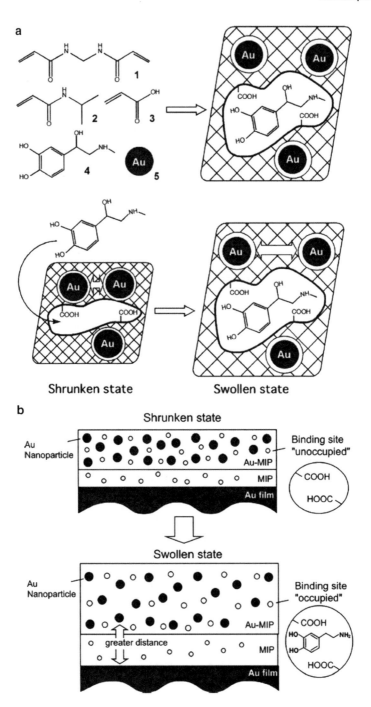

Fig. 12 (**a**) Schematic representation of the preparation of a molecularly imprinted polymer with immobilized Au nanoparticle and the detection of an analyte upon selective swelling of the MIP. Reprinted with permission from [123]. Copyright (2004) American Chemical Society. (**b**) Schematic representation of Au-MIP/MIP-coated SPR sensor chip. 'Reprinted with permission from [124]. Copyright (2005) American Chemical Society

Fig. 13 (**a**) Schematic representation of the MIP-Au composite sensor particles. (**b**) TEM images of the sensor particles [125]

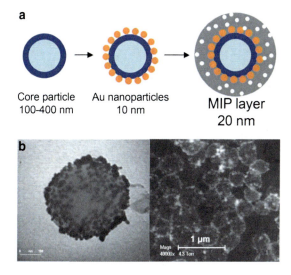

obtained with these composite particles, compared to conventional Raman measurements on single plain MIP particles.

3.1.4 Electrospun Nanofibers Including MIPs

Electrospinning is a method allowing creation of polymer fibers with diameters in the range between a few tens of nanometers to a few micrometers, starting from a solution of preformed polymer. MIP nanoparticles have been included into nanofibers by electrospinning [126, 127]. In another case, the nanofibers were directly produced by electrospinning and polymerizing an MIP-precursor solution [128]. Such MIP fibers can then be used, for example, for the preparation of affinity separation materials [129] or as affinity layers in biosensors [127, 130].

3.2 MIP Nanowires and Nanotubes Obtained by Nanomolding

MIP nanowires have been synthesized using anodized alumina as a sacrificial nanoporous inorganic template for the imprinting of small molecules and proteins [131–135]. It was possible to render these nanowires magnetic by inclusion of superparamagnetic $MnFe_2O_4$ nanocrystallites, thus facilitating their separation from a suspension [133]. Of particular interest for biosensing applications will be if these nanofilaments are attached to a flat surface. Our group has recently shown that layers of surface-bound MIP nanofilaments, in addition to molecular recognition, allow one to fine-tune the surface properties towards superhydrophobicity or superhydrophilicity [134, 135]. Figure 14 shows scanning electron microscope images of a porous alumina

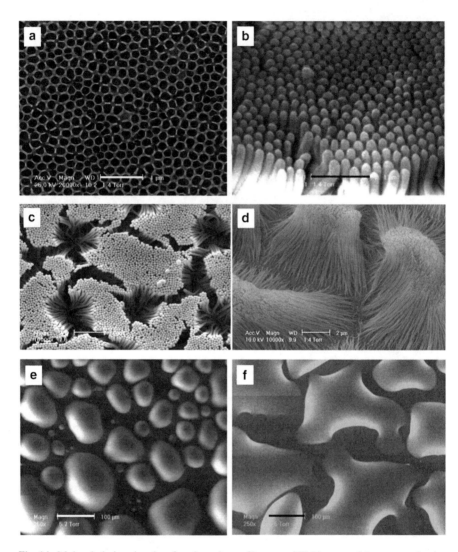

Fig. 14 Molecularly imprinted surface-bound nanofilaments. SEM images of the porous alumina template (**a**) and MIP nanofilaments after nanomolding (**b**). SEM images of 200 nm × 2 μm MIP (**c**) and of 50 nm × 5 μm nanofilaments (**d**). ESEM images of water microdroplets condensed on a surface with 200 nm × 1 μm nanofilaments (**e**) and on a flat surface consisting of the same polymer (**f**) [134, 135]. *Scale bars*: 1 μm (**a**, **b**), 100 μm (**c**, **d**)

template, a layer of surface-bound poly(divinylbenzene) nanofilaments, and water microdroplets condensed on such a nanostructured surface compared to a chemically identical flat surface. It can be clearly seen that the nanostructuring results in a more hydrophobic surface since the contact angles of the droplets are considerably higher than those on the flat surface. More recently we have shown that these nanofilaments can be patterned on a surface by projection photolithography [76]. Using a similar

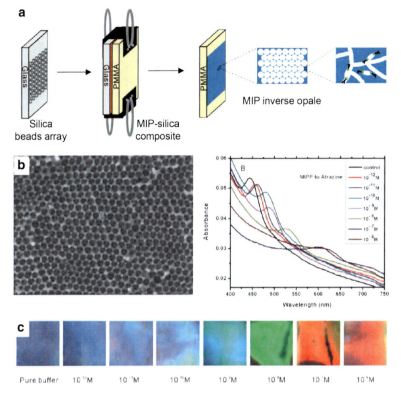

Fig. 15 (**a**) Schematic representation of the preparation of an MIP photonic crystal. (**b**) SEM image of an MIP photonic polymer film. (**c**) Color changes induced in the MIP photonic polymer by the rebinding of atrazine at different concentrations. (**d**) Optical response of the MIP photonic polymer after incubation with atrazine at different concentrations. Reproduced with permission from [139]. Copyright 2008 WILEY-VCH Verlag GmbH & Co

nanomolding process on nanoporous alumina, molecularly imprinted sol–gel nanotubes have been generated. However, the alumina template was not dissolved, and the resulting composite membranes could be used for the separation of the target molecule, estrone. The authors showed that, owing to the very thin imprinted sol–gel layer, release of the imprinting template was facilitated [136]. Xie et al. reported the molecular imprinting of 2,4,6-trinitrotoluene (TNT) in the walls of highly uniform silica nanotubes. Silica nanotubes were produced by nanomolding in a porous alumina membrane by a sol–gel imprinting process, whereby acid–base pairing interactions were established between the template TNT and the functional monomer 3-aminopropyltriethoxysilane (APTES). Nanotubes were separated by centrifugation, and binding properties were studied by UV–visible spectroscopy of TNT in the resultant floating liquid [137].

3.2.1 MIP Photonic Crystals

Another nice example of nanostructuring an MIP layer is the work published by Wu et al. [138, 139] who developed a label-free optical sensor based on molecularly imprinted photonic polymers. Photonic crystals were prepared by self-assembly of silica nanospheres. The space between the spheres was then filled with MIP precursor solution. After polymerization, the silica was dissolved, leaving an MIP in the form of a 3D-ordered interconnected macroporous inverse polymer opal (Fig. 15). The authors were able to detect traces of the herbicide atrazine at low concentrations in aqueous solution [139]. Analyte adsorption into the binding sites resulted in a change in Bragg diffraction of the polymer characterized by a color modification (Fig. 15).

4 Conclusion

In conclusion, micro and nanofabrication methods have a great potential for the development of sensors and microchips based on molecularly imprinted polymers, since they allow the patterning and structuring of MIPs on transducer surfaces. It has been shown that, because of their stability, MIPs can be easily integrated in a number of standard microfabrication processes. Thereby, the possibility of photopolymerizing MIPs is a particular advantage. In addition to specific molecular recognition properties, nanostructured MIPs and MIP nanocomposites allow for additional interesting properties in such sensing materials, for example, amplification of electromagnetic waves by metal nanoparticles, magnetic susceptibility, structural colors in photonic crystals, or others. These materials will therefore find applications in particular for chemical and biochemical detection, monitoring, and screening.

References

1. Arshady R, Mosbach K (1981) Makromol Chem 182:687–692
2. Wulff G, Sarhan A (1972) Angew Chem Int Ed 11:341
3. Haupt K (2003) Anal Chem 75:376A–383A
4. Zimmerman SC, Lemcoff NG (2004) Chem Commun 5–14
5. Bossi A, Bonini F, Turner APF, Piletsky SA (2007) Biosens. Bioelectron. 22:1131–1137
6. Fischer L, Müller R, Ekberg B, Mosbach K (1991) J Am Chem Soc 113:9358–9360
7. Schweitz L, Andersson LI, Nilsson S (1997) Anal Chem 69:1179–1183
8. Vlatakis G, Andersson LI, Müller R, Mosbach K (1993) Nature 361:645–647
9. Andersson LI (1996) Anal Chem 68:111–117
10. Haupt K, Mayes AG, Mosbach K (1998) Anal Chem 70:3936–3939
11. Surugiu I, Danielsson B, Ye L, Mosbach K, Haupt K (2001) Anal Chem 73:487–491
12. Piletsky SA, Parhometz YP, Lavryk NV, Panasyuk TL, El'skaya AV (1994) Sens Actuators B 18–19:629–631

13. Kriz D, Ramström O, Svensson A, Mosbach K (1995) Anal Chem 67:2142–2144
14. Jakusch M, Janotta M, Mizaikoff B, Mosbach K, Haupt K (1999) Anal Chem 71:4786–4791
15. Dickert FL, Hayden O (2002) Anal Chem 74:1302–1306
16. Ayela C, Vandevelde F, Lagrange D, Haupt K, Nicu L (2007) Angew Chem Int Ed 46:9271–9274
17. Andersson LI (2000) Analyst 125:1515–1517
18. Crescenzi C, Bayoudh S, Cormack PAG, Klein T, Ensing K (2001) Anal Chem 73:2171–2177
19. Pichon V, Haupt K (2006) J Liq Chromatogr Relat Technol 29:989–1023
20. Liu JQ, Wulff G (2008) J Am Chem Soc 130:8044–8054
21. Hedin-Dahlstrom J, Rosengren-Holmberg JP, Legrand S, Wikman S, Nicholls IA (2006) J Org Chem 71:4845–4853
22. Alexander C, Smith CR, Whitcombe MJ, Vulfson EN (1999) J Am Chem Soc 121:6640–6651
23. Allender CJ, Brain KR, Heard CM, Pellett MA (1997) Proc Int Symp Controlled Release Bioact Mater 24:585–586
24. Norell MC, Andersson HS, Nicholls IA (1998) J Mol Recogn 11:98–102
25. Byrne ME, Park K, Peppas NA (2002) Adv Drug Deliv Rev 54:149–161
26. Byrne ME, Hilt JZ, Peppas NA (2008) J Biomed Mater Res A 84A:137–147
27. Lueking A, Cahill DJ, Müllner S (2005) Drug Discov Today 10:789–794
28. Espina V, Mehta AI, Winters ME, Calvert V, Wulfkuhle J, Petricoin EF, Liotta LA (2003) Proteomics 3:2091–2100
29. Takahashii K, Seio K, Sekime M, Hino O, Esashi M (2002) Sens Actuators B 83:67–76
30. Bates AK, Rotschild M, Bloomstein TM, Fedynyshyn TH, Kunz RR, Liberman V, Switkes M (2001) IBM J Res Dev 45:605–614
31. Heise C, Bier FF (2005) Top Curr Chem 261:1–27
32. Schena M, Shalon D, Davis RW, Brown PO (1995) Science 270:467–470
33. Xia Y, Rogers JA, Paul KE, Whitesides GM (1999) Chem Rev 99:1823–1848
34. Lalo H, Vieu C (2009) Langmuir 7:7752–7758
35. Ayela C, Nicu L, Soyer C, Cattan E, Bergaud C (2006) J Appl Phys 100:054908–054908–09
36. Reculusa S, Heim M, Gao F, Mano N, Ravaine S, Kuhn A (2011) Adv Funct Mater 21:691–698
37. Malitesta C, Losito I, Zambonin PG (1999) Anal Chem 71:1366–1370
38. Pietrzyk A, Suriyanarayanan S, Kutner W, Chitta R, D'Souza F (2009) Anal Chem 81:2633–2643
39. Hutchins RS, Bachas LG (1995) Anal Chem 67:1654–1660
40. Spurlock LD, Jaramillo A, Praserthdam A, Lewis J, Brajtertoth A (1996) Anal Chim Acta 336:37–46
41. Panasyuk TL, Mirsky VM, Piletsky SA, Wolfbeis OS (1999) Anal Chem 71:4609–4613
42. Cheng ZL, Wang EK, Yang XR (2001) Biosens Bioelectron 16:179–185
43. Ulyanova YV, Blackwell AE, Minteer SD (2006) Analyst 131:257–261
44. Du D, Chen S, Cai J, Tao Y, Tu H, Zhang A (2008) Electrochim Acta 53:6589–6595
45. Huang HC, Huang SY, Lin CI, Lee YD (2007) Anal Chim Acta 582:137–146
46. Gomez-Caballero A, Goicolea MA, Barrio RJ (2005) Analyst 130:1012–1018
47. Cai D, Ren L, Zhao HZ, Xu CJ, Zhang L, Yu Y, Wang HZ, Lan YC, Roberts MF, Chuang JH, Naughton MJ, Ren ZF, Chiles TC (2010) Nat Nanotechnol 5:597–601
48. Lynch M, Mosher C, Huff J, Nettikadan S, Johnson J, Henderson E (2004) Proteomics 4:1695–1702
49. Leïchlé T, Nicu L (2008) J Appl Phys 104:111101–111101–16
50. Vancura C, Li Y, Lichtenberg J, Kirstein K-U, Hierlmann A, Josse F (2007) Anal Chem 79:1646–1654
51. Shekhawat G, Tark S-H, Dravid VP (2006) Science 311:1592–1595

52. Belaubre P, Guirardel M, Garcia G, Leberre V, Dagkessamanskaia A, Trévisiol E, François JM, Pourciel JB, Bergaud C (2003) Appl Phys Lett 82:3122–3124
53. Vandevelde F, Leïchlé T, Ayela C, Bergaud C, Nicu L, Haupt K (2007) Langmuir 23:6490–6493
54. Piner RD, Zhu J, Xu F, Hong SH, Mirkin CA (1999) Science 283:661–663
55. Lewis A, Kheifetz Y, Shambrodt E, Radko A, Khatchatryan E, Sukenik C (1999) Appl Phys Lett 75:2689–2691
56. Taha H, Marks RS, Gheber LA, Rousso I, Newman J, Sukenik C, Lewis A (2003) Appl Phys Lett 83:1041–1043
57. Hyun J, Kim J, Craig SL, Chilkoti AJ (2004) J Am Chem Soc 126:4770–4771
58. Demers LM, Ginger DS, Park SJ, Li Z, Chung SW, Mirkin CA (2002) Science 296:1836–1838
59. Sokuler M, Gheber LA (2006) Nano Lett 6:848–853
60. Belmont A-S, Sokuler M, Haupt K, Gheber LA (2007) Appl Phys Lett 90:193101–193101–3
61. Ionescu RE, Marks RS, Gheber LA (2005) Nano Lett 5:821–827
62. Abe K, Suzuki K, Citterio D (2008) Anal Chem 80:6928–6934
63. Blanchard AP, Kaiser RJ, Hood LE (1996) Biosens Bioelectron 11:687–690
64. Mannerbro R, Ranlöf M, Robinson N, Forchheimer R (2008) Synth Met 158:556–560
65. Van Oosten CL, Bastiaansen CWM, Broer DJ (2009) Nat Mater 8:677–682
66. Schmidt RH, Mosbach K, Haupt K (2004) Adv Mater 16:719–722
67. Schmidt RH, Belmont A-S, Haupt K (2005) Anal Chim Acta 542:118–124
68. Lieberzeit PA, Rehman A, Najafi B, Dickert FL (2008) Anal Bioanal Chem 391:2897–2903
69. Sener G, Ozgur E, Yilmaz E, Uzun L, Say RI, Denizli A (2010) Biosens Bioelectron 26:815–821
70. Kawata S, Sun HB, Tanaka T, Takada K (2001) Nature 412:697–698
71. Conrad PG, Nishimura PT, Aherne D, Schwartz BJ, Wu D, Fang N, Zhang X, Roberts MJ, Shea KI (2003) Adv Mater 15:1541–1544
72. Henry OYF, Piletsky SA, Cullen DC (2008) Biosens Bioelectron 23:1769–1775
73. Huang HC, Lin CI, Joseph AK, Lee YD (2004) J Chromatogr A 1027:263–268
74. Hilt JZ, Byrne ME, Peppas NA (2006) Chem Mater 18:5869–5875
75. Guillon S, Lemaire R, Linares AV, Haupt K, Ayela C (2009) Lab Chip 9:2987–2991
76. Linares AV, Cordin AF, Gheber LA, Haupt K (2011) Small 7:2318–2325
77. Granlund T, Nyberg T, Roman LS, Svensson M, Inganäs O (2000) Adv Mater 12:269–273
78. Kim E, Xia Y, Whitesides GM (1995) Nature 376:581–584
79. Zhao X-M, Xia Y, Whitesides GM (1996) Adv Mater 8:837–840
80. Blümel A, Klug A, Eder S, Scherf U, Moderegger E, List EJW (2007) Org Electron 8:389–395
81. Gates BD, Xu Q, Stewart M, Ryan D, Willson CG, Whitesides GM (2005) Chem Rev 105:1171–1196
82. Lalo H, Cau JC, Thibault C, Marsaud N, Severac C, Vieu C (2009) Microelectron Eng 86:1428–1430
83. Yan M, Kapua A (2001) Anal Chim Acta 435:163–167
84. Voicu R, Faid K, Farah AA, Bensebaa F, Barjovanu R, Py C, Tao Y (2007) Langmuir 23:5452–5458
85. Lalo H, Ayela C, Dague E, Vieu C, Haupt K (2010) Lab Chip 10:1316–1318
86. Balla T, Spearing SM, Mo KA (2008) J Phys D Appl Phys 41:174001–174001–10
87. Guo LJ (2007) Adv Mater 19:495–513
88. Forchheimer D, Luo G, Montelius L, Ye L (2010) Analyst 135:1219–1223
89. Ohtsu M (2009) Nanophotonics and nanofabrication. Wiley VCH
90. Girard C, Dereux A (1996) Rep Prog Phys 59:657–699
91. Espanet A, Ecoffet C, Lougnot DJ (1999) J Polym Sci A Polym Chem 37:2075–2085
92. Espanet A, Santos GD, Ecoffet C, Lougnot DJ (1999) Appl Surf Sci 138:87–92
93. Ecoffet C, Espanet A, Lougnot DJ (1998) Adv Mater 10:411–414

94. Fuchs Y, Linares AV, Mayes AG, Haupt K, Soppera O (2011) Chem Mater 23:3645–3651
95. Soppera O, Jradi S, Ecoffet C, Lougnot DJ (2007) Proc SPIE 6647:66470I
96. Chegel V, Whitcombe MJ, Turner NW, Piletsky SA (2009) Biosens Bioelectron 24:1270–1275
97. Ishi-i T, Nakashima K, Shinkai S (1998) Chem Commun 9:1047–1048
98. Zimmerman SC, Wendland MS, Rakow NA, Zharov I, Suslick KS (2002) Nature 418:399–403
99. Pérez N, Whitcombe MJ, Vulfson EN (2001) Macromolecules 34:830–836
100. Vaihinger D, Landfester K, Kräuter I, Brunner H, Tovar GEM (2002) Macromol Chem Phys 203:1965–1973
101. Carter SR, Rimmer S (2004) Adv Funct Mater 14:553–561
102. Biffis A, Graham NB, Siedlaczek G, Stalberg S, Wulff G (2001) Macromol Chem Phys 202:163–171
103. Hunt CE, Pasetto P, Ansell RJ, Haupt K (2006) Chem Commun 16:1754–1756
104. Wulff G, Chong B-O, Kolb U (2006) Angew Chem Int Ed 45:2955–2958
105. Carboni D, Flavin K, Servant A, Gouverneur V, Resmini M (2008) Chem Eur J 14:7059–7065
106. Hoshino Y, Kodama T, Okahata Y, Shea KJ (2008) J Am Chem Soc 130:15242–15243
107. Pérez N, Whitcombe MJ, Vulfson EN (2000) J Appl Polym Sci 77:1851–1859
108. Bompart M, Haupt K (2009) Aust J Chem 62:751–761
109. Lu C, Zhou W, Han B, Yang H, Chen X, Wang X (2007) Anal Chem 79:5457–5461
110. Gai Q-Q, Qu F, Liu Z-J, Dai R-J, Zhang Y-K (2010) J Chromatogr A 1217:5035–5042
111. Li Y, Li X, Chu J, Dong C, Qi J, Yuan Y (2010) Environ Pollut 158:2317–2323
112. Gonzato C, Courty M, Pasetto P, Haupt K (2011) Adv Funct Mater 21:3947–3953
113. Sondi I, Siiman O, Koester S, Matijevic E (2000) Langmuir 16:3107–3118
114. Mitchell GP, Mirkin CA, Letsinger RL (1999) J Am Chem Soc 121:8122–8123
115. Lin CI, Joseph AK, Chang CK, Lee YD (2004) J Chromatogr A 1027:259–262
116. Diltemiz SE, Say RI, Büyüktiryaki S, Hür D, Denizli A, Ersöz A (2008) Talanta 75:890–896
117. Li HB, Li YL, Cheng J (2010) Chem Mater 22:2451–2457
118. Lee E, Park DW, Lee JO, Kim DS, Lee BH, Kim BS (2008) Colloids Surf A 313–314:202–206
119. Kan XW, Zhao Y, Geng ZR, Wang ZL, Zhu JJ (2008) J Phys Chem C 112:4849–4854
120. Lee HY, Kim S (2009) Biosens Bioelectron 25:587–591
121. Choong CL, Bendall JS, Milne WI (2009) Biosens Bioelectron 25:652–656
122. He C, Long Y, Pan J, Li K, Liu F (2008) J Mater Chem 18:2849–2854
123. Matsui J, Akamatsu K, Nishiguchi S, Miyoshi D, Nawafune H, Tamaki K, Sugimoto N (2004) Anal Chem 76:1310–1315
124. Matsui J, Akamatsu K, Hara N, Miyoshi D, Nawafune H, Tamaki K, Sugimoto N (2005) Anal Chem 77:4282–4285
125. Bompart M, De Wilde Y, Haupt K (2010) Adv Mater 22:2343–2348
126. Chronakis IS, Jakob A, Hagström B, Ye L (2006) Langmuir 22:8960–8965
127. Piperno S, Tse Sum Bui B, Haupt K, Gheber LA (2011) Langmuir 27:1547–1550
128. Chronakis IS, Milosevie B, Frenot A, Ye L (2006) Macromolecules 39:357–361
129. Yoshimatsu K, Ye L, Lindberg J, Chronakis IS (2008) Biosens Bioelectron 23:1208–1215
130. Yoshimatsu K, Ye L, Stenlund P, Chronakis IS (2008) Chem Commun 17:2022–2024
131. Yang H-H, Zhang S-Q, Tan F, Zhuang Z-X, Wang X-RJ (2005) J Am Chem Soc 127:1378–1379
132. Li Y, Yang H-H, You Q-H, Zhuang Z-X, Wang X-R (2006) Anal Chem 78:317–320
133. Li Y, Yin X-F, Chen F-R, Yang H-H, Zhuang Z-X, Wang X-R (2006) Macromolecules 39:4497–4499
134. Vandevelde F, Belmont A-S, Pantigny J, Haupt K (2007) Adv Mater 19:3717–3720
135. Linares AV, Vandevelde F, Pantigny J, Falcimaigne-Cordin A, Haupt K (2009) Adv Funct Mater 19:1–5

136. Yang H-H, Zhang S-Q, Yang W, Chen X-L, Zhuang Z-X, Xu J-G, Wang X-R (2004) J Am Chem Soc 126:4054–4055
137. Xie C, Liu B, Wang Z, Gao D, Guan G, Zhang Z (2008) Anal Chem 80:437–443
138. Wu Z, Hu X, Tao CA, Li Y, Liu J, Yang C, Shen D, Li G (2008) J Mater Chem 18:5452–5458
139. Wu Z, Tao CA, Lin CX, Shen DZ, Li GT (2008) Chem Eur J 14:11358–11368

Top Curr Chem (2012) 325: 111–164
DOI: 10.1007/128_2010_94
© Springer-Verlag Berlin Heidelberg 2010
Published online: 29 October 2010

Immuno-Like Assays and Biomimetic Microchips

M.C. Moreno-Bondi, M.E. Benito-Peña, J.L. Urraca, and G. Orellana

Abstract Biomimetic assays with molecularly imprinted polymers (MIPs) are bound to be an alternative to the traditional immuno-analytical methods based on antibodies. This is due to the unique combination of advantages displayed by the artificial materials including the absence of animal inoculation and sacrifice, unnecessary hapten conjugation to a carrier protein for stimulated production, the possibility of manufacturing MIPs against toxic substances, excellent physicochemical stability, reusability, ease of storage, and recognition in organic media. If the selectivity and affinity of MIPs are increased, many more immuno-like assays will be developed using radioactive, enzymatic, colorimetric, fluorescent, chemiluminescent, or electrochemical interrogation methods. This chapter provides a comprehensive comparison between the bio- and biomimetic entities and their usage.

Keywords Biomimetic assay · Immuno-like assay · Microchip · Molecularly imprinted polymer

Contents

M.C. Moreno-Bondi (✉), M.E. Benito-Peña, and J.L. Urraca
Department of Analytical Chemistry, Faculty of Chemistry, Universidad Complutense, 28040
Madrid, Spain
e-mail: mcmbondi@quim.ucm.es

G. Orellana
Department of Organic Chemistry, Faculty of Chemistry, Universidad Complutense, 28040
Madrid, Spain

1 General Principles of MIP Immuno-Like Assays

Antibodies, or immunoglobulins (Igs), are γ-globulin proteins folded into well defined three-dimensional structures synthesized by living organisms, e.g., mice, rabbits, or goats, or by living cells, e.g., hybridoma or *E. coli* cells, in response to the presence of a foreign substance known as the antigen. The main functions of antibodies are to neutralize toxins and viruses and to opsonize bacteria, making them easier to phagocytize [1].

Antibodies are extremely versatile weapons of the immune system, but also useful tools for science in general, as well as for laboratory and clinical medicine, agriculture, and biotechnology. They were a great discovery for analytical chemistry as antibodies can be easily applied to sample analysis without any analyte enrichment, purification, or pre-treatment usually required with standard methods based on liquid or gas chromatography and mass spectrometry (MS).

The use of antibodies has allowed the development of a technique known as immunodiagnostics, especially attractive for the direct analysis of intrinsically complex samples such as blood, serum, urine, food, etc. An immunoassay is a test that uses antibody–antigen complexes as a means of generating a measurable result. The specific interaction between epitopes and paratopes lies at the heart of every immunoassay. The role of antibodies in immunoassays is based on the observation that, in a system containing the determinant and a specific antibody, the distribution of the former between its antibody-bound and antibody-free forms is quantitatively related to the total analyte concentration.

The first immunoassays were introduced in the 1960s by Berson and Yalow [2] for the analysis of insulin and by Ekins for thyroxine determinations [3] using radioactive tracers; since then, their application has continued to grow and is currently worth billions of dollars annually. Nowadays, antibodies can be produced against almost any molecule. Even antibodies against toxic molecules or molecules that are metabolized in vivo can be produced using recombinant antibodies in combination with in vitro evolutive methods [4], something impossible a few years ago. In spite of the benefits of antibodies for assay or sensor development, there are also drawbacks in their production and application, as the former requires usage of laboratory animals, increasing the production costs and raising social concerns, and the latter refers to the biomolecule instability at moderately high temperatures or in organic solvents.

One of the main goals of bioorganic chemistry in the past few years has been the development of fully synthetic materials able to mimic the binding properties of antibodies but prepared by chemical means, thus avoiding the problems associated with in vivo production. To that end, some of the antibody mimics that have attracted a larger attention are the so-called molecularly imprinted polymers (MIPs) since these materials also bear specific binding sites within a significantly rigid insoluble polymeric structure. MIPs are produced with the tools of organic chemistry and are highly stable under harsh conditions. Assays based on MIPs have received the name of Molecular Imprinting Sorbent Assays (MIAs) or MIP immuno-like assays (MIP-ILAs) and have been reviewed elsewhere [5–12].

Molecular imprinting allows generation of artificial recognition sites by forming a polymer structure around a target molecule used as a template. The resulting materials display a high affinity for the imprinted molecule so that they have found applications not only as recognition elements for assay or sensor development but also as enzyme mimics for catalysis or stationary phases for chromatography, as reviewed in other chapters of this book. In principle, the synthetic procedure for MIP preparation allows for an almost unlimited number of chemical structures to be imprinted, including the ability of imprinting more than one template molecule into the same polymer.

Similarly to the immunoassay definition, in an MIA or MIP-ILA the interaction of the target analyte with the imprinted polymer binding sites will provide a measurable result that allows quantification of the former.

1.1 Antibody Structure and Function

There have been three major hypotheses to explain the generation of antibody diversity: the instructive hypothesis, the directive hypothesis, and the selective hypothesis [13]. Breinl and Haurowitz [14] introduced the instructional theory that was further developed by Pauling [15]. This hypothesis suggests that the function of the antigen is to serve as a template on which the complementary structure of antibody molecules are formed, i.e., the antigens themselves act as templates for the antibody production. Pauling suggested that the amino-acid composition or amino-acid order in the antibody polypeptide chain is not determined by the antigen. The amino acids in the active site of the antibody were irrelevant as the selectivity arose from the influence of the antigen molecule that causes the polypeptide chain to assume configurations complementary to surface regions of the antigen (template). MIPs are synthesized in the presence of a template as proposed in the instructional theory. This theory was influential for over 20 years but it was discarded after it was recognized that changes in the protein structure cannot be translated into changes in the DNA necessary to retain the alteration of the antigen during clonal proliferation [13].

The directive hypothesis also regarded the antigen to be responsible for generating the molecules necessary for recognition, suggesting that it occurs at the DNA level. This hypothesis was also discarded in favor of the selective theory.

The selective hypothesis was first postulated by Ehrlich in 1900 [16] and defends the case that the antigen encounters a variety of different Igs generated by the immune system and only one of them fits the antigen. Therefore, only the cells producing this type of antibody respond to the antigenic stimulus. The progress in molecular biology in the 1950s and 1960s focused attention on the selective model and Jerne and Burnet, independently, introduced the idea of clonal selection, i.e., each lymphocyte produces one type of Ig only and the antigen selects and stimulates cells carrying such Ig type. Burnet received the Nobel Prize in 1960 for the clonal selection theory that has been accepted since 1967 [17–19].

All Ig molecules have the same basic structure. They are heterodimers with a similar architecture that consists of four polypeptide chains: two identical "heavy chains" (H) with a molecular weight of approximately 55–70 kDa each, equivalent to about 450–600 aminoacid residues, and two identical "light chains" (L) with a molecular weight of approximately 24 kDa and about 220 residues each one [20, 21]. They also contain a 3–12% by mass carbohydrate as an oligosaccharide side chain depending on the Ig class. In the antibody, each L chain is bound to an H chain by one or more disulfide bonds (–S–S–), and the two H chains are bound together by disulfide bonds in the "hinge" region to form a Y-shape with a central axis of symmetry (Fig. 1). Some Ig H chains (μ and ε) do not contain a hinge region.

The enzymatic cleavage of the Ig with papain results in two type of fragments: the "Fab" (fragment antigen binding) one that contains the N-terminal end of a heavy chain together with a linked light chain, and the "Fc" (crystallizable) fragment [22, 23].

Each H and L chain is formed by homologous segments, containing about 70–110 amino acids, that form independently folded domains. The light chain can be

Fig. 1 Schematic representation of an antibody molecule (human subclass IgG$_1$). The homologous domains within the heavy (H) and light (L) chains are indicated, and the hypervariable segments within the variable (V) regions are shown. Inter- and intra-chain disulfide bridges are depicted by *solid lines*

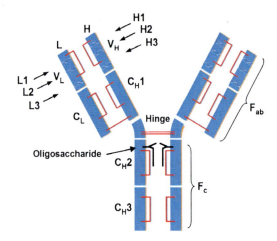

divided into a variable, V_L, and a constant, C_L, region and the heavy chain into V_H, C_{H1}, hinge, C_{H2}, and C_{H3} regions. As mentioned above, there is no hinge region in some H chains, but they contain an additional C_{H4} region. The N-terminal domain of both H and L chains that corresponds to the variable region is different in antibodies of different specificity. Within the end of each V region, three segments show hypervariability and correspond to the Ig complementary-determining regions (CDR). The V_L and V_H domains associate non-covalently and form a β-barrel structure which place a unique combination of about 50 aminoacid residues distributed in the six CDRs loops (termed L1, L2, L3, H1, H2, H3, Fig. 1) close to each other to form the combining site or paratope of an antibody. This region interacts with the antigenic determinant or epitope, i.e., the group of amino acids or chemical groups in the surface of the molecule, generally a protein, that generates the immune response and are recognized by a certain antibody [20–23]. A certain antigen may have many epitopes.

There are five classes of Igs, or isotypes, in mammals that differ in the sequence of heavy chains and in their carbohydrate contents. The five classes are known as IgG, IgM, IgA, IgD, and IgE and contain different heavy chains known as γ, μ, α, δ, and ε. In contrast, there are only two types of light chains (λ and κ) that are the same for all the Ig classes. IgG is the most abundant class of Igs in the serum of mammals and is synthesized predominantly by the plasma cells after secondary exposure to an antigen. They bear two identical paratopes able to bind specifically its complementary antigen [22].

Igs are produced by plasma cells derived from the B-lymphocytes, although their synthesis also involves cooperation of the so-called antigen presenting cells (APC), a class of phagocytic leukocytes, and the T-lymphocytes. The antigen is "captured" by the APCs that retain fragments on its surface. The surface receptor of the T-helper cells recognizes the antigen and provides help to the B-cells which recognize the antigen by their specific antigen receptors, the Igs. Then the B-cells are stimulated and proliferate dividing into antibody forming cells that secrete antibodies [13].

Immunoassays require the availability of antibodies with reproducible, specific and high-affinity interactions with the antigen. Antibodies are produced after immunization of an animal with the selected antigen. If the target analyte is a small molecule which is not immunogenic by itself (called a hapten) it must be coupled to a carrier protein, such as bovine serum albumin (BSA), keyhole limpet hemocyanin (KLH), or ovalbumin before injection. The animal is bled and the isolated antiserum, or purified IgG fraction, can be used for further immunoassay development. The antiserum will contain a mixture of several hundred of different antibodies [22, 23] with varying strengths of binding and selectivities which are known as polyclonal antibodies.

Monoclonal antibodies are produced by cells that arise from a single antibody-producing cell. For immunization the animal, usually a mouse, is injected with a specific antigen. The mouse immune system develops antibodies against the antigen, and the antibody producing white blood cells (B-cells) are removed from the mouse spleen. These cells are immortalized by fusion with immortal myeloma cells, a rapidly multiplying white blood cell cancer, so that they will grow and divide indefinitely to yield a range of B-lymphocyte clones (hybridomas), each of which produces it own antibody. The interesting clones are selected and expanded in tissue culture [13, 22, 23]. Monoclonal antibodies are preferable reagents because they are standard and eternal but they involve important production costs and can be overspecific and not detect variants [23].

Most MIPs show a heterogeneous distribution of binding sites and can be considered as "polyclonal" in their nature. In non-covalent imprinting, the amorphous material contains binding sites which are not identical because they may have different cross-linking density or accessibility. Moreover, the monomer (M) and the template molecule (A) may form complexes of different stoichiometry (M_nA) in the pre-polymerization mixture [5]

$$nM + A \rightleftharpoons M_nA, \tag{1}$$

$$\beta_n = \frac{[M_nA]}{[M]^n[A]} \tag{2}$$

that will be characterized by different association constants (β_n).

In theory, the use of stoichiometric non-covalent or covalent imprinting yields a homogeneous binding site population ("monoclonality") after template removal from the material [24].

1.2   Immunoassay Formats

Immunoassays can be classified according to different criteria. The particular type selected has a strong influence on the assay performance with regard to precision and sensitivity. The main criteria include [3, 22, 23]: (1) labeled or unlabeled assay formats, with different type of labels; (2) competitive or non-competitive immunoassays, and (3) homogeneous or heterogeneous immunoassays. These classifications can also be extended to MIP-ILAs (Fig. 2):

(a) The *direct assay* formats are based on the measurement of antigen–antibody binding without any labeling compound and usually in real time. This method depends on achieving a high ratio of specific to non-specific binding. The antigen–antibody binding leads to physical or bulk effects that determine changes in the local density, dielectric constant, or refractive index that can be monitored by means of electrical, gravimetric, or optical transducers. In this way, the transducer converts the binding event into an electronic signal that can be easily measured. Techniques such as surface plasmon resonance (SPR), reflectometric interference spectroscopy (RIfS), the quartz crystal microbalance (QCM), cantilever-based methods, or isothermal titration calorimetry (ITC) have all been used to monitor direct assays.

Indirect methods require the use of labeled compounds (tracers) or coupled reactions to detect binding events. The first indirect assays were based on

CLASSIFICATION OF (BIO)MIMETICS ASSAYS ACCORDING TO WHETHER A TRACER IS USED OR NOT

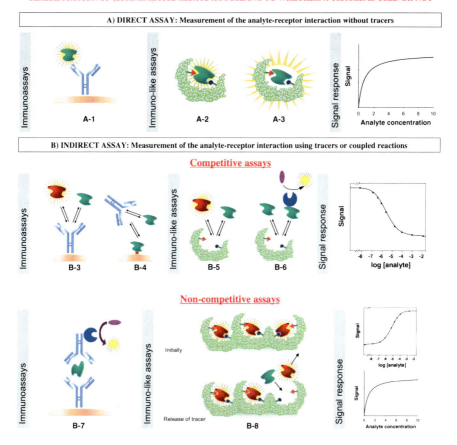

CLASSIFICATION OF (BIO)MIMETICS ASSAYS ACCORDING TO WHETHER THE ANTIBODY-ANALYTE COMPLEX IS SEPARATED FROM THE FREE ANALYTE OR NOT

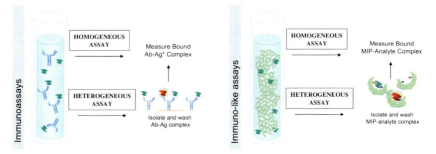

Fig. 2 Classification of (bio)mimetic assays

radioactive tracers and received the name of radioimmunoassays (RIA). Non-isotopic labels such as fluorophores, fluorescence quenchers, enzymes (mainly horseradish peroxidase and alkaline phosphatase), lanthanide ions, chemilumines-cent compounds, latex particles, or liposomes have also been used [22, 23].

The MIP-ILAs based on a direct format will be described elsewhere. ILAs based on indirect methods will be reviewed in this chapter. Like immunoassays, the first MIP-ILA, reported in 1993 by the Mosbach group [25] for the analysis of theo-phylline and diazepan, was based on a radioactive tracer. Since this work, many other indirect MIP-ILAs have been developed using non-isotopic tracers such as fluorophores, enzymes, or electroactive probes.

(b) *Competitive* immunoassays, also called "limited reagent" or "saturation" assays [3, 22, 23], are those in which a limited amount of antibody is used that is insufficient to bind all of the antigen. For instance, a labeled analyte competes with the analyte for a limited number of antigen-binding sites. After incubation of the mixture with the antibody, the bound or the free fraction of the tracer is measured and correlated with the analyte concentration in the sample. Thus, in a competitive immunoassay, a decrease in the signal from the labeled species means that more of the unlabeled antigen (analyte) is present in the sample, i.e., the signal is inversely related to the analyte concentration. The sensitivity of the assay increases with decreasing antibody concentration and is affected by the affinity of the antibody for its antigen, the non-specific binding, the specific activity of the labeled antigen, and the experimental error in measuring the fraction of bound and unbound antigen. The major drawbacks of these assays include limited sensitivity and working range, slow reaction kinetics, lower precision, and development of a negative endpoint.

Competitive immunoassays can be based on different formats [3, 22, 23]: (1) an anti-analyte antibody is immobilized on a solid support and the analyte in the sample competes with an added labeled analyte for a limited number of antibody binding sites; (2) a solid phase is coated with an antibody that binds to the anti-analyte antibody and the assay is carried out with the analyte and the labeled derivative mixture; (3) the immobilized antigen approach consists of the competi-tion of the analyte in the sample with the immobilized antigen for binding to labeled antibody molecules.

Competitive MIP-ILAs are based on the first configuration: the analyte competes with a labeled derivative for the specific binding sites of the imprinted polymer prepared in the form of beads or thin films.

In *non-competitive* or "reagent excess" immunoassays [26], an excess of immu-noreagent (antibody or antigen) is used so that all the analyte forms an immunocom-plex that is further quantified and related to the analyte concentration in the sample. These assays are also known as immunometric assays and their advantages over competitive assays include higher sensitivity, precision, and analyte working range.

Some configurations used in non-competitive immunoassays include the follow-ing two. First, two-site (sandwich type) immunoassays, used for the determination of macromolecular antigens, where simultaneous binding of two antibodies to the antigen is allowed without steric hindrance. A "capture" antibody is immobilized onto a solid phase and incubated (in excess) with the sample. After washing, the amount of bound analyte is quantified by adding an excess of labeled anti-analyte ("detection") antibody. In this way, the signal will be directly proportional to the analyte concentration in the sample. This assay is suitable for analytes with a molecular weight larger than ca. 1,000 D. Second, immunoassays for the determination of specific antibodies can be carried out using two different approaches. In the first, the solid support is coated with an excess of antigen and, after incubation with the sample, the antibody bound to the solid support is quantified using a labeled anti-immunoglobulin.

According to Ekins [27], the fundamental difference between competitive and non-competitive methods is based solely on the approach adopted to detect the antibody occupancy from which the analyte concentration in the system is deduced. Competitive assays rely on the indirect measurement of occupancy by observation of unoccupied sites. In this case the amount of antibody must be kept small to minimize errors in the indirect estimate of the occupied sites. Non-competitive assays rely on direct measurement of binding site occupancy so that the use of large amounts of antibody is advantageous.

In agreement with the above definitions, MIP-ILAs, based on the direct measurement of the analyte binding, can be considered as non-competitive, as they measure the binding site occupancy in the polymeric material and an excess of MIP binding sites is used in this configuration. To quantify binding site occupancy by the analyte, the MIP can be labeled and its signal will be modified upon the analyte binding, or the analyte itself may be responsible for the generated signal, directly or after coupling to a "developer" reaction. The two-site approach (capture MIP–analyte–detection MIP) seems difficult to implement with imprinted polymers due to steric hindrance. Nevertheless, as the imprinting of macromolecules becomes successful, this configuration may be a matter of future research.

The conventional competitive or non-competitive assays do not allow continuous detection so that, for on-line measurements, the so-called displacement assays are usually applied. Typically, the antibody is immobilized onto a solid support and packed in a column, and the corresponding antigen is labeled. The antibody binding sites are saturated with labeled antigens and the sample, containing the free (unlabeled) analyte, is injected into the column, resulting in displacement of the bound labeled analyte, as the affinity of the antibody for the labeled analyte is usually much lower than its affinity for the unlabeled analyte. The displaced labeled analyte is eluted and detected at the outlet of the column and the measured signal is directly proportional to the analyte concentration in the sample.

This measuring scheme, with slight modifications, has also been applied using MIPs in the same way as their biological counterparts. The imprinted polymer is packed in a column and the mobile phase containing a constant concentration of fluorophore-labeled or an enzyme-labeled analyte is pumped through the sorbent until a stable baseline is achieved. When the sample containing both the analyte and the labeled analyte, in the same concentration as in the loading step, are injected into the system, the probe is displaced from the MIP increasing its concentration in the mobile phase. The presence of the analyte is detected as a peak above the baseline, followed by a trough that evidences the elution of the bound analyte as it is displaced by the labeled analyte from the specific binding sites. Following Ekins definition [27], this assay could be classified as non-competitive as the binding site occupancy is measured directly by considering that the signal increase corresponds to the amount of labeled analyte ("probe") displaced upon analyte binding.

(c) Immunoassays can also be classified as *heterogeneous* or *homogeneous* depending on whether the antibody-bound analyte needs to be separated from the free analyte before quantification or not, respectively. In homogeneous immunoassays, the signal is modulated as a result of the immunoreaction and the conjugate formation can be monitored directly. These assays are well suited to automation and provide high turnaround time. The calibration curve is stable from several days to weeks with the corresponding savings in technical time and reagent expenses.

In heterogeneous assays the bound and free label forms must be separated by different means, such as precipitation of antibody or coupling of the antibody to a solid support. Heterogeneous immunoassays are more versatile as inclusion of a separation step eliminates most of the interferences before quantification. However, these assays are also more labor-intensive and time-consuming. Any of these methods can be performed in either competitive or non-competitive format.

Most MIP-ILAs are based on a heterogeneous format, as the MIP-bound analyte is separated from the free form before measurement. However, the introduction of soluble imprinted polymers or MIPs that yield stable suspensions has allowed the development of homogeneous ILAs [28, 29].

1.3   Advantages and Disadvantages of MIPs in Immuno-Like Assays

The main *advantages* of MIPs as alternative recognition elements to antibodies in analytical applications include:

–   Antibodies require in vivo production with the corresponding expenses associated to animal growing and the social concern about experiments with animals.
–   The production of antibodies for small molecules is difficult to achieve as they must be coupled to a carrier protein before inoculation. Moreover, such conjugation may change the structural properties of the antigen exposed to the immune system and the obtained antibodies will be directed against a structure slightly different from the target compound.
–   As indicated above, the rising of antibodies against toxic or immunosuppressant compounds is not straightforward; however, any compound may be imprinted provided it can be dissolved in the polymerization mixture.

- Antibodies display limited thermal and chemical stability. Their binding site may be altered or blocked during immobilization processes. However, the binding site of MIPs is very stable and robust, even under harsh conditions.
- Antibody recognition in organic media is poor due to partial or total denaturalization but MIPs perform very well in non-aqueous solvents broadening the application of MIP-ILAs only to analytes soluble in organic media.
- Unlike antibodies, MIPs can easily be reused for many assays without losing their binding characteristics.
- MIPs can be stored in the dry state at ambient temperature for several years without losing their recognition features.

The main *limitations* associated with the application of MIPs, instead of antibodies or other biomimetic receptors, in ILAs are summarized as follows:

- Large (stoichiometric) amounts of analyte are required to manufacture MIPs. This is an important issue if the (analyte) template is expensive, difficult to isolate, or toxic. In principle, the analyte could be recovered after template extraction although this process can also be cumbersome.
- Imprinting of high molecular weight templates, proteins, or whole cells still remains challenging.
- Initially, MIP performance in aqueous samples was characterized by significant non-specific binding but, in the past years, materials that show selective recognition in aqueous samples have been developed [30–34].
- The use of enzymatic tracers for MIP-ILAs has been limited due, on the one hand, to the poor performance of these biomolecules in organic media where many MIPs work and, on the other, to the limited accessibility of the large protein macromolecules to the binding sites of these hydrophobic and highly cross-linked materials. The use of imprinted microspheres or films with binding sites at or close to the surface can overcome the problem of binding site accessibility [9].
- In combination with optical measurement techniques, the large size of the MIP particles can give rise to light scattering problems. The use of imprinted membranes may avoid this limitation.
- The time required to reach static equilibrium is higher than for homogeneous immunoassays. Like in the case of solid-phase immunoassays, this behavior can be related to the diffusion dependence of interfacial reaction kinetics, directly correlated with the ratio of the solution phase volume to the volume of the reactive interface.

The equilibrium phenomena involved in the interaction between the target molecule (A) and a homogeneous population of imprinted binding sites (M) can be simplified as follows [24]:

$$M + A \rightleftharpoons MA, \tag{3}$$

$$K = [MA]/[M][A] = k_2/k_{-1}, \tag{4}$$

k_2 and k_{-1} being the second-order and first-order rate constants, respectively, of the association and dissociation processes described by (3).

The association rate constants (k_2) for solution-phase antibody systems are on the order of 10^7–$10^8 \ M^{-1} \ s^{-1}$ [23] but those for reactions on synthetic solid phases and cell surfaces are two to four orders of magnitude slower, mainly as a consequence of the sluggish diffusion and slower mass transfer of the reactants to the interaction sites. The dissociation rate constant for heterogeneous systems (k_{-1}) is on the order of 10^{-4}–$10^{-5} \ s^{-1}$, up to two orders of magnitude slower than for solution-phase systems, and for solid phase immunoassays, it is attributed to multivalent interactions and to surface coagulation or aggregation via translational diffusion in the presence of extensive cross-linking at the interface [23]. Most of the MIA-based assays described in the literature require equilibration times on the order of hours, and more synthetic efforts are required to reduce this analysis time.

– One of the most challenging tasks in MIP-ILAs, as well as for MIP-based sensors, is the difficulty of transforming the binding event into a measurable signal. Such a signaling process requires preparation of novel monomers with responsive functionalities, synthesis of labeled analyte derivatives, conjugation of binding sites with transducers, or generation of conformational changes upon analyte binding.

1.4 Binding Site and Heterogeneity

Different techniques have been applied to study the protein–protein, protein–ligand and, in particular, MIP–ligand interactions. They may serve to estimate or determine the binding constant and the number of independent binding sites (N) of a ligand-to-receptor (MIP or antibody) interaction. The range of affinity constants that can be calculated depends on the sensitivity of the assay and, in those cases where the separation of the bound and free species is a step of the assay, perturbation of the equilibria in the separation step will also be important [22]. Direct non-separation techniques such as spectroscopic techniques (e.g., SPR or fluorescence polarization) can be used as well as indirect separation techniques such as radiolabeling [22].

The affinity of an MIP for the ligand will depend not only on the intrinsic binding process but also on the conditions under which the recognition is carried out, such as solvent, pH, ionic strength, temperature, etc. The time required to reach equilibrium will be influenced by the MIP format (monolith, film, beads, etc.) because the surface area will be significantly different. In any case, to calculate the binding isotherms, the MIP is incubated with the ligand and, after equilibrium is attained, the amount of bound vs the concentration of free ligand is plotted and fitted to different models.

The simplest model is the Langmuir isotherm or its derivative, the Scatchard equation, the most widely used mathematical approach to describe quantitatively the multiple equilibria that take place when an antibody, or an MIP, binds reversibly to a ligand molecule (antigen or analyte) [35]. This model focuses on the individual binding sites in the binder and applies the law of mass action to each of them (M) considering that the affinity of each site for the ligand is not affected by the extent of the occupancy of the other sites, i.e., the binding sites are independent and non-interacting. The equilibrium (5) taking place can be described by [5, 22–24]:

$$M + A \; \rightleftharpoons \; MA, \tag{5}$$

$$K_a = \frac{B}{F(N - B)}, \tag{6}$$

where B represents the moles of analyte bound per mass of binder, F is the molar concentration of free analyte, N is the total concentration of binding sites (in mol per mass of binder), and K_a is the affinity constant (in L mol^{-1}). Therefore, $N-B$ is the number of free (unoccupied) binding sites in the binder.

The total ligand concentration (T) will be given by:

$$T = B + F. \tag{7}$$

Rearranging (6) we obtain the well-known Scatchard equation:

$$\frac{B}{F} = K_a(N - B) = K_a N - K_a B. \tag{8}$$

A plot of B/F vs B for various ligand concentrations and a constant amount of binder is known as the Scatchard plot and should yield a straight line (8) if the polymer has a single class of binding sites. Figure 3 shows a simulation of the binding of a ligand to different binder concentrations. The slope is $-K_a$ and the x-axis intercept corresponds the total concentration of binding sites (N). As for antigen binding to monoclonal antibodies, imprinted polymers containing a homogeneous distribution of binding sites, such as covalent MIPs, yield a straight line.

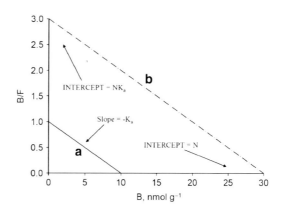

Fig. 3 Simulated Scatchard plots for the binding of a ligand to various concentrations of identical and independent binding sites (N). (a) $N = 10^{-8}$ mol g^{-1}, (b) $N = 3 \times 10^{-8}$ mol g^{-1}; $K_a = 108$ M^{-1}

The presence of non-specific binding, defined as low-affinity non-saturable binding of the ligand to the solid phase involved in the experiment, distorts the linear behavior. The binding model that incorporates non-specific binding can be written as

$$\frac{B}{F} = \frac{K_a N}{1 + K_a F} + n,$$ (9)

where n is a constant and accounts for the extent of non-specific binding which is proportional to the free ligand concentration. At very high concentrations B/F becomes equal to n and the plot will be a straight line parallel to the x-axis.

As mentioned in Sect. 1.1, a curvilinear plot is obtained in non-covalently prepared MIPs indicating that the polymer contains a heterogeneous distribution of binding sites with a range of affinities for the imprint molecule. Using the Scatchard model, the binding sites can be grouped in m different classes, each one with N_i individual sites with the same affinity for the ligand. The concentration of ligand bound to each class of binding sites will be given by (10):

$$B_i = \frac{K_{a_i} N_i F}{1 + K_{a_i} F}.$$ (10)

The total concentration of bound ligand is given by (11):

$$B_n = \sum_{i=1}^{m} \frac{K_{a_i} N_i F}{1 + K_{a_i} F}.$$ (11)

sites ($m = 2$) are present in the binder ("bis-Langmuir" model):

$$B = \frac{K_{a_1} N_1 F}{1 + K_{a_1} F} + \frac{K_{a_2} N_2 F}{1 + K_{a_2} F},$$ (12)

and including the non-specific binding component, (12) becomes

$$B = \frac{K_{a_1} N_1 F}{1 + K_{a_1} F} + \frac{K_{a_2} N_2 F}{1 + K_{a_2} F} + nF.$$ (13)

The Scatchard plot for this model is a hyperbola (Fig. 4) and the asymptotes (14) can be used for the graphical estimation of the binding parameters:

$$y = K_{a_1} N_1 - K_{a_1} x \quad \text{and} \quad y = K_{a_2} N_2 - K_{a_2} x.$$ (14)

While Scatchard plots may be useful to display the binding data, they are not a useful way to analyze them. It is usually preferred to fit the raw binding data with a nonlinear regression. Selection of the dependent and independent variables

Fig. 4 Scatchard plot depicting the binding of a fluorescent labeled penicillin to a pencillin G-imprinted polymer [36]. $N_1 = 0.024$ μmol g^{-1}, $K_{a1} = 2.4 \times 10^2$ M^{-1}; $N_2 = 0.125$ μmol g^{-1}, $K_{a2} = 7 \times 10^2$ M^{-1}

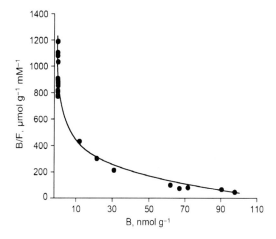

is important for a statistically correct regression; therefore, the total ligand concentration should be used as the independent variable, using the experimental data (fluorescence, counts, etc.) as the dependent one.

As mentioned in Sect. 1.1, the heterogeneity of the binding sites in non-covalent imprinting is due, to a large extent, to the formation of complexes of different stoichiometry between the monomer and the template (M_nA) in the pre-polymerization mixture. Taking into account that in the final MIP the binding site distribution reflects the complexes formed before polymerization, the strongest binding sites will be those with the highest number of interactions while the weakest association sites will correspond to those of the free monomer that was not involved in any complex with the template.

The data shown in Fig. 4 correspond to the binding of a fluorescent labeled penicillin G (PAAP) to a penicillin G-imprinted polymer in a mixture of acetonitrile–water (99:1, v/v) [36]. The experimental data were fitted in this case to the bis-Langmuir model. At low analyte concentrations, corresponding to the left-hand side of the plot, the calculated affinity constant is attributed to the presence of a small population of strong binding sites within the material ($K_{a_1} = 2.4 \times 10^4$ M^{-1} and $N_1 = 0.024$ μmol g^{-1}). However, at high analyte concentrations (right-hand side of the plot), the response is attributed to the presence of a large amount of weak binding sites ($K_{a_2} = 7 \times 10^2$ M^{-1} and $N_2 = 0.125$ μmol g^{-1}). Higher affinity constants have been obtained in radioimmuno-like assays (RILAs). For instance, Anderson [33] has optimized an RILA for the analysis of S-propanolol in organic or aqueous buffer media, the results being fitted to the two-site model for high-low affinity binding sites. The apparent K_a values were found to be 2.5×10^8 and 2.4×10^5 M^{-1}, with binding site populations of 0.63 and 28 μmol g^{-1}, respectively. These values are higher than those obtained for the PAAP binding but, in the latter case, the polymer was imprinted against penicillin G and not its fluorescent derivative used for the binding assay.

Nevertheless, it must be pointed out that the two-site model lacks physical significance if we consider that there is probably a much broader distribution of binding sites in the polymer. Umpleby et al. [37] have developed the "affinity spectrum" (AS) model supposing that MIPs contain a heterogeneous continuum of binding sites and therefore a continuous distribution of affinity constants should be expected. The AS represents an improvement over Scatchard plots and it can be applied to the characterization and comparison of MIPs prepared by different synthetic procedures.

It has been pointed out [5] that, for preparation of non-covalent MIPs for ILAs, a good design strategy is to reduce the amount of template in the polymerization mixture, i.e., employ template-to-monomer ratios of 1:1000 instead of the usual 1:4 or 1:10. Under a monomer excess, the formation of complexes of higher stoichiometry (M_nA) is favored in detriment of those with a lower one. The amount of free monomer binding sites will increase compared to the typical 1:4 formulations, but considering that MIP ILAs use only the binding sites with strongest affinity for the probe, the features of the rest of the polymer are not as important as for other MIP applications.

1.5   Experimental Approach to Immuno-Like Assays

There are some critical aspects in the optimization of immunoassays that should also be considered in immunolike-like assays [38]: (1) the non-specific binding; (2) MIP batch-to-batch variations; (3) the assay format; (4) the analyte binding conditions to the imprinted material; (5) the length of incubation.

The polymer can be produced in different formats (bulk, beads, film, etc.) that can be applied to MIP-ILA development. A non-imprinted polymer (NIP) or a control polymer (CP), prepared in the absence of template or in the presence of an unrelated molecule, respectively, must also be synthesized to account for non-specific binding. The template must be completely removed from the material before the analysis because the analyte and the tracer will compete for the strongest binding sites that are supposed to be the most selective for MIP-ILAs and they must be empty before the analysis. Binding to the NIP or the CP is considered to be entirely non-specific and the solvent composition must be optimized to minimize this interaction.

As has been pointed out [5], it is important to distinguish between the polymerization template, the analyte, and the labeled analyte (tracer) that will compete for the polymer binding sites. The bound or free fraction of the tracer can be monitored and correlated to the analyte concentration.

The format chosen for the MIP-ILA will have a strong influence on the sensitivity and selectivity of the assay. In direct assays, the analyses are carried out in a shorter time because the number of steps necessary for the assay is reduced; however, indirect MIP-ILAs are more sensitive.

Label selection will also affect assay sensitivity and, sometimes, selectivity. Radioactive tracers, luminescent labels, or enzymes have been applied for MIP-ILAs development and will be described in more detail in the next sections.

Ansell [5] has described a procedure for optimization of the amount of polymer to be used in competitive ILA when the probe and the analyte are chemically identical, as in the case of RILAs. After optimizing the binding of the probe to the MIP, the MIP concentration should be such that an excess of binding sites, compared to the amount of probe, is available and the solvent strength should be such that only 50–80% of the probe binds to the polymer and only to the high affinity binding sites. The amount of MIP required to achieve such binding depends on the quality of the binding sites. If the imprinting is poor and the number of high quality binding sites is low, more polymer will be necessary for the assay.

When the label and the analyte are not identical, as happens in the case of competitive fluoroimmuno-like or enzyme-immunolike assays, we have found it useful to follow the experimental approach suggested by Ekins [22] for the optimization of the antibody concentration in competitive immunoassays.

First of all, the lowest concentration of label to be used in the assay has to be optimized to maximize sensitivity while minimizing the assay price. If the free label in solution is to be measured after incubation with the polymer, it should be determined with a relative standard deviation of less that 5% [5]. It is important to use a reliable separation system to separate the free label from the MIP-bound one.

The experiment consists in the incubation of the labeled antigen (1) with increasing concentrations of MIP or NIP/CP, and (2) with the same set of concentrations but in the presence of a constant concentration of analyte sufficient to displace significantly the probe from the polymer [22]. Each point in the dilution curve should be determined at least by duplicate and preferably between five and ten replicates so that we can obtain not only the optimum polymer concentration for the assay but also an estimate of the standard deviation of the response at each point along the curves. The curve representing the difference (% bound label) of the bound label in the absence and in the presence of the analyte will render the optimum MIP concentration to be used in the assay. Using this approach we have optimized the amount of polymer required for the development of a fluoroimmuno-like assay for penicillin analysis [36].

Initially, it is necessary to select carefully the assay incubation time. With that aim, different batches of MIP particles (2.5 mg mL^{-1}) were incubated for between 0 and 18 h at room temperature in polypropylene tubes on a shaking table with 250 nmol L^{-1} of PAAP in the absence and in the presence of 166 µg mL^{-1} of penicillin G (in acetonitrile-water, 99:1), with 2 mL of a fluorescently labeled penicillin (PAAP) [36]. As shown in Fig. 5, binding to the CP was very low in comparison to that of the MIP and 7 h were required to achieve equilibria.

To calculate the optimum amount of polymer to be used in the competitive assay, 2 mL of PAAP (250 nmol L^{-1}) were incubated with increasing concentrations of MIP or CP (0–10 mg mL^{-1}) in the absence and in the presence of a constant concentration of penicillin G. After centrifugation (5 min, 14,000 rpm) the fluorescence supernatant was measured and the % PAAP bound was plotted as a function of the polymer concentration. The largest differences in the fraction of PAAP bound to the polymer, in the presence and in the absence of the analyte, corresponded to an MIP concentration of 2.5 mg mL^{-1} so that this value was selected for

Fig. 5 Kinetics of PAAP
binding to: (*inverted
triangles*) a CP (in the
absence of penicillin G) and
to a penicillin G-imprinted
MIP in the absence (*black
circles*) and in the presence
(*red circles*) of 426 µmol L^{-1}
of the antibiotic;
[PAAP] = 250 nmol L^{-1},
[Polymer] = 2.5 mg mL^{-1},
binding solvent: acetonitrile–
water (99:1, v/v), T = 25 °C
(n = 9)

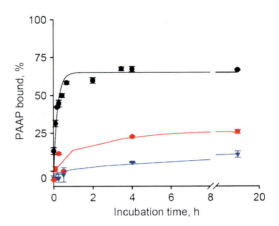

Fig. 6 Optimization of the
concentration of imprinted
polymer in immuno-like
assays. The plot represents
the differences (%) observed
in the binding of 250 nmol L^{-1}
PAAP to the polymer when
incubated in the absence and in
the presence of 426 µmol L^{-1}
of penicillin G. Binding
solvent: acetonitrile–water
(99:1, v/v), T = 25 °C,
incubation time 7 h (n = 9)

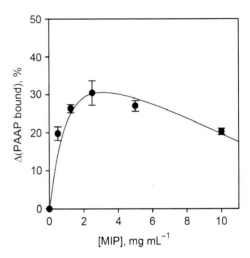

the competitive assay (Fig. 6). The amount of probe bound to the MIP at this polymer concentration, in the absence of penicillin G was 67% and that bound to the CP was <5%. This value is consistent with the range proposed for RILAs (50–80% of the probe binds to the MIP) [5].

The assay solvent must be selected to maximize the differences between the amount of probe bound to the MIP and to the NIP (or the CP), considered to be entirely non-specific. When the assay is optimized in aqueous solutions, a small amount of nonionic surfactant such as Triton X-100 (0.5%, w/v) or miscible organic solvents, such as ethanol, may be added to increase polymer wettability. The additive may also help to reduce hydrophobic interactions, especially for MIPs based on ethylenglycoldimethacrylate (EGDMA) or divinylbenzene (DVB) [5].

Competitive assays are usually carried out only with the MIP, although it is advisable to check that no competition is observed when the NIP/CP is used instead.

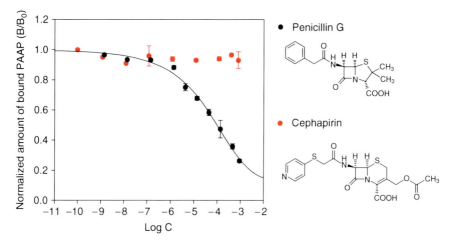

Fig. 7 Dose response curve ($n = 9$) for penicillin G (*black circle*) and for cephapirin (*red circle*) in acetonitrile–water (99:1) for imprinted polymer (MIP, 2.5 mg L^{-1})/labeled β-lactam antibiotic (PAAP, 250 nmol L^{-1}), adapted from [36]

Once the optimum concentration of MIP and label, as well as the incubation time and the solvent nature, have been selected, the assay is carried out in the absence and in the presence of increasing concentrations of the analyte or the interferent in a concentration range as wide as possible. A control measurement must be taken only with the label, without the MIP and the analyte. At least three replicates of each point should be measured.

The amount of label bound to the MIP in the absence of the analyte is known as B_0 and B is the amount of label bound to the MIP in the presence of each concentration of analyte. The plot of the ratio B/B_0 as a function of the log[analyte], or the log [interferent], is a sigmoid curve such as the one shown in Fig. 7 for the penicillin G assay described above. As the concentration of penicillin G increases in the sample, the amount of bound PAAP decreases as does the B/B_0 ratio. Another β-lactam antibiotic not derived from penicillin, such as cephapirin, did not show any cross-reactivity (Fig. 7).

1.6 Analytical Characteristics

MIP ILAs offer interesting practical advantages over immunoassays; however, their acceptance still depends on the demonstration of their quality and validity in comparison with traditional methods. Much effort is still needed in this field so that this section will focus on some aspects related to MIP ILAs development.

Analytical methods for trace analysis are usually classified in two categories depending on the purpose of the analysis and the confidence required in the results: routine methods (screening and surveillance) and regulatory methods (confirmatory and reference). MIP-ILAs could be used in routine analysis as an alternative to immunoassays. Screening methods can be quantitative or semiquantitative, should be rapid, and should permit a high throughput of samples at low cost. Moreover, they should be sensitive enough to eliminate false negatives, i.e., where an analyte which is present is not detected. Surveillance methods are similar to screening methods but yield quantitative results and their specificity is better, yet their response is slower that the screening ones [39].

The performance characteristics of any analytical method involve the evaluation of the following parameters: calibration range, limit of detection, precision, trueness, specificity, recovery, and robustness [40].

(a) *Calibration*

As with any analytical technique, generation of a reproducible standard curve with minimal error is critical. An assay calibration consists of several steps during which the value of the primary standard is transferred to the calibrators used in the final assay [22]. Immunoassay optimization is usually difficult due to protein heterogeneity and matrix effects and these factors, heterogeneity and matrix effects, will also affect MIP based assays [22].

In a competitive assay the system contains the analyte (A), a labeled derivative of it (A*) and a constant concentration of polymer binding sites (M) so that the following equilibria take place [23]:

$$M + A \ \rightleftharpoons \ MA \ , \tag{15}$$

$$M + A^* \ \rightleftharpoons \ MA^* \ . \tag{16}$$

Considering that B^* and F^* are the concentrations of bound and free labeled analyte, respectively, $(M - B - B^*)$ represents the concentration of non-occupied binding sites. The affinity constants for the analyte and the labeled analyte will be given by

$$K_a = \frac{B}{F(N - B - B^*)}, \tag{17}$$

$$K_a^* = \frac{B^*}{F * (N - B - B^*)}. \tag{18}$$

Dividing (17) and (18) and assuming that the labeled analyte and the analyte have identical binding properties (i.e., $K_a = K_a^*$), for instance when a radiolabel tracer is used, results in

$$\frac{B^*}{F^*} = \frac{B}{F} = r \tag{19}$$

and

$$\frac{B^*}{T^*} = \frac{B}{T} = y,$$ (20)

where T^* is the total labeled analyte concentration. Using (19) and (20) the average concentration of analyte bound and free can be calculated for each (total) analyte level.

In the case when $K_a \neq K_a^*$, for instance when fluorescent labeled analyte derivatives or non-related fluorescent dyes are used as competitors, the following general expression can be formulated [23]:

$$\frac{B^*}{F^*} = \frac{y}{1-y}.$$ (21)

There are several ways to represent the dose-response curves in MIP ILAs. Plots of $1/B^*$ vs T, B_0^*/B^* vs T, B^*/B_0^* vs T, and F^*/B^* vs T, where T is the (total) analyte concentration, etc., can be represented and fitted to two different models, namely the mechanistic (or theoretical, based on assay theory) or the empirical based on the mathematical fitting of the measured responses [41].

Mechanistic models are based on the Law of Mass Action, taking into account concentrations of reagents, binder affinity for the ligand, and assuming that the reaction is at equilibrium as was described in Sect. 1.4. These models are very useful to gain information on the assay system but, in practice, the mechanistic models cannot be adequately defined or the assumptions fully met, so that the resulting models are complicated and not useful for routine analysis.

The *empirical models* are based on mathematical functions that mimic the distribution of the standards measured in the assay. They can be based on point to point (interpolation) methods or regression methods. The most widely used empirical models that have been applied to MIP-ILAs include the log-logit model and the four-parameter logistic model.

The logit plot is a mathematical transformation of the standard data to linearize the curve:

$$\text{logit } z = \ln\frac{z}{1-z} = a + b\ln(T).$$ (22)

The assay response (z) must be expressed as the bound fraction, or as B^*/B_0^*, so that it has values between 0 and 1. The plot of logit z vs $\ln(T)$ yields a straight line. This model has been widely used for the fitting of many competitive immunoassays and also for the graphical representation of MIP-ILAs data. The bound and free concentrations of the tracer must be corrected for the non-specific binding (n); this value can be obtained by measuring the concentration of bound tracer in the presence of a large excess of unlabeled analyte. Systems that show asymmetrical calibration curves may not fit the log-logit calibration plot, particularly (1) in polymers with a large heterogeneity of binding sites, (2) for assays with very low concentrations of MIP, (3) when the labeled analyte is different from the analyte, or (4) in immuno-metric-like assays [22].

The 4-parameter logistic model is considered the most versatile one for fitting the dose-response curves in immunoassays [42]. The fitting equation is given by [22, 23, 41]

$$Y = d + \frac{a - d}{1 + \left(\frac{T}{c}\right)^b} , \tag{23}$$

where Y is the assay response and can be expressed as B^*, F^*, B_0^*/B^*, B^*/T^*, B^*/F^* or F^*/T^*. In this model the values of B, T^* and B_0^* are used without correction for the non-specific binding. The parameter a represents the response at zero analyte concentration, b corresponds to the slope of the log-logit model, c corresponds to the analyte concentration at the true mid-point of the curve (50% maximum binding, i.e., is the concentration of competing ligand which displaces 50% of the specific binding of the tracer, IC_{50}), and d is the expected response when T is extremely high, i.e., represents the non-specific binding. Unlike the log-logit model that requires good estimates of the non-specific binding and the response at zero dose, the parameters a and d of the four logistic model can take values that ensure the best fit to the data [23]. This equation has also been recommended for limited reagent ILAs [41].

The main problem of this fitting is that the curve is essentially symmetrical about IC_{50}. The excess reagent ILAs curves can be treated as inverted rectangular hyperbolas and, to accommodate the asymmetry of this curve, a fifth logistic equation is suggested [41]. Due to the sigmoidal shape of the calibration curve it is of paramount importance to use enough levels of reference standards (calibration standards) in order to define the upper and lower limit of quantification (linear range of the calibration curve), the maximum response, and the limit of detection of the test [43].

Like in the case of immunoassays, the MIP-ILAs will be governed by the Law of Mass Action when working at equilibrium so that the reagents are under equilibrium conditions and subject to temperature fluctuations. Shaking can also affect the local concentration of reagents and the reaction rate; these factors must be controlled to improve the assay precision.

The *sensitivity* of the assay generally refers to the minimum amount of analyte required to produce a change in the analytical response significantly different from that obtained in the absence of analyte ("zero dose analyte") [22, 23]. It is frequently expressed as the minimum detectable dose (MDD) or detection limit of the assay, although other definitions refer to the slope of the dose-response curve [44]. When empirical fitting approaches are used, the value usually given refers to the dose of analyte that displaces 10% of the tracer, i.e., the 90% binding level [23, 45]. Sensitivity is a function not only of the successful competition between the tracer and the analyte for the good recognition sites but also of the inherent sensitivity with which the probe can be detected.

(b) *Precision* is the closeness of agreement between independent test results obtained under stipulated conditions and is usually specified in terms of the standard deviation or relative standard deviation. Precision of the method must be evaluated by performing repetitive measurements of the entire method and not of the final determination step. Precision may be expressed as repeatability and reproducibility [43].

Repeatability is defined as the value below which the absolute difference between two single results, obtained with the same method on an identical test sample under the same conditions (including the same analyst), may be expected to lie within a probability of 90%.

Reproducibility is closeness of agreement between independent results obtained with the same method on identical test material but under different conditions (different operators, different apparatus, different laboratories, and/or at different intervals of time). Reproducibility is usually specified in terms of standard deviation or relative standard deviation.

The response of an MIP-ILA is nonlinear and the variance is non-uniform so it would be desirable to establish the precision profiles by calculating the standard deviation, or the percentage of the coefficient of variation (%CV) vs concentration. This shows the interval of concentrations where precision is maximum (usually given as 20–80% inhibition). In immunoassays, and by extension in ILAs, this range is considered as the *assay working range* [45].

(c) *Trueness* of the assay is defined as the difference between the experimentally determined value and the true value (where known) [40]. Trueness is stated quantitatively in terms of "bias," with smaller bias indicating greater trueness. The bias is the systematic component of error and the precision is the random component [38]. It is typically determined by comparing the response of the method to a certificate reference material with the known value assigned to the material [39]. If not available, the results obtained can be compared with those obtained by other well established reference methods. If none of the materials/methods described above is available, bias can be investigated by spiking and recovery tests [38, 40]. A test material is analyzed by the assay under validation in its original state and after the addition ("spiking") of a known mass of the analyte. The difference between the two results as a proportion of the mass added is called the surrogate recovery or sometimes the marginal recovery.

(d) *Specificity* or *selectivity* is a measure of the extent to which other substances affect the determination of the analyte. Such effects are known as interferences [38]. In MIP-ILAs it may be established by assaying a number of structurally related analytes and determining their cross-reactivity in the assay. The strategy usually consists of the evaluation of the dose-response curve for the test compound and for the analyte and the comparison of them with a test of similarity ("parallelism"). The cross-reactivity is usually given as a percentage:

$$\%\text{Cross} - \text{reactivity} = 100 \times (\text{IC}_{50}(\text{analyte})/\text{IC}_{50}(\text{interferent})). \quad (24)$$

These studies will help to identify which analytes may interfere in the assay or give false positives, or whether degradation products are still being detected [23, 45]. Cross-reactivity values below 0.001% are not a major concern; higher percentages must be appraised for every assay. If a significant interference is observed it will be necessary to include a clean-up step before the assay.

(e) The *recovery* of the method is the fraction or percentage of the actual amount of substance that can be obtained after one or more manipulative stages of the method [38]. Recovery evaluation can be carried out using· spiked blank samples or with matrix-spiked samples.

Studies of *matrix effects* are usually carried out by preparing standard curves on the matrix and comparing their parallelism with the curve prepared with the measuring buffer/solvent. A shift to the right (loss of sensitivity) or to the left (increase of sensitivity) whilst keeping parallelism may indicate the need for performing measurements by preparing the standard curve on the matrix. Alternatively, artificial samples can be prepared mimicking the effect of the matrix. The evaluation of the matrix effect depends on every system. In case this effect affects the interaction between the analyte and the MIP, sample preparation methods must be employed before the assay [45].

(f) The *robustness*, or *ruggedness*, of an analytical method is a measure of its capacity to remain unaffected by small but deliberate variations in the main parameters of the method. It provides an indication of the reliability of the assay during routine applications so that it is important to identify which steps are particularly sensitive to small changes in the experimental conditions of the MIP-ILA in order to modify the method in such a way that these operations are more robust.

Method validation is usually carried out in several steps (Fig. 8). First, the sensitivity and specificity of the assay are evaluated. Second, bias is estimated by comparison with reference methods using reference materials, analysis of samples of known composition or of spiked samples. Finally, the new method must be

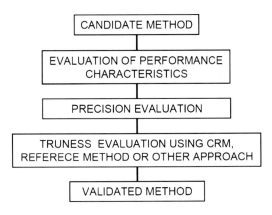

Fig. 8 Process validation chart [38]

applied to real sample analysis. Not many MIP ILAs have been applied to real sample analysis and the most important ones are described in the next sections. The goal of *method validation* is to ensure that the developed method provides data that are fit for purpose [41]. Collaborative studies, also known as collaborative trials or inter-laboratory method performance studies, are regarded as the ultimate evaluation of a method [41, 43]. They require a minimum number of laboratories and test materials to be included in the collaborative trial. To the best of our knowledge, MIP-ILAs have not been included so far in such studies and it would be of paramount importance to invest more efforts in this regard to demonstrate the applicability of these new materials to replace their biological counterparts.

2 Applications of Molecularly Imprinted Polymers in Immuno-Like Assays

Affinity assays using antibodies (immunoassays), enzymes, or nucleic acids have been established in research, clinical diagnosis, and industry for many years. However, for portable, easy to use, cost-effective devices, new sensitive receptors and measuring principles are needed.

MIPs are molecular recognition materials that can be tailored towards a wide variety of target molecules. These human-made materials are relatively easy to prepare and have found use as biomimetic receptors. The application of MIPs to ILAs requires not only high affinity highly selective binding sites but also a sensitive way to signal the binding event. Several analytical detection modes have already been used in combination with MIPs for ILAs development, mainly optical, enzymatic, radiochemical, or electrochemical. Similarly to the immunoassay classification outlined above, the ILAs included in this section have been separated into different groups depending on the transduction technique.

2.1 Radioimmuno-Like Assays

As in RIA, high sensitivity RILAs can be obtained through amplification of the binding reaction between the MIP and the target by means of radiolabels which act as "tracers." Obviously, as the analyte itself is not radioactive, the measuring principle is based on an indirect assay, typically a competitive one.

2.1.1 Heterogeneous Assays

The Mosbach group described in 1993 the first assay in which polymers imprinted with either theophylline or diazepam replaced antibodies in competitive RILA experiments [25]. Unfortunately, recognition did not take place in aqueous media (human serum) and it was necessary to change the media, from water to an organic solvent, prior to analysis. The labeled and unlabeled drug competes for the limited specific binding sites of the polymer and, after batch incubation, the MIP was removed by centrifugation and the radioactivity present in the supernatant was measured. The dynamic range in blood was 14–224 μM of theophylline. Moreover, this RILA was equivalent, in terms of cross-reactivity, to a commercial enzyme-multiplied immunotest performed with monoclonal antibodies.

Since this seminal work, many ILAs have been developed for determination of different compounds such as pesticides, drugs, and toxins. Table 1 [34, 46–56] collects a comparison of the analytical figures-of-merit obtained in the quantification of some analytes using immunoassays and ILAs. In a later work [30] (Table 2), using anti-morphine and anti-enkephalin MIPs, the same authors confirmed the ability of these materials to behave biomimetically in ILAs. The anti-morphine polymer imprints showed high affinity for binding morphine, with K_d values as low as 10^{-7} M, and levels of selectivity similar to those of antibodies (Table 2). Preparation of imprints against enkephalin was greatly facilitated by the linkage of anilide derivative to the peptide due to improved solubility of the template system obtained in the pre-polymerization mixture. Free enkephalin was efficiently recognized by this polymer ($K_d \sim 10^{-7}$ M) showing only a weak affinity for two D-amino acid-containing analogs of enkephalin. Enantioselective recognition of the L-enantiomer of phenylalanylglycine anilide, a truncated analog of the N-terminal end of enkephalin, was observed. In addition, it was possible to carry out this assay directly in an aqueous environment.

Andersson et al. [57] have reported a molecularly imprinting polymer for determination of the total concentration of cyclosporin A and its metabolites using a radioligand binding assay. The assay method comprises an extraction step of hemolyzed whole blood with diisopropyl ether followed by competitive RILA in the organic phase employing the MIP and [^{3}H]-cyclosporin. The assay measured the parent drug and the AM1, AM9, and AM4N metabolites with equal response.

The Haupt group [58] have described a competitive assay for 17-β-estradiol using the radioligand [2,4,6,7-^{3}H]-β-estradiol and MIPs prepared with methacrylic acid (MAA) and EGDMA in acetonitrile as bulk and spherical formats. The imprinted polymers were highly specific to β-estradiol since the CPs bound virtually no radioligand. The bulk polymer was then employed to screen structurally related endocrine disrupting chemicals such as α-estradiol, estrone, and ethynylestradiol with a cross-selectivity of 14%, 5%, and 0.7%, respectively. These results were compared to those obtained with a bioassay using stably transfected yeast cells in culture bearing the human estrogen receptor. The receptor was activated by several estrogen-like compounds and, to a lesser extent, by some structurally related species.

Further radiolabeled ligand-binding assays have been reported for a wide variety of analytes, namely morphine and leu-enkephaline [30], (S)-propanolol [33], atrazine [47], 17-β-estradiol [59], 2,4-dichlorophenoxyacetic acid (2,4-D) [32], yohimbine and corynanthine [60], and corticosterone/cortisol [29]. Other assays are summarized in Table 3.

Table 1 Comparison between the figures of merit of selected immunoassays and immuno-like assays for the analysis of environmental and food samples

Template	Monomer/crosslinker	Tracer	Dynamic range	Detection limit	Sample	Ref.
Atrazine	–	HRP-simazine derivative	0.24–3.64 µg L⁻¹	0.11 µg L⁻¹	Water	[46]
	MAA/EGDMA	[ring-¹⁴C]atrazine	1–1,000 µmol L⁻¹	0.2 µmol L⁻¹	Water	[47]
2,4-D	–	Goat anti-rabbit IgG-HRP	0.01–1,000 µg L⁻¹	0.008 µg L⁻¹	Water	[48]
	PVA	2,4-D Tobacco peroxidase	0.01–500 µg L⁻¹	0.004 µg L⁻¹	Water	[48]
Clenbuterol	–	Clenbuterol-HRP	5.0–40 nmol L⁻¹	1.2 nmol L⁻¹	Urine	[50]
	MAA/EGDMA	–	1–50 µg L⁻¹	0.3 µg L⁻¹	Urine	[51]
Fenbendazole	–	HRP-fenbendazol derivative	1–20 µg L⁻¹	0.5 µg L⁻¹	Milk	[52]
	MAA/EGDMA	–	–	57 µg L⁻¹	Liver	[53]
Penicillin G	–	PAAP	6–191 µg L⁻¹	2.4 µg L⁻¹	Water	[54]
	MAM/2P/EGDMA	PAAP	250–2,700 µg L⁻¹	83 µg L⁻¹	Water	[34]
Zearalenone	–	Zearalenone-HRP derivative	0.020–0.420 µg L⁻¹	0.007 µg L⁻¹	Cereals	[55]
	1-ALPP/TRIM	PMRA	16,400–8,200 µg L⁻¹	8.200 mg L⁻¹	–	[56]

PAAP [$S,5R,6R$]-3,3-Dimethyl-7-oxo-6-[(pyren-1-ylacetyl)amino]-4-thia-1-azabicyclo-[3.2.0] heptane-2-carboxylic acid, *PMRA* 2,4-Dihydroxy-*N*-pyren-l-ylmethylbenzamide, *2P* 1-(4-Vinylphenyl)-3-(3-trifluromethylphenyl) urea, *PVA* Poly(vinyl alcohol), *1-ALPP* 1-allylpiperazin

Table 2 Cross-reactivity of opiates to the binding of [^{3}H]morphine to the anti-morphine MIP and anti-morphine antibodies. Cross-reactivities are expressed as the molar ratio (%) of morphine to ligand that produces a 50% inhibition of radiolabeled morphine binding. Information on MAb1–4 in [30]

Ligand	Cross-reactivity (%)					
	Buffer	Toluene	MAb1	MAb2	MAb3	MAb4
Morphine	100	100	100	100	100	100
Codeine	25	4.7	18	104	36	<0.1
Normorphine	9.9	8.3				
Hydromorphone	15	6.0		112	9.8	
Heroin	8.3	2.3				
Naloxone	0.4	<0.1	<0.5	0.1	0.7	0.1
Naltrexone	0.3	<0.1		<<0.1	0.2	<<0.1
Leu-enkephalin	<<0.1					
Met-enkephalin	<<0.1			<<0.1		<<0.1

Table 3 Examples of radio immuno-like assays

Template	Monomer/crosslinker	Porogen	Tracer	Dynamic range	Ref.
Theophyllin	MAA/EGDMA	Chloroform	[^{3}H]Theophyllin	2–6 nmol L^{-1}	[61]
Morphine	MAA/EGDMA	Acetonitrile	[^{3}H]Morphine	1–1,000 µmol L^{-1}	[30]
Atrazine	MAA/EGDMA	Dichloromethane	[ring-U-^{14}C] Atrazine	0.1–100 µmol L^{-1}	[47]
2,4-D	4-VP/EGDMA	Methanol–water	[^{14}C]2,4-D	0.01–10 mg L^{-1}	[32]
17β-Estradiol	MAA/EGDMA	Acetonitrile	[2,4,6,7-^{3} H] β-Estradiol	0.01– 1,000 µmol L^{-1}	[59]

2.1.2 Homogeneous Assays

These formats are highly popular with immunoassays (IAs) because there is no need for physical separation of the solid-phase and the supernatant: reagents are simply added and measurements recorded. The scintillation proximity assay (SPA) applied to MIPs was first developed by Ye and Mosbach who described the principle of using a "universal" scintillation reporter embedded in molecularly imprinted microspheres. The polymer containing the scintillation reporter is imprinted against a β-adrenergic antagonist, S-propanolol. When tritium-labeled S-propanolol binds to the MIP, its β radiation triggers the nearby reporter to emit long wavelength fluorescence that can be directly quantified. When used in competitive-assay mode, the fluorescence signal decreases due to the non-labeled analyte competing for the limited number of binding sites. Several parameters, such us the amount of MIP and reaction kinetics, were optimized using the above-mentioned radioligand. This MIP-based SPA has the potential of providing a very high sample throughput. The assay seems to work properly in the 1–1,000 ng mL^{-1} analyte range although real samples were not evaluated [62]. Figure 9 depicts the principle of using scintillation reporters embedded in molecularly imprinted microspheres.

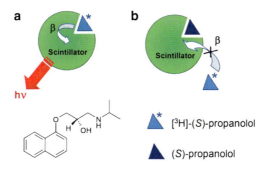

Fig. 9 Schematic representation of the scintillation proximity assay of (*S*)-propanolol using imprinted microspheres. The light green area represents the aromatic "antenna" element. (**a**) The bound, tritium-labeled (*S*)-propranolol triggers the scintillator to generate a fluorescent light. (**b**) When the tritium-labeled (*S*)-propranolol is displaced by the unlabeled (*S*)-propranolol, it is too far away from the scintillator antenna to transfer efficiently the radiation energy; therefore, no fluorescence can be generated (as described in [62])

Furthermore, the description of soluble imprinted microgels by the Wulff group [63] may provide a step towards homogeneous RILAs. Obviating the need for separation does not entirely depend on the use of a soluble antibody: a binding matrix that stays in stable suspension may serve as well.

Finally, we must point out that nowadays the widely extended RILAs are basically used for MIP characterization purposes. In a general way, the use of RILAs has been reduced drastically in the last few years. The literature shows that they have been replaced by other tracers with fluorescence, chemiluminescence or absorbance detection often combined with labeled enzymes. Simplicity in their use, easy access to these techniques, and the lack of radioactive compounds are the main reasons for such substitution.

2.2 Enzyme Immuno-Like Assays

Enzymes are perhaps the most versatile and popular class of labeling substrates for IAs and they have a sure future in this field [64, 65]. They were introduced in 1971 as an alternative to isotopes in immunoassays [66].

The most common enzyme label in IAs and ILAs is horseradish peroxidase (HRP) due to its high turnover number, the sensitivity of its colorimetric and luminometric assays, its suitability for diverse conjugation procedures, and relative small molecular size (40 kDa compared to 100 kDa for alkaline phosphatase). The use of labeled enzymes as tracers allows qualitative and quantitative assay procedures that are not dependent on instrumentation. Thus absorbance, luminescent, electrochemical, or multistage assay systems could be performed (Fig. 10) [23].

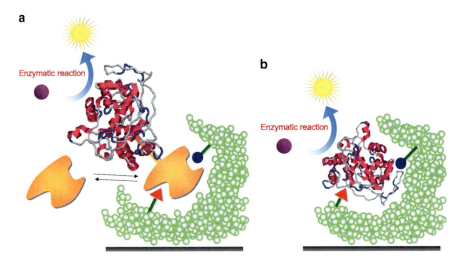

Fig. 10 Schematic representation of the two approaches mainly used in EzILAs. (**a**) The enzyme conjugate and the analyte compete for the selective binding sites of the polymer; finally, a substrate is converted into a product that generates a chemical signal (e.g., fluorescence, absorbance, electrochemical) at a rate which is proportional to the amount of bound enzyme and hence to the concentration of analyte in the sample. (**b**) Direct assay where the analyte is the enzyme which is quantified by a coupled enzymatic reaction

Table 4 Examples of enzyme immuno-like assays

Template	Monomers	Porogen	Enzyme	Dynamic range	Detection limit	Ref.
2,4-D	VPY, TRIM	Methanol–water	TOP	40–600 mg L^{-1}	–	[67]
2,4-D	VPY, TRIM	Methanol–water	TOP	1–200 mg L^{-1}	–	[67]
Epinephrine	APBA	Water	HRP	0.001–0.1 mol L^{-1}	–	[69]
Epinephrine	APBA, TBA	Water	HRP	–	8 μmol L^{-1}	[70]
Atrazine	MAA, EGDMA	Acetonitrile	HRP	–	0.25 μmol L^{-1}	[70]
Microperoxidase	APBA	Water	–	0.001–10 mg L^{-1}	–	[73]

TOP Tobacco peroxidase, *HRP* Horseradish peroxidase, *APBA* 3-Aminophenylboronic acid monohydrate, *TBA* 3-Thiopheneboronic acid, *VYP* 4-vinylpyridine, *TRIM* Trimethylolpropane trimethacrylate

However, the use of enzyme labels in ILAs seems to be less practical for a number of reasons. First, they often work only in aqueous buffers as enzymes are more susceptible to inactivation than fluorescent or chemiluminescent compounds under harsh conditions such as organic solvents. Additionally, enzymes are invariably high molecular weight species that diffuse slowly into the MIP cavities and may have a greater tendency to increase the non-specific interactions of some other labels. Accordingly, few enzyme ILAs have been reported to date (Table 4) and, in all cases, rely on the observation of binding sites unoccupied by the analyte using an enzyme-labeled template.

Surugiu et al. [67] have introduced an Enzyme Immuno-Like Assays (EzILA) for the herbicide 2,4-dichlorophenoxyacetic acid (2,4-D). The label was a 2,4-D conjugate with the tobacco peroxidase (TOP) enzyme, which allows for both colorimetric and chemiluminescent detection. In this case, the polymer imprinted with 2,4-D was synthesized in the form of microspheres. In contrast, despite its higher binding capacity for radiolabeled 2,4-D, a conventional MIP prepared by bulk polymerization showed only weak binding of the 2,4-D-TOP tracer.

The same authors developed flow injection capillary [68] and imaging [49] chemiluminescence ELISA-like MIP assays based on competition of 2,4-D-TOP with the same 2,4-D herbicide. In the former, a glass capillary was coated with the imprinted polymer and mounted in a flow system; in the imaging format, microtiter plates were coated with MIP beads fixed in place by poly(vinyl alcohol) as a glue. Calibration curves corresponding to analyte concentrations of $0.01-100$ μg mL^{-1} (imaging format) and 0.5 ng mL^{-1}–50 μg mL^{-1} (flow injection format) were produced. When compounds structurally related to 2,4-D were tested in the competitive assay, cross-reactivities of 75%, 32%, and 10% for 2,4,5-trichlorophenoxyacetic acid, 4-chlorophenoxyacetic acid, and phenoxyacetic acid were obtained, respectively. A clear difference between the binding of 2,4-D to the imprinted polymers and CPs was also observed. The assay was applied to the analysis of 2,4-D-spiked water samples at different concentration levels. The use of either flow injection or imaging format enables consecutive and/or simultaneous measurements of a larger number of samples leading to an increase of the sample throughput or the development of multisensors for simultaneous determination of different analytes.

Piletsky et al. [69, 70] have developed a novel technique for coating microtiter plates with MIPs for the specific recognition of epinephrine. To that end, 3-aminophenylboronic acid was polymerized in the presence of epinephrine with oxidation of the monomer by ammonium persulfate. This process led to the grafting of a thin polymer layer onto the polystyrene surface of the microplates. The polymer affinity was determined by an enzyme-linked assay using a conjugate of horseradish peroxidase and norepinephrine (HRP-N) Such microplates, coated with the epinephrine-imprinted polymer, were tested using competition between the free ligand and HRP-N. The enzymatic reaction of HRP with hydrogen peroxide and 2,2′-azino-bis(3-ethylbenzothiazoline-6-sulfonic acid) (ABTS) was used to determine the concentration of conjugate adsorbed on the polymer by measuring the absorbance of the generated product at 450 nm. It was found that imprinting yielded increased affinity of the polymer towards HRP-N and epinephrine. The effect of the buffer pH and concentration on the polymer affinity was analyzed to demonstrate that both factors modulate the strength of the electrostatic and reversible covalent interactions. The assay sensitivity is in the range of $1-100$ μM, a figure that can probably be increased with further optimization of the polymer composition and assay performance. The dissociation constant for the MIP-epinephrine complex is 9.2 μM. This value is in the data range found for the natural adenylate-cyclase coupled β-adrenergic receptor [71]. Moreover, the high stability of the polymers and good reproducibility of the measurements make MIP coatings an attractive alternative to traditional antibodies or receptors employed in ELISAs.

In a similar approach, Wang and co-workers [72] have described a direct competitive ELISA-like MIP assay for the determination of estrone in natural waters. Polymer films of controlled thickness were directly prepared on the well surface of 96-well plates by a room-temperature oxidative chemical process mediated by an ionic liquid, 1-butyl-3-methylimidazolium hexafluorophosphate. The use of the ionic liquid allowed a reduction of the cracking and shrinking of the imprinted films, facilitating the access of the HRP-labeled conjugate to the binding sites. The IC_{50} and the detection limit were 200 ± 40 μg L^{-1} and 8.0 ± 0.2 μg L^{-1}, respectively, showing a better performance than other MIP-based assays [71]. Compared to methods that use polyclonal antibodies against estrone, the detection limit of the MIP assay was 100-fold higher although the analysis time was considerably reduced to 80 min and the films could be reused more than 50 times without loss of sensitivity. The assay was applied to the analysis of estrone-spiked lake and river water at three concentration levels (100, 200, and 400 μg L^{-1}) and validated by HPLC, although the statistical comparison of the reported results shows significant differences for some samples.

Another issue is related to the use of enzymes in molecular imprinting technology as protein templates (Fig. 10). This approach has been pioneered by the Piletsky group [73] in the development of an ILA based on microplate wells coated with MIPs for determination of several enzymes. The MIP is synthesized by chemical oxidation of 3-aminophenylboronic acid (APBA). As described above, the APBA-based polymer is tightly grafted to the surface of the plate by aromatic ring electron-pairing interactions. A few proteins (horseradish peroxidase, lactoperoxidase, microperoxidase, hemoglobin HbA_0, and cytochrome C), differing in mass and charge, were selected as templates. The quantity of bound enzyme was measured as a function of its peroxidase activity using ABTS as substrate and measuring the optical density at 450 nm. The resulting K_d values are shown in Table 4. From the rebinding experiments, it was evident that the size (shape) of the molecule and its charge affect the imprinting efficiency. For a small protein such as microperoxidase, the dissociation constant was rather high (1.5 μM). Large proteins, such as lactoperoxidase (M_r 77,000), hemoglobin A_0 (M_r 65,000), and horseradish peroxidase (M_r 44,000), exhibited much lower K_d values (36, 56, and 82 nM, respectively). The protein charge also significantly affects the formation of binding sites and the polymer affinity. Thus, highly positively charged proteins such as cytochrome c (pI 10.6) are not able to form imprints in the APBA polymer, which has a net positive charge.

In conclusion, the easy preparation of the MIPs, their high stability, and their ability to recognize small and large proteins, and to discriminate macromolecules with small variations in charge, make this approach attractive and broadly applicable in biotechnology, assays, and sensors. Up to now, several studies on the imprinting of macromolecules for the analysis of proteins have been performed and this topic has been extensively reviewed [74–76].

2.3 Absorbance Immuno-Like Assays

The major problem for the application of absorption measurements to MIP-based direct or indirect assays is the dispersion of light at low wavelengths and the absorption at the same wavelength of other species. For these reason is necessary to develop optical dyes with high absorption in the lowest energy part of the electromagnetic spectrum.

Greene and Shimizu [77] have reported a displacement assay using a colorimetric dye (N,N-dimethyl-N,N-(7-nitro-2,1,3-benzoxadiazol-4-yl)-1,2-ethanediamine) with absorption maximum at 460 nm. This assay allowed the determination of seven different aromatic amines. Nevertheless, although the dye seems to be more sensitive in the polymer than in solution, after thorough optimization of the assay conditions, the displacement is only significant for the mmol L^{-1} amine concentration range.

2.4 Fluoroimmuno-Like Assays

Fluorometry is a superior optical technique in terms of sensitivity and specificity. Merits of fluoroimmunoassays (FIAs) and fluoroimmuno-like assays (FILAs) include the stability and freedom from hazards of fluorescent labels compared to radioactive tracers, the moderate cost of analysis, the wide availability of the equipment needed, and the potential high sensitivity. In general, the sensitivity of fluorescence measurements is 10- to 1,000-fold higher than the absorption counterparts.

An indication of the potential sensitivity of fluorometry is that single-molecule detection has been based almost exclusively on the use of fluorescent labeled compounds. In addition, fluorometric determinations can combine several parameters simultaneously ("multidimensional" techniques), such as excitation and emission wavelengths, fluorescence lifetime, and polarization, providing additional specificity and versatility to the analytical measurements.

2.4.1 Heterogeneous Assays

The heterogeneous FILAs can be based on one of the following detection schemes: (1) the analyte itself is fluorescent and its binding to the MIP can be monitored directly; (2) the analyte is not fluorescent so that it is necessary to synthesize a fluorescent analog capable of competing with the analyte for the polymer binding sites, upon the measuring principle of competitive or displacement assays; (3) the polymer is labeled with a fluorescent reporter that will modify its emission intensity, lifetime, and/or emission wavelength upon analyte binding. If the analyte itself cannot interact strongly enough with the fluorescent polymer, an external non-related quencher or modifier can be added to monitor the fluorescence change upon analyte binding.

Fluoroimmuno-Like Assays Based on Direct Measurements

In this case, the MIP performs as the selective recognition element and quantification is generally achieved by measuring the specific emission of the fluorescent analyte after its release from the polymer binding sites. Table 5 [78–85] collects some recent examples of such type of assays.

For instance, Karube et al. [86] have described the development of a sensitive system, in combination with HPLC, for the analysis of the fluorescent hormone β-estradiol. In their HPLC method, the MIP is used as the stationary phase. The polymers were prepared using MAA and EGDMA and the direct measurement of the β-estradiol fluorescence was carried out. They compared performance of the direct assay with that of the competitive mode, based on the displacement by the analyte of a fluorescent compound from the specific binding sites in the imprinted polymer. Unfortunately, the reporter compounds used in this approach (β-estradiol dansylate, boc-L-tryptophan, 11-((5-dimethylamino-naphthalene-1-sulfonyl) amino)undecanoic acid, or 6-(N-(7-nitrobenz-2-oxa-1,3-diazol-4-yl)amino)hexanoic acid) did not compete efficiently with the template for the polymer binding sites. This behavior was attributed to their larger size in comparison with β-estradiol, to the differences in their 3D structure, and/or to the lower strength of their interactions with the binding cavities. The detection based on direct interrogation allowed determination of the analyte in the 0.1–4 µM range, with a detection limit of 100 nM, good reproducibility, and excellent selectivity.

Wandelt et al. [87] have reported the application of a FILA for nucleotide monitoring. The assays were based on the fluorescence quenching of 3-diphenyl-6-vinyl-1H-pyrazole[3,4-b]quinoline (PAQ), by guanosine 30,50-cyclic monophosphate (cGMP), adenosine 30,50-cyclic monophosphate (cAMP), and cytidine 30,50-cyclic monophosphate (cCMP). Although no analytical characterization of the polymers was reported, the authors carried out a rebinding study to ascertain the applicability of the MIPs for such application.

The main limitation of direct FILAs is the limited availability of fluorescent target molecules. The preparation of a fluorescent derivative is time-consuming and requires the presence in the molecule of an appropriate functional group. On the other hand, derivatization may compromise MIP selectivity.

Fluoroimmuno-Like Assays Based on Indirect (Competitive or Displacement) Schemes

In these assays the target is not fluorescent itself but a labeled analog is prepared to compete with the analyte for the binding sites of the imprinted polymer. Most of the labeled analytes described in the literature for competitive or displacement ILAs are

Table 5 Analytical characteristics of fluoroimmuno-like assays

Template	Monomers	Tracer	Measuring solvent	Dynamic range	Detection limit	Ref.
Methotrexate	MAA/EGDMA	Intrinsic fluorescence of oxidized methotrexate	Water-ACN (95:5, v/v)	4–500 ng mL^{-1}	0.82 ng mL^{-1}	[78]
Chloramphenicol	VPY/DAM	Dansylated chloramphenicol	Acetonitrile	8–100 µg mL^{-1}	8 µg mL^{-1}	[79]
2,4-Dichlorophenoxyacetic acid	VPY/EGDMA	CMMC	20 mM PB pH 7.0, 0.1% TX-100 Acetonitrile	0.1–50 µmol L^{-1} 0.1–10 µmol L^{-1}	100 nmol L^{-1}	[80]
Pinacolyl methyl phosphonate	MAA/DVB	4MU	Toluene	5–30 nmol	5 nmol	[82]
β-Estradiol	MAA/EGDMA	Intrinsic fluorescence of beta-estradiol	Water-ACN (97.5:2.5, v/v)	4–80 µg L^{-1}	1.12 µg L^{-1}	[82]
Sialic acid	AA/EGDMA	OPA	100 mM sodium borate, pH 10	0.5–10 µmol L^{-1}	0.5 µmol L^{-1}	[83]
Triazine	MAA/EGDMA	DTAF	Ethanol	0.01–100 mmol L^{-1}	0.01 mM	[84]
Digoxin	MAA/EGDMA	Digoxin-FITC	Acetonitrile	0.0–600 µg L^{-1}	0.017 µg L^{-1}	[85]

MAA Methacrylic acid, *EGDMA* Ethylenglycoldimethacrylate, *DAM* Diethylaminoethylaminoethylmethacrylate, *VPY* 4-Vinylpiridine; *DVB* Divinylbenzene, *CMMC* 7-Carboxymethoxy-4-methylcoumarin, *AA* Allylamine, *4MU* 4-Methylumbelliferone, *OPA* o-Phthalic dialdehyde + beta-mercaptoethanol, *APBA* 3-Aminophenylboronic acid, *TBA* 3'-Thiopheneboronic acid, *DTAF* 5-[(4,6-Dichlorotriazin-2-yl)amino]fluorescein, *FITC* Fluorescein isothiocyanate radioactive analogs but fluorescent tags are also known [10, 25, 88]. However, it is not uncommon that the analyte cannot easily be derivatized with a fluorescent label or the labeled derivative is not able to bind specifically to the MIP in competition with the unlabeled analyte [80, 86].

The use of fluorescent derivatives of the target in ILAs are not as widespread as the direct methods, because the selective sites of the polymer material are not always able to accommodate the bulkier fluorescent probe suitably. Anyhow, some indirect FILAs have been described in the literature based on different approaches (Fig. 11):

1. The analyte itself or an analog of it is tagged with a fluorophore and the polymer is imprinted with the unmodified analyte [36, 89, 90].
2. The competitor is a fluorescent species more or less unrelated to the analyte structure, and the polymer is imprinted with the unmodified analyte [80, 91].
3. The analyte or analog is labeled with a fluorophore and the polymer is imprinted with it.

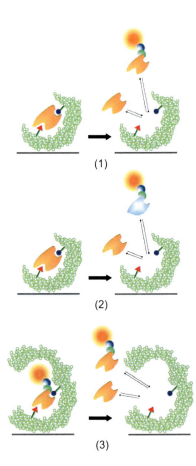

Fig. 11 Schematic representation of the two approaches mainly used in FILAs. (1) The analyte or analog of it is labeled with a fluorophore and the polymer is imprinted with the native analyte. (2) The probe is a fluorescent species unrelated to the analyte structure and the polymer is imprinted with the native analyte. (3) The analyte or its analog is labeled with a fluorophore and the polymer is imprinted with it

The first work in this field was probably that of Piletsky et al. [84] that described a competitive FILA for the analysis of triazine using the fluorescent derivative 5-[(4,6-dichlorotriazin-2-yl)amino]fluorescein. The fluorescence of the supernatant after incubation was proportional to the triazine concentration and the assay was selective to triazine over atrazine and simazine. The same fluorescent triazine derivative was applied to competitive assays using atrazine-imprinted films [70]. To this end an oxidative polymerization was performed in the presence of the template, the monomer(s) 3-thiopheneboronic acid (TBA) or mixtures of 3-amino-phenylboronic acid (APBA) and TBA (10:1) in ethanol–water (1:1 v/v) where the template is more soluble. The polymers were grafted onto the surface of polystyrene microplates. The poly-TBA polymers yielded a detection limit of 8 μM atrazine whereas for the poly-TBA-APBA plates it was lowered to 0.7 μM after 5 h of incubation. However, a 10–20% decrease in the polymer affinity was observed after 2 months.

Karube and coworkers [92] have also developed an MIP-HPLC method for the analysis of chloramphenicol (CAP) based on the competitive displacement of a chloramphenicol-methyl red (CAP-MR) conjugate by the antibiotic from the polymer binding sites during the HPLC. The best polymers were obtained using (diethylamino)ethylmethacrylate (DAEM) as functional monomer. The mobile phase contained CAP-MR and the injection of CAP, and to a lesser extent triamphenicol (TAM), resulted in a concentration-dependent conjugate displacement. The linear response range was $3–1,000$ μg mL^{-1}, the sample measuring time was 5 min, and the assay was applied to the detection of CAP in serum samples. In a further work [89], the polymer was imprinted inside an LC column. The column capacity was significantly lower (linear range for CAP: $0–30$ μg mL^{-1}; detection limit: 3 μg mL^{-1}) than that obtained with the bulk-polymerized beads but the ability to separate CAP and TAM improved.

A competitive flow-through FILA assay for the analysis of CAP has been described by Suárez-Rodríguez and Díaz-García [93] using dansylated chloramphenicol. The measuring scheme was similar to that by Karube et al. [92] but, in this case, the polymer was packed in a flow-through cell instead of an LC column. The detection limit was 8 μg mL^{-1} and the working range was up to 100 μg mL^{-1} CAP. The sample throughput was five to six samples per hour.

Haupt et al. [80, 94] have developed a fluorescent competitive assay for 2,4-dichlorophenoxyacetic acid (2,4-D) in organic and aqueous solvents using 7-methoxycarbonyl-4-methylcoumarin (CMMC), a non-related probe with some structural similarities to the target analyte. Initially, fluorescein isothiocyanate was coupled to the carboxyl group of 2,4-D via ethylenediamine or 1,6-diaminohexane spacers but no competition with 2,4-D was obtained so that CMMC was selected for further experiments. In the latter case, measurements were performed by incubating for 2 h a constant concentration of the fluorescent probe (640 nM) with increasing concentrations of the analyte in 20 mM, pH 7.0 sodium phosphate buffer containing 0.1% of Triton X-100, and measuring the fluorescent emission of the supernatant. Binding of CMMC to the MIP, evaluated from radioactive competitive assays with ^{14}C-2,4-D, was just 6% due to the small resemblance of the probe to the analyte.

The detection limit was 0.1 μM 2,4-D, comparable to that of the radioassay, and the response range was 0.1–50 μM in buffer. Cross-reactivity of the polymer towards 2,4-dichlorophenoxyacetic acid methyl ester (2,4-D-OMe) was lower than that reported for immunoassays.

Piletsky et al. [91] have also applied a non-related fluorescent probe, rhodamine B, to the detection of L-phenylalaninamide (L-Phe-NH$_2$), D-phenylalaninamide (D-Phe-NH$_2$), L-phenylalanine (L-Phe), and L-tryptophane (L-Trp). The imprinted polymer was packed in a chromatographic column and saturated with the dye until no change was detected in the concentration eluted from the column. The injection of analyte aliquots into the eluent dye solution originates the displacement of the fluorescent dye from the polymer binding sites and the corresponding variation in the peak areas. The polymer showed an association constant (K_d) of about 60 μM for the template L-Phe-NH$_2$ and of about 133 μM for rhodamine B. Surprisingly, the CP also displayed a high affinity for the template ($K_d = 83$ μM) and showed a different response for the aminoacid enantiomers that was not observed in the MIP.

It must be underlined that, for the development of a successful FILA based on the use of non-related tracers, the latter should also show sufficient affinity for the specific binding sites of the imprinted polymer; otherwise the assay will not be selective. For instance, in order to facilitate the competition between the labeled derivative and the analyte, Moreno-Bondi et al. have developed a FILA for the analysis of penicillins [34, 36] using novel fluorescently labeled β-lactam antibiotics with a close resemblance to the analyte (Fig. 12) [95].

The highly fluorescent competitors (emission quantum yields of 0.4–0.95) were molecularly engineered to contain pyrene or dansyl fluorescent tags while keeping intact the 6-aminopenicillanic acid moiety for efficient competition with penicillin G (PenG) for the polymer binding sites. A library of six polymers imprinted with PenG was synthesized and the interaction of the labeled antibiotics and the MIPs were evaluated using competitive binding assays with radiolabeled PenG. The fluorescent tagged antibiotic [2S,5R,6R]-3,3-dimethyl-7-oxo-6-[(pyren-1-ylacetyl) amino]-4-thia-1-azabicyclo[3.2.0]heptane-2-carboxylic acid (PAAP) and a PenG-imprinted polymer prepared with MAA and TRIM (10:15 molar ratio) in acetonitrile provided the best sensitivity for PenG analysis. Molecular modeling experiments showed that recognition of the fluorescent analogs of PenG by the MIP was due to a combination of size and shape selectivity demonstrating the importance of the choice of label and tether chain for the assay success. The dynamic range of the FILA was 3–890 μM with a detection limit of 0.32 μM PenG. Cross-reactivity was observed for some antibiotics derived from 6-aminopenicillanic acid, particularly amoxicillin, ampicillin, and penicillin V, but not oxacillin, cloxacillin, dicloxacillin, or nafcillin. Other antibiotics, such as chloramphenicol, tetracycline, or cephapirin did not compete with PAAP.

In order to overcome the limitations of the methacrylate-based MIPs for selective recognition in aqueous samples, Urraca et al. have prepared PenG-imprinted polymers using an urea-based functional monomer to target the single oxyanionic groups in the template molecule [34]. This polymer has shown excellent recognition in aqueous samples [90] and has been applied to the development of the first

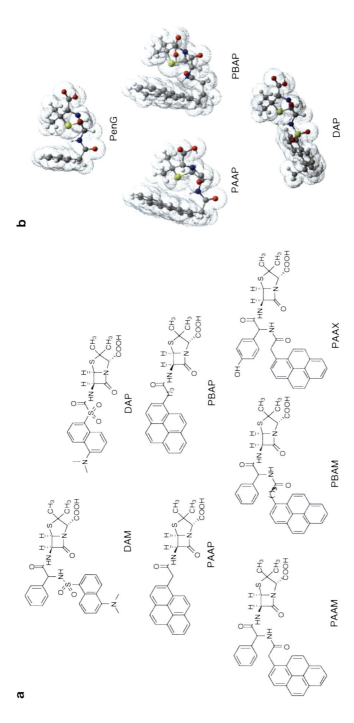

Fig. 12 (a) Chemical structures of the fluorescent tracers synthesized for β-lactam antibiotic analysis: **PAAP**: [2S,5R,6R]-3,3-dimethyl-7-oxo-6-[(pyren-1-yl)acetyl) amino]-4-thia-1-azabicyclo[3.2.0]heptane-2-carboxylic acid; **PBAP**: [2S,5R,6R]-3,3-dimethyl-7-oxo-6-[(4-pyren-1ylbutanoyl)amino]-4-thia-1-azabicyclo[3.2.0] heptane-2-carboxylic acid; **PAAM**: [2S,5R,6R]-3,3-dimethyl-7-oxo-6-[((2R)-2-phenyl-2-[(pyren-1-yl)acety)amino]ethanoyl]amino)-4-thia-1-azabicyclo[3.2.0] heptane-2-carboxylic acid; **PBAM**: [2S,5R,6R]-3,3-dimethyl-7-oxo-6-(((2R)-2-phenyl-2-[(pyren-1-ylbutanoyl)amino]ethanoyl]amino)-4-thia-1-azabicyclo [3.2.0]heptane-2-carboxylic acid; **PAAX**: [2S,5R,6R]-6-[[(2R)-2-amino-2-(4-hydroxyphenyl)ethanoyl]amino)-3,3-dimethyl-7-oxo-4-thia-1-azabicyclo[3.2.0]hep-tane-2-carboxylic acid; **DAM**: [2S,5R,6R]-6-[[(5-(dimethylamino)-1-naphthyl]sulfonyl]amino-2-phenylethanoyl]amino-3,3-dimethyl-7-oxo-4-thia-1-azabicyclo [3.2.0]heptane-2-carboxylic acid, **DAP**: [2S,5R,6R]-6-[[(5-(dimethylamino)-1-naphthyl]sulfonyl]amino-3,3-dimethyl-7-oxo-4-thia-1-azabicyclo[3.2.0]heptane-2-carboxylic acid. (b) AM1-energy-minimized structures of PenG, PAAP, PBAP, and DAP

automated molecularly imprinted sorbent based assay for the analysis of penicillin-type β-lactam antibiotics. The pyrene-labeled antibiotics shown in Fig. 12 were evaluated for the assay optimization and PAAP was again the tracer that provided the highest sensitivity for PenG analysis. The analyte and a constant concentration of PAAP were allowed to compete for the binding sites of the MIP, which was packed into a reactor. After application of a desorbing solution, the fluorescence of the labeled derivative eluted from the sorbent was measured and related to the analyte concentration in the sample. The detection limit was decreased to 0.197 μM and the dynamic range was 0.680–7.21 μM PenG in 0.1 M pH 7.5 acetonitrile–HEPES buffer (40:60, v/v) solutions. Cross-reactivity was observed with other antibiotics of the penicillin family such as ampicillin (71%), oxacillin (66%), penicillin V (56%), amoxicillin (13%), and nafcillin (46%). The automatic MIA has been successfully applied to the direct analysis of PenG in spiked urine samples with excellent recoveries (mean value 92%) and results statistically comparable to those obtained by HPLC-DAD.

Furthermore, the same authors have also developed an automated fluoroimmunoassay for penicillins using the same pyrene-labeled antibiotics [54]. Protein A/G covalently bound to an azlactone-activated polymer support was used for the oriented capture of antibody-PAAP/antibody-PENG immunocomplexes. Upon desorption from the immunosupport, the emission signal generated by the antibody/PAAP complexes is related to the antibiotic concentration in the sample. The detection limit for PenG was 6.44 nM (2.4 ng mL^{-1}) and the dynamic range was 16.1–512.7 nM (6.0–191 ng mL^{-1}).

The generic nature of the antiserum was shown by good relative cross-reactivities with penicillin type β-lactam antibiotics such as amoxicillin (50%), ampicillin (47%), and penicillin V (145%), and a lower response to the isoxazolyl penicillins such as oxacillin, cloxacillin, and dicloxacillin. No cross-reactivity was obtained for cephalosporin type β-lactam antibiotics (cephapirin), chloramphenicol, or fluoroquinolones (enrofloxacin and ciprofloxacin).

Table 6 summarizes a comparison of the IC$_{50}$ and cross-reactivity for three types of assays for PenG analysis, two of them based on MIPs and the third one on the

Table 6 IC$_{50}$ values and cross-reactivities (CR) in competitive fluoroimmuno-like MIP assays and an immunoassay; tracer: PAAP. Measuring conditions as in [34, 36, 54]

Analyte	MIP1 assay [36]		MIP2 assay [34]		Immunoassay [54]	
	IC$_{50}$ (μmol L^{-1})	CR (%)	IC$_{50}$ (μmol L^{-1})	CR (%)	IC$_{50}$ (μmol L^{-1})	CR (%)
PenG	53	100	1.8	100	0.1	100
AMOX	180	29	13.9	13	0.2	50
AMPI	140	37	2.5	71	0.2	47
PENV	130	40	3.2	56	0.05	145
OXA	250	21	2.7	66	0.5	12
CLOX	930	6.0	6.7	27	1.5	3.9
DICLOX	5,400	0.9	11.6	16	4.0	1.5
NAFCI	1,700	3.0	4.1	46	nc	nc
CEPHA	nc	nc	23.1	8	nc	nc

nc No competition

specific antibody. Although in the case of MIP1 the analyte required extraction from the aqueous samples into an almost purely organic media (ACN–water 99:1), the three fluoroligand-binding assays were equivalent in terms of cross-reactivity confirming the potential of MIPs as antibody mimics in ILAs. Moreover, MIP2 demonstrates that efficient binding can be achieved in an aqueous environment and the three recognition entities show remarkable group specificity. Only low cross-reactivities were obtained for oxacillin, cloxacillin, dicloxacillin, and nafcillin while a non-related 6-APA antibiotic, cephapirin, did not react. These results demonstrate the importance of shape-recognition and functional group distribution for the MIPs synthesized and tested.

More recently, new formats of MIPs have been developed with the aim of increasing selectivity and capacity against the template molecule for sorbent assays and drug delivery. In this regard, submicrometer-sized MIP particles are particularly well suited [96, 97]. During the past few years, MIP nanoparticles have been produced by several methods such as precipitation or emulsion polymerization [98]. Surface-imprinted core-shell nanoparticles can be an attractive alternative since imprinted nanoparticles can be prepared by grafting MIPs to or from the surface of supporting preformed-core nanomaterials (e.g., silica nanoparticles, magnetite nanoparticles, quantum dots, or carbon nanotubes).

In this way, Lu and coworkers [99] have prepared core-shell particles selective to 2,4-D, via RAFT controlled/living polymerization, to allow growth of uniform MIP shells with adjustable thicknesses (Fig. 13). Moreover, a competitive assay was accomplished between 2,4-D and the structurally related fluorophore 7-carboxy-4-methylcoumarin (CMMC). A typical competitive curve is obtained in buffered solution. The useful concentration range for the detection of 2,4-D ranged from 50 nM to 20 μM. The detection limit was around 10 nM, making possible its application to real samples.

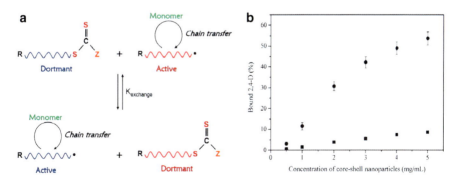

Fig. 13 (**a**) Schematic representation of the general RAFT polymerization mechanism. (**b**) Binding of 2,4-D to imprinted (*circles*) and control (*squares*) core-shell nanoparticles. The points represent mean values of five determinations (Adapted with permission from [99]. Copyright 2007 American Chemical Society)

2.4.2 Homogeneous Assays

Like in RILAs, an advantage of fluorescence detection is the possibility of developing homogeneous FILAs using direct or indirect (competitive or displacement) approaches. The fluorescence polarization immunoassay (PFIA) and their homolog fluorescence polarization immuno-like assay (PFILA) are two of the most widely used procedures in homogeneous fluoroassays. Both are based on the principle that fluorescence polarization gives a direct measure of the bound/free ratio of the labeled analyte (tracer) without the need for their separation [23, 28].

In this way, Hunt et al. [28] carried out, for the first time, an indirect PFILA where 2,4-D competes with 7-carboxy-4-methylcoumarin (CMMC) for the binding sites of an MIP microgel. Dynamic light scattering at 90° of the hydrogel suspension in methanol demonstrated that the MIP particle size was between 140 mm and 220 nm. The results showed that the assay is selective for 2,4-D with an IC_{50} value about 10 μM, approximately 30 times lower than that for phenoxyacetic acid, although 2,4-dichlorobutyric acid competes strongly.

Prahl and coworkers [100] have developed a direct homogeneous FILA assay applying steady-state and time-resolved fluorescence anisotropy measurements to get some insight of the MIP polymerization process and template rebinding. These authors compared the fluorescence anisotropy of polyurethanes imprinted with anthracene and in the absence of the template molecule during polymerization. The anisotropy of the polymer solutions increased during polymerization, probably reflecting the increase of viscosity of the local environment around the fluorophore. They found that MIPs and NIPs had the same steady-state anisotropy and observed that analytes rebound to the polymer had a shorter fluorescence lifetime and correlational time than those measured in the initially imprinted materials. This would suggest a shorter distance and a tighter binding between anthracene and the polymer after rebinding, although further confirmation of this hypothesis is required.

An example of direct characterization of hydrogels by homogeneous FILA has been reported by Aburto et al. [101]. The hydrogel was synthesized using cross-linking of chitosan with glutaraldehyde in the presence of dibenzothiophene sulfone (DBTS) as template. In this case, the double role of the template molecule is very interesting, because after the study of cross-linker ratio (2.0 moles of glutaraldehyde per mole of glucosamine), swelling and temperature (50 °C), the template was used as ligand and as fluorescent species for probing the high affinity binding sites. For the first time, these assays have been used to identify the molecular recognition conformation based on the ligands and not on the polymers.

With respect to the drawbacks associated with FILAs, synthesizing a fluorescent monomer or labeled derivatives is time-consuming and may not render the desired output. In fact, sometimes the magnitude of the fluorescence change is smaller than expected, or the derivatized monomer/analyte does not show the same fluorescence characteristics as the isolated label. It may also be possible that, due to heterogeneity of the binding sites, only in a small fraction of binding sites is the orientation of the fluorophore adequate to signal the binding event.

It should be preferable to design assays based on fluorescence enhancement upon analyte binding rather than on fluorescence quenching, because this is usually less sensitive to small concentration changes. With this aim, new fluorophores must be designed and synthesized that can lead to the preparation of highly sensitive and selective MIPs with a broad response range to the target analyte.

In this context, Wang and coworkers [102] have synthesized molecularly imprinted silica beads embedded with CdSe quantum dots (QDs) for pentachloro-phenol (PCP) analysis in water samples. This approach avoids the interference of autofluorescence and light scattering from the sample matrices, allowing the detection of trace levels of PCP (86 nM). The molecular recognition sites were formed on the surface of a silica layer deposited onto Mn-doped ZnS QDs. In the recognition process, the MIP- and NIP-coated QDs interact by acid-base pairing with the acidic hydroxyl group of PCP, although the RTP quenching efficiency was just twofold higher in the imprinted than in the non-imprinted material. The authors suggest that the sensing mechanism is based on a charge transfer from the QDs to the PCP species that yields a quenching of the room temperature phosphorescence signal of the QDs. The polymer showed cross-reactivity to other phenols such as 2,4-dichlorophenol (DCP, pK_a 7.9), 2,4-dinitrophenol (DNP, pK_a 4.0), and phenol (Phe, pK_a 10.0), although there was no correlation between recognition and phenol acidity, as would happen in the absence of imprinting effect. In fact, the Stern–Volmer quenching constant (K_{SV}) of DNP was much lower than that of PCP (pK_a 4.9), indicating an efficient imprinting effect. The method was applied to the analysis of PCP in spiked river water samples, with recoveries of 93–106%.

2.5   Chemiluminescence Immuno-Like Assays

Methods based on chemiluminescent labels are other non-isotopic IAs and ILAs that continue to be the focus of active research. Most common approaches under this category are direct measurements of the chemiluminescence accomplished by the action of the imprinted polymer (e.g., catalytic effect) or by indirect methods, such as competitive CILAs.

Chemiluminescence (CL) and electrogenerated chemiluminescence (ECL) are commonly used for detecting the hydrogen peroxide product of enzymatic reactions. These techniques are characterized by simple optical instrumentation: no excitation light source is required and therefore no additional apparatus is necessary to remove the scattered excitation light. These methods also have advantages in selectivity and sensitivity over traditional absorbance, fluorescence, and reflectance optical techniques, as the background is greatly reduced [103].

One of the earliest investigations on MIPs as CILAs materials was carried out by Lin and Yamada using an imprinting of DNS-L-Phe [104] and Cu(II)/1,10-phenan-throline [105]. In this case, a direct flow-through CILA based on a polymer containing a ternary complex with an efficient catalyst for the decomposition of hydrogen peroxide has been reported. During the hydrogen peroxide decomposition, the generated superoxide radical ion reacts with 1,10-phenanthroline (template) and gives a chemiluminescent emission in the presence of copper. Although the assay displays a fast response and a relatively low detection limit, it was not satisfactory enough compared with other versions yet it was a good approximation to this kind of assay.

Since 2004, Zhang et al. [51, 106–109] have described several chemiluminescent ILAs with drug-imprinted MIPs synthesized from MAA and EGDMA as monomers and acetonitrile as porogen. The MIP-adsorbed analyte (e.g., salbutamol, terbutaline, or hydralazine) enhances the CL intensity of the reaction of luminol with ferricyanide in a micro-flow-through sensor. Other CL systems have also been used. In all cases the recognition of the analytes took place in aqueous media such as urine and serum. Under the experimental conditions, low detection limits have been obtained even using direct introduction of the sample in the flow-assay system without previous treatment and manipulation of the sample (e.g., 16 ng mL^{-1} for salbutamol, 4 ng mL^{-1} for terbutaline, or 0.6 ng mL^{-1} for terbutaline). Unfortunately, the authors do not always report the MIP cross-reactivity with potential urine interference nor do they provide any information about the binding behavior to NIPs. Moreover, the mechanism of the chemiluminescence enhancement is not clearly understood.

Fei et al. and Nie et al. [110, 111] have reported two assays for indapamide and indomethacin, respectively, also based on luminol as chemiluminescent reagent. Again MAA and EGDMA were used as monomers and acetonitrile as porogen. Excellent detection limits were achieved: 10^{-8} M for indapamide and indomethacin in urine and 10^{-9} M for epinephrine in serum. The authors describe the cross-reactivity for other contaminants potentially present in urine such as antibiotics (norfloxacin), ions (Fe^{2+}, Ca^{2+}, K^+), or glucose.

He et al. [112] have developed a complex design for determination of trimethoprim in urine samples. The chemiluminescence intensity was linearly related to the concentration of trimethoprim over the 5.0×10^{-8} to 5.0×10^{-6} g mL^{-1} range with a 4.8% relative standard deviation. The detection limit is as low as 2×10^{-8} g mL^{-1} and the assay seems to be robust to the presence in urine of the most common organic molecules such as glucose, galactose, glycine, uric acid, or ascorbic acid. Also, after a study of the different conditions, where the adsorption time and washing solvent play a critical role, there is a clear difference between the imprinted and the non-imprinted polymer.

Feng et al. [113] have described a direct assay for the determination of 2,4-dichlorophenol (2,4-DCP) with an online system that combines CL detection and molecularly imprinted solid-phase extraction. Before actually collecting the CL signal, a MISPE procedure is first carried out, and after evaporation the eluent is reconstituted in 2 mL of water and introduced into the CL system. Several parameters of the assay were optimized, namely the binding capacity, time response, and tracer concentration (luminol, H_2O_2). The calibration curve was linear from 10^{-7} to 2×10^{-5} g mL^{-1}, with a limit of detection of 1.8×10^{-8} g mL^{-1}. The selectivity was measured against several ions being over 1,000 for K^+, Na^+, Cl^-, SO_2^{4-}, NO_3^-, and NO_2^-, over 100 for Cr^{3+}, and over 50 for ClO_4^-, NH_4^+, Mg^{2+}, and Ca^{2+}. The retention and the applicability of this procedure were tested with the analysis of tap and river water. The recovery in these samples was 72–98% for the MIP and 33–39% for the CP. Nevertheless, this selectivity was reduced drastically for other compounds with similar structure such as 4-chlorophenol (4-CP), where the retention was 48% for the imprinted polymer and 30% for the NIP.

An example of a CILA using optical fibers has been described by Wang et al. [114] for the analysis of 6-mercaptopurine (6-MP). The template, 6-MP, was oxidized to a strong fluorescent compound by H_2O_2 in alkaline solution. Upon optimization of the H_2O_2 and NaOH concentrations and of the assay temperature, the sensor showed a linear response in the 1.0×10^{-8} to 6.0×10^{-6} g mL^{-1} range with a detection limit of 3.0×10^{-9} g mL^{-1}. Cross reactivity to metal ions, amino acids, and carbohydrates was tested and the sensor was applied to the analysis of 6-MP in spiked serum.

Fang and co-workers [115] have reported a flow injection chemiluminescence assay for the detection of maleic hydrazide (MH). The imprinted poly(MAA-co-EDMA) polymer selectively retained the herbicide that was further treated with alkaline luminol–potassium periodate to produce a strong CL signal. Upon reaction, the absorbed MH was destroyed and removed by the flowing solution. The polymer was reused up to ten times and the linear response was between 3.5×10^{-4} and 5.0×10^{-2} mg mL^{-1} with a detection limit of 6.0×10^{-5} mg mL^{-1}. The CILA flow injection system was used for the analysis of potato and onion samples spiked with 1.0, 2.0, and 4.0×10^{-3} mg mL^{-1} of MH. Recoveries ranged from 98% to 103% (RSD 2.3%), demonstrating the successful application of the method.

More recently, a new CILA has been developed for detection of the (free) amino acid proline in maize and honey samples [116]. With this aim, three different imprinted polymers were prepared by suspension polymerization using proline, hydroxyproline, and dansylproline as templates. The assay was performed by monitoring the CL response of the dansyl-proline derivative in the presence of bis (2,4,6-trichlorophenyl)oxalate (TCPO), H_2O_2, and imidazole, which enhances the CL signal. The recognition properties of the polymer microspheres and the analytical characteristics of the developed CILA were evaluated using a 96-well plate platform. The results showed that the best selectivity and sensitivity were obtained using hydroxyproline as template for the MIP preparation. Under optimized conditions, the CILA showed a detection limit of 3×10^{-7} M and a linear response between 1×10^{-6} M and 4×10^{-5} M. Proline concentration in enriched maize and corn samples, following derivatization with dansyl chloride, have been evaluated with recoveries ranging between 93% and 104% . The cross-reactivity study was limited to other aminoacids and selectivity was strongly dependent on the imprinted template.

2.6   Electrochemical Immuno-Like Assays

The electrochemical detection of ILAs has not been very much used but it is of interest because it has the potential to be applicable to the development of low cost miniaturisable electrodes. To that aim, electrochemical transducers are employed in combination with MIP recognition materials for example, MIP films immobilized on ion-selective electrodes (ISEs), self-assembled monolayers (SAMs), etc.

Although multistep affinity assays with redox-labeled targets have been described (Wang et al. [117]), most of the assays use enzyme-labeled species in conventional indirect formats (competitive, non-competitive). Direct EILAs based on multistep electrochemical affinity assays have also been developed with excellent results. In all these cases the MIP is used to extract the analyte from the sample and, after elution, the analyte is carried on to the electrochemical flow-through cell for being detected.

This principle was used by Pingarrón et al. for determination of fenbendazole antibiotic in beef liver [53]. The fenbendazole-imprinted MIP exhibited a good selectivity for benzimidazoles with respect to other veterinary drugs. Using square-wave voltammetry (SWV) at a cylindrical carbon fiber microelectrode the limit of detection was 57 μg L^{-1}, a value under the MRL allowed by the European Union regulations. The same authors have described the analysis of sulfamethazine using the same principle and combining a further electrode surface preconcentration of the analyte at a Nafion-coated glassy carbon electrode (GCE). Oxidative SWV of the accumulated sulfamethazine was employed for its quantification [118].

Electrochemical binding assays have also been applied for testing computational predictions which render the highest stabilization energy for the pre-polymerization mixture of several formulations traditionally used in non-covalent MIPs [119]. The batch binding assays and voltammetric detection confirm the theoretically best monomer–porogen solvent mixture for preparation of a recognition material for the dopamine metabolite homovanillic acid.

Mazzotta et al. [120] have developed a sensor for ephedrine monitoring in the 0.5–3 mM range. The sensor was based on an MIP modified with a polypyrrole (PPY) film, potentiostatically grown on a glassy carbon electrode using an unconventional cell configuration. The specificity of the sensor was evaluated in the presence of other analytes such as dopamine, uric acid, and aspartic acid, uric acid exhibiting the strongest interference. However, the working range of the sensor was limited to high concentrations, so that a previous sample clean up and pre-concentration was required for direct sensing applications.

A potentiometric sensor for the determination of hydroxyzine in tablets and biological fluids has been reported by Javanbakht et al. [121]. This is probably the first carbon paste electrode described in the literature based on MIPs and potentiometric detection. The polymer was prepared with a very general composition, MAA as functional monomer, EDMA as the cross linker, and chloroform as porogen. Response range was found to be 0.01–100 mM, with a moderate response time (30–40 min). The method was applied to the analysis of hydroxyzine in tablets, spiked human serum, and human urine.

Kubo et al. [122] have covered the sensing area of a gold electrode lattice with a polyimide layer which supports an atrazine selective MIP prepared with MAA and EDMA and functional monomer and cross linker, respectively. The detection limit was 50 nM (11 ppb) with a working range up to 15 μM atrazine. Other herbicides such as simazine or ametryn could be detected as well but with lower sensitivity. In spite of being robust, even after 5 days of use, the developed MIP-based chips were not applied to the analysis of real samples.

In a different approach, Thoelen et al. [123] have developed an MIP-based sensor for L-nicotine analysis using impedance spectroscopy in a basic electro-chemical cell and monitoring the time-resolved capacitance. Signal enhancement was achieved using a conjugated polymer, poly(2-methoxy-5-(3,7-dimethylocty-loxy))-1,4-phenylene vinylene (OC1C10-PPV), to immobilize the non-conducting MIP particles onto the electrode surface. The adsorption isotherms of the MIP and NIP polymers showed clear differences in the millimolar concentration range, yet the authors reported a linear response of 2–5 nM. For concentrations higher than 5 nM the impedance change was similar for MIP and NIP. Therefore, the applica-tion of this sensor for real sample analysis is very limited.

3 Biomimetic Microchips

It has been demonstrated that MIPs possess recognition and binding properties close to those of antibodies and enzymes, especially in organic media, and display higher stability than their natural counterparts. These properties suggest that MIPs can be suitable as recognition elements for chemical (biomimetic) sensors or biomimetic chips.

Conventional microarrays are usually obtained by depositing picoliters or nano-liters of a solution containing the recognition element onto a suitable surface using inject printing or contact deposition techniques. However, these techniques cannot be directly applied in combination with MIPs due to the problems associated with solvent evaporation, viscosity, or wettability of the polymer substrate.

Byrne et al. [124] have shown the possibility of creating imprinted polymer ordered micropatterns, of a variety of shapes and dimensions, on polymer and silicon substrates using iniferters and photopolymerization. They applied this approach to the recognition of D-glucose using copolymer networks containing poly(ethylene glycol) and functional monomers such as acrylic acid, 2-hydro-xyethyl methacrylate, and acrylamide.

Hong and coworkers [125] have developed a compact, fast, low cost, and disposable microfluidic biochip with on-chip MIP-based recognition. The sensing scheme is based on the reaction of the anesthetic propofol initially retained in the MIP film with the color reagent (quinoneimine dye) and the released product is optically monitored at 655 nm. The experimental results showed that the developed biomimetic chip could successfully detect propofol in the 0.25–10 mg L^{-1} range with a limit of detection of 0.25 mg L^{-1}. The authors claimed that the new platform compared well with chromatographic methods usually applied to the detection of this drug, although no comparative data were included in the study.

Significant advances have been provided by the Haupt group in this field, [126]. Biomimetic microchips were produced by depositing MIP microdot arrays on surfaces. MIP precursors with fluorescein as the template were deposited with a microcantilever array (Fig. 14). A sacrificial polymeric porogen, poly(vinylacetate), was used to obtain porous dots. Fluorescence microscopy revealed the specific binding of fluorescein to the MIP dots. With another MIP selective to the herbicide 2,4-D, competitive binding assays were performed using a coumarin derivative as fluorescent probe [89].

More recently, Haupt and coworkers [127] have demonstrated the possibility of producing mimetic microbiochips by wafer-scale production using patterned MIPs (Fig. 15). In this approach standard photolithographic equipment was used to create microbiochips by patterning an MIP layer after spin coating the precursor solution. For proof of the concept, a dansyl-L-phenylalanine fluorescent probe was selected as target analyte and MIP patterns were imprinted with the non-fluorescent BOC-L-Phe using 2-carboxyethyl acrylate (CEA) and TRIM as functional monomer and crosslinker, respectively. The mixture of diethyleneglycoldimethylether (diglyme) and poly(vinyl acetate) at 1% was used as porogen.

The pattering process was optimized for leading a 400-nm thick MIP membrane and features were engraved by a 12-s UV exposure. Multiplexed chips were also successfully prepared using the same process and their features were spatially

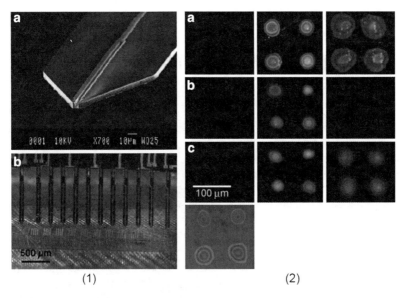

(1)   (2)

Fig. 14 (**1a**) Scanning electron micrograph of the tip of a silicon cantilever showing the integrated channel. (**1b**) Photograph of the cantilevers during deposition of MIPs on a glass slide. (**2**) Fluorescence microscopy images of microdot matrices of the non-imprinted control polymer (*left*), positive control with the copolymerized template (*middle*), and imprinted polymer (*right*) synthesized with DMSO–diglyme–1% PVAc as the porogen (**2a**) directly after deposition, (**2b**) after washing and elution of the template, and (**2c**) after re-incubation in a solution of 30 μM fluorescein. *Bottom row*: white-light image of control dots. Reprinted with permission from [126]. Copyright 2007 American Chemical Society

Fig. 15 Four-inch silicon substrate with multiplexed MIPs. Features defining one chip are repeated on the entire wafer. *Inset*: A, MIP templated with dansyl-L-Phe; B, MIP templated with boc-L-Phe, C, non-imprinted control polymer. [127] Reproduced by permission of The Royal Society of Chemistry

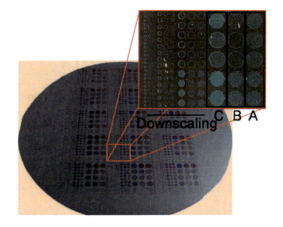

resolved with a resolution of 1.5 μm. Finally, to demonstrate the specific binding properties of the patterned MIP, incubation with 10 μM dansyl-L-Phe was performed and the emission of the bonded probe was imaged by fluorescence microscopy. The authors also suggested the possibility of chip regeneration. These results confirm the applicability of MIPs as valuable recognition elements for the development of mimetic biochips for a broad range of analytical applications.

Nevertheless, the major issue in the miniaturization of this kind of assays continues to be the strong non-selective interactions with the polymers. This problem is often avoided, or at least minimized, on large scale experiments such as chromatographic separations. However, it is not easy to solve in the case of microsensors or microchips and direct or indirect assays, where competition can be strongly affected by unspecific binding.

4 Conclusions

The main limitation towards the widespread application of MIPs as selective recognition elements in ILAs is that their performance is still not as brilliant as their biological counterparts. Their sensitivity and selectivity in *aqueous* media is, in most cases, lower than that of the antibodies. In fact there are not many examples so far of ILAs for the analysis of complex clinical, environmental or food samples. Moreover, the number of MIP-based *validated* methods described in the literature is still limited and this issue should be addressed before these materials can be considered as practical alternatives to the biological recognition entities.

However, MIPs offer definitive advantages over the antibodies: they display excellent performance in organic samples, their chemical stability is far superior and, in principle, they can be precisely tailored to the target species with low production costs, especially for the analysis of small molecules, for which antibody production is more complicated. The synthesis of novel monomers with strong binding with the target molecule, favoring stoichiometric imprinting, will also allow the development of materials with a homogeneous distribution of high affinity binding sites that can be pivotal to the application of ILAs.

The use of radiolabels in ILAs is being abandoned in favor of other tracers, particularly those based on fluorescence measurements. With the decrease in the cost of time-resolved emission and fluorescence anisotropy instrumentation, fluorescence ILAs will be based not only on steady-state measurements but also on fluorescence lifetime determinations or polarization measurements. Time-resolved measurements avoid the problems associated with tracer photobleaching or instrumental drifts due to source or detector aging. Therefore, these techniques hold enormous promise for the future of ILAs. Moreover, the application of Förster resonance energy transfer (FRET) or fluorescence anisotropy measurements to detecting the binding events with outstanding sensitivity are also almost unexplored fields. Nano ILAs will surely be based on the modern fluorescence lifetime imaging microscopy (FLIM) with laser diode excitation and confocal arrangement.

The use of enzymatic tracers for ILAs has not yet been exploited to a large extent, probably due to the limitations associated with the effective competition of the analyte and its labeled derivative for the polymer binding sites. The availability of nanoparticles and microgels that can be used in homogeneous assays and allow a much better accessibility of the tracer to the selective sites in the MIP will certainly help to extend their application to highly sensitive detection of analytes.

One of the most promising applications of MIPs is the fabrication of chips with microarrays containing several selective polymers against different targets. Fabrication of such platforms will extend the application and the commercial impact of such materials to many different areas where multianalyte capability is a must. Nevertheless, further efforts are still needed to optimize fabrication of localized polymer structures, avoiding problems related to long term stability of the biomimetic material, MIP deposition, and transducer coupling.

The ingenuity of the human being is challenging Nature in many areas. Artificial receptors for analytical applications will certainly take the lead in the years to come for the sake of a cleaner environment, a safer food, and more affordable medical diagnostics.

Acknowledgments The authors' work on MIPs has been funded by the European Marie Curie Programme (MRTN-CT-2006-033873), the Spanish Ministry of Science and Innovation (grants CTQ2006-15610-C02 and CTQ2009-14565-C03), the Madrid Regional Government (ref. S-0505/AMB/0374), the ESF, the ERDF, and Complutense University (UCM-BSCH GR58-08-910072).

References

1. Levinson WE (2004) Medical microbiology and immunology: examination and board review, 8th edn. McGraw-Hill, New York
2. Yallow RS, Berson SA (1959) Nature 184:1648–1649
3. Ekins RP (1960) Clin Chim Acta 5:453–459

4. Barbas CF, Burton DR, Scott JK (2004) Phage display: a laboratory manual. Cold Spring Harbor Laboratory Press, Cold Spring Harbor
5. Ansell RA (2005) Applications of MIPs as antibody mimics in immunoassays. In: Yan M, Ramstrom O (eds) Molecularly imprinted polymers: science and technology. Marcel Dekker, New York
6. Ansell RA, Kriz D, Mosbach K (1996) Curr Opin Biotechnol 7:89–94
7. Alexander C, Andersson H, Andersson LI, Ansell RI, Kirsch N, Nicholls I, O'mahony J, Whitcombe MJ (2006) J Mol Recognit 19:106–180
8. Haupt K (2003) Chem Commun 2:171–178
9. Ye L, Haupt K (2004) Anal Bioanal Chem 378:1887–1897
10. Piletsky S, Turner APF (2002) New materials based on imprinted polymers and their application in sensors. In: Ligler FS, Rowe-Taitt CA (eds) Optical biosensors: present and future. Elsevier, Amsterdam
11. Piletsky S, Turner APF (2008) Imprinted polymers and their application in sensors. In: Ligler FS, Rowe-Taitt CA (eds) Optical biosensors: today and tomorrow, 2nd edn. Elsevier, Amsterdam
12. Zimmerman SC, Lemcoff NG (2004) Chem Commun 1:5–14
13. Rolt I, Brostoff J, Male D (1987) Immunology. Gower Medical Publishing Ltd, London
14. Breinl F, Haurowitz F (1930) Z Physiol Chem 192:45–57
15. Pauling L (1940) J Am Chem Soc 62:2643–2657
16. Ehrlich P (1900) Proc R Soc Lond 66:424–448
17. Burnet FM (1959) The clonal selection theory of acquired immunity. Cambridge University Press, Cambridge
18. Hodgkin PD, Heath WR (2007) Nat Immunol 8:1019–1026
19. Silverstein AM (1989) A history of immunology. Academic Press, San Diego
20. Pinchuk G (2001) Schaum's outline of immunology. McGraw-Hill, Blacklick, OH
21. Mikkelsen SR (2004) Bioanalytical chemistry. Wiley, Hoboken, NJ
22. Price CP, Newman DJ (eds) (1997) Principles and practise of immunoassay, 2nd ed. MacMillan Reference Ltd., London
23. Diamandis EP, Christopoulos TK (eds) (1996) Immunoassay. Academic Press, San Diego
24. Sellergren B (2001) The non-covalent approach to molecular imprinting. In: Sellergren B (ed) Molecularly imprinted polymers. Man made mimics of antibodies and their applications in analytical chemistry. Elsevier, Amsterdam
25. Vlatakis G, Andersson LI, Muller R, Mosbach K (1993) Nature 361:645–647
26. Miles LEM, Hales CN (1968) Nature 219:186–189
27. Ekins RP (1998) Clin Chem 44:2015–2030
28. Hunt CE, Pasetto P, Ansell RJ, Haupt K (2006) Chem Commun 16:1754–1761
29. Ramstrom O, Ye L, Mosbach K (1996) Chem Biol 3:471–477
30. Andersson LI, Muller R, Vlatakis G, Mosbach K (1995) Proc Natl Acad Sci USA 92:4788–4792
31. Bengtsson H, Roos U, Andersson LI (1997) Anal Commun 34:233–235
32. Haupt K, Dzgoev A, Mosbach K (1998) Anal Chem 70:628–631
33. Andersson LI (1996) Anal Chem 68:111–117
34. Urraca JL, Moreno-Bondi MC, Orellana G, Hall AJ, Sellergren B (2007) Anal Chem 79:4915–4923
35. Segel LA (1991) Biological kinetics. Cambridge University Press, New York
36. Benito-Peña E, Moreno-Bondi MC, Aparicio S, Orellana G, Cederfur J, Kempe M (2006) Anal Chem 78:2019–2027
37. Umpleby RJ II, Bode M, Shimizu KD (2000) Analyst 125:1261–1265
38. Prichard FE (1996) Trace analysis. A structured approach to obtaining reliable results, LGC. The Royal Society of Chemistry, Cambridge
39. Crosby NT, Day JA, Hardcastle W, Holcombe DG, Treble RD, Prichard FE (1995) Quality in the analytical chemistry laboratory. Wiley, Chichester, New York

40. Thomsom M, Ellison SLR, Wood R (2002) Pure Appl Chem 74:835–855
41. Little JA (2004) Chromatographia 59:S177–S181
42. Gee SJ, Hammock BD, Van Emon JM (1996) Environmental immunodetection analysis for detection of pesticides and other chemicals. A user's guide. Noyes Publications, Westwood, New Jersey
43. Krotzky AJ, Zeeh B (1995) Pure Appl Chem 67:2065–2088
44. Ekins R, Edwards P (1997) Clin Chem 43:1824–1831
45. Marco MP, Gee S, Hammock BD (1995) Trends Anal Chem 14:415–425
46. Herranz S, Ramón-Azcón J, Benito-Peña E, Marazuela MD, Marco MP, Moreno-Bondi MC (2008) Anal Bioanal Chem 391:1801–1812
47. Siemann M, Andersson LI, Mosbach K (1996) J Agric Food Chem 44:141–145
48. Kaur J, Boro RC, Wangoo N, Singh KR, Suri CR (2008) Anal Chim Acta 607:92–99
49. Surugiu I, Danielsson B, Ye L, Mosbach K, Haupt K (2001) Anal Chem 73:487–491
50. Ji X, He Z, Ai X, Yang H, Xu C (2006) Talanta 70:353–357
51. Zhou H, Zhang Z, He D, Hu Y, Huang Y, Chen D (2004) Anal Chim Acta 523:237–242
52. Brandon DL, Bates AH, Binder RG, Montague WC Jr, Whitehand LC, Barker SA (2002) J Agric Food Chem 50:5791–5796
53. Guzmán-Vázquez de Prada A, Loaiza OA, Serra B, Morales D, Martínez-Ruiz P, Reviejo AJ, Pingarrón JM (2007) Anal Bioanal Chem 388:227–234
54. Benito-Peña E, Moreno Bondi MC, Orellana G, Maqueira A, van Ameronguen A (2005) J Agric Food Chem 53:6635–6642
55. Urraca JL, Benito-Peña E, Pérez Conde C, Moreno Bondi MC, Pestka JJ (2005) J Agric Food Chem 53:3338–3344
56. Navarro-Villoslada F, Urraca JL, Moreno-Bondi MC, Orellana G (2007) Sens Actuators B 121:67–73
57. Senholdt M, Siemann M, Mosbach K, Andersson LI (1997) Anal Lett 30:1809–1821
58. Tse Sum Bui B, Belmont AS, Witters H, Haupt K (2008) Anal Bioanal Chem 390:2081–2088
59. Ye L, Cormack PAG, Mosbach K (1999) Anal Commun 36:35–38
60. Berglund J, Nicholls IA, Lindbladh C, Mosbach K (1996) Bioorg Med Chem Lett 6:2237–2242
61. Svenson J, Nicholls IA (2001) Anal Chim Acta 435:19–24
62. Ye L, Surugiu I, Haupt K (2002) Anal Chem 74:959–964
63. Biffis A, Graham NB, Siedlaczek G, Stalberg S, Wulff G (2001) Macromol Chem Phys 202:163–171
64. Enguall E, Perlmann PG (1971) Immunochemistry 8:871–874
65. Van Weeman BK, Schuurs AHWM (1971) FEBS Lett 15:232–236
66. Nakane PK, Pierce GB (1966) J Histochem Cytochem 14:929–931
67. Surugiu I, Ye L, Yilmaz E, Dzgoev A, Danielsson B, Mosbach K, Haupt K (2000) Analyst 125:13–16
68. Surugiu I, Svitel J, Ye L, Haupt K, Danielsson B (2001) Anal Chem 73:4388–4392
69. Piletsky SA, Piletska EV, Chen B, Karim K, Weston D, Barrett G, Lowe P, Turner APF (2000) Anal Chem 72:4381–4385
70. Piletsky SA, Piletska EV, Bosi A, Karim K, Lowe P, Turner APF (2001) Biosens Bioelectron 16:701–707
71. Mukherjee C, Caron MG, Mullikin D, Lefkowitz R (1975) J Mol Pharmacol 12:16–31
72. Wang S, Xu ZX, Fang GZ, Zhang Y, Liu B, Zhu HP (2009) J Agric Food Chem 57:4528–4534
73. Bossi A, Piletsky SA, Piletska EV, Righetti PG, Turner APF (2001) Anal Chem 73:5281–5286
74. Bergmanna NM, Peppas NA (2008) Prog Polym Sci 33:271–288
75. Fukusaki I, Saigo A, Kajiyama SI, Kobayashi A (2000) J Biosci Bioeng 90:665–668
76. Bossi A, Bonini F, Turner APF, Piletsky SA (2007) Biosens Bioelectron 22:1131–1137
77. Greene NT, Shimizu KD (2005) J Am Chem Soc 127:5695–5700
78. Chen S, Zhang Z (2008) Spectrochim Acta A Mol Biomol Spectrosc 70:36–41
79. Suárez-Rodríguez JL, Díaz-García ME (2001) Anal Chim Acta 16:955–961

80. Haupt K, Mayes AG, Mosbach K (1998) Anal Chem 70:3936–3939
81. Malosse L, Buvat P, Dominique A, Siove A (2008) Analyst 133:588–595
82. Bravo JC, Fernández P, Durand JS (2005) Analyst 130:1404–1409
83. Piletsky SA, Piletskaya EV, Yano K, Kugimiya A, Elgersma AV, Levi R, Kahlow U, Takeuchi T, Karube I (1996) Anal Lett 29:157–170
84. Piletsky SA, Piletskaya EV, Elskaya AV, Levi R, Yano K, Karube I (1997) Anal Lett 30:445–455
85. Paniagua González G, Fernández Hernando P, Durand Alegría JS (2008) Biosens Bioelectron 23:1754–1758
86. Rachkov A, McNiven S, El'skay A, Yano K, Karube I (2000) Anal Chim Acta 405:23–29
87. Wandelt B, Sadowska M, Cywinski P, Hachulka K (2008) Mol Cryst Liq Cryst 486:257–270
88. Al-Kindy S, Badia R, Suarez-Rodrguez JL, Daz-Garca ME (2000) Crit Rev Anal Chem 30:291–309
89. McNiven S, Kato M, Levi R, Yano K, Karube I (1998) Anal Chim Acta 365:69–74
90. Urraca JL, Moreno-Bondi MC, Hall AJ, Sellergren B (2007) Anal Chem 79:695–701
91. Piletsky SA, Terpetschnig E, Anderson HS, Nichols IA, Wolfbeis OS (1999) Fresenius J Anal Chem 364:512–516
92. Levi R, McNiven S, Piletsky SA, Cheong SH, Yano K, Karube I (1997) Anal Chem 69:2017–2021
93. Suárez-Rodríguez JL, Díaz-García ME (2001) Biosens Bioelectron 16:955–961
94. Haupt K (1999) React Funct Polym 41:125–131
95. Orellana G, Aparicio S, Moreno-Bondi MC, Benito-Peña E (2005) Spanish Patent 2,197,811
96. Lavignac N, Allender CJ, Brain KR (2004) Anal Chim Acta 510:139–145
97. Ansell RJ (2004) J Chromatogr B 804:151–165
98. Yoshimatsua K, Reimhultc K, Krozerc A, Mosbacha K, Sodeb K, Yea L (2007) Anal Chim Acta 584:112–121
99. Lu CH, Zhou WH, Han B, Yang HH, Chen X, Wang XR (2007) Anal Chem 79:5457–5461
100. Chen YC, Wang Z, Yan M, Prahl SA (2006) Luminescence 21:7–14
101. Aburto J, Borgne S (2004) Macromolecules 37:2938–2943
102. Wang HF, He Y, Ji TR, Yan XP (2009) Anal Chem 81:1615–1621
103. Fahnrich KA, Pravda M, Guilbault GG (2001) Talanta 54:531–559
104. Lin JM, Yamada M (2000) Anal Chem 72:1148–1155
105. Lin JM, Yamada M (2001) Analyst 126:810–815
106. Xiong Y, Zhou H, Zhang Z, He D, He C (2007) Spectrochim Acta A Mol Biomol Spectrosc 66:341–348
107. He D, Zhang Z, Zhou H, Huang Y (2006) Talanta 69:1215–1220
108. Yan X, Zhou H, Zhang Z, He D, He C (2006) J Pharm Biomed Anal 41:694–700
109. Zhou H, Zhang Z, He D, Xiong Y (2005) Sens Actuators B 107:798–804
110. Fei N, Jiuru L, Weifen N (2005) Anal Chim Acta 545:129–136
111. Nie F, Lu J, He Y, Du J (2005) Talanta 66:728–733
112. He Y, Lu J, Liu M, Du J (2005) Analyst 130:1032–1038
113. Feng QZ, Zhao LX, Yan W, Ji F, Wei YL, Lin JM (2008) Anal Bioanal Chem 391:1073–1079
114. Wang L, Zhang Z (2008) Talanta 76:768–771
115. Fang YJ, Yan SL, Ning BA, Liu N, Gao ZX, Chao FH (2009) Biosens Bioelectron 24:2323–2327
116. Chang PP, Zhang ZJ, Yang CY (2010) Anal Chim Acta 666:70–75
117. Wang J, Ibanez A, Chatrathi MP (2002) Electrophoresis 23:3744–3749
118. Guzmán-Vázquez de Prada A, Reviejo AJ, Pingarrón JM (2006) J Pharm Biomed Anal 40:281–286
119. Dineiro Y, Menendez MI, Blanco-Lopez MC, Lobo-Castanon MJ, Miranda-Ordieres AJ, Tunon-Blanco P (2006) Biosens Bioelectron 22:364–371
120. Mazzotta E, Picca RA, Malitesta C, Piletsky SA, Piletska EV (2008) Biosens Bioelectron 23:1152–1156

121. Javanbakht M, Fard S, Mohammadi A, Abdouss M, Ganjali M, Norouzi P, Safaraliee L (2008) Anal Chim Acta 612:65–74
122. Kubo I, Shoji R, Fuchiwaki Y, Suzuki H (2008) Electrochemistry 76:541–544
123. Thoelen R, Vansweevelt R, Duchateau J, Horemans F, D'Haen J, Lutsen L, Vanderzande D, Ameloot M, vandeVen M, Cleij T, Wagner P (2008) Biosens Bioelectron 23:913–918
124. Byrne ME, Oral E, Hilt JZ, Peppas NA (2002) Polym Adv Technol 13:798–816
125. Hong CC, Chang PH, Lin CC, Hong CL (2010) Biosens Bioelectron 25:2058–2064
126. Vandevelde F, Leïchlé T, Ayela C, Bergaud C, Nicu L, Haupt K (2007) Langmuir 23:6490–6493
127. Guillon S, Lemaire R, Linares AV, Haupt K, Ayela C (2010) Lab Chip 9:2987–2991

Top Curr Chem (2012) 325: 165–266
DOI: 10.1007/128_2010_92
© Springer-Verlag Berlin Heidelberg 2010
Published online: 3 December 2010

Chemosensors Based on Molecularly Imprinted Polymers

Subramanian Suriyanarayanan, Piotr J. Cywinski, Artur J. Moro, Gerhard J. Mohr, and Wlodzimierz Kutner

Contents

S. Suriyanarayanan
Institute of Physical Chemistry, Polish Academy of Sciences, Kasprzaka 44/52, 01–224, Warsaw, Poland

P.J. Cywinski (✉)
Department of Physical Chemistry, University of Potsdam, Karl-Liebknecht-Str 24-25, 14476 Potsdam-Golm, Germany

Institute of Physical Chemistry, Friedrich-Schiller-University Jena, Lessingstrasse 10, 07743 Jena, Germany
e-mail: Piotr.Cywinski@uni-potsdam.de

A.J. Moro
Institute of Physical Chemistry, Friedrich-Schiller-University Jena, Lessingstrasse 10, 07743 Jena, Germany

G.J. Mohr
Fraunhofer Institution for Modular Solid State Technologies EMFT, Workgroup Sensor Materials, Josef-Engert-Strasse 9, D-93053 Regensburg, Germany

W. Kutner (✉)
Faculty of Mathematics and Natural Sciences, School of Science, Cardinal Stefan Wyszynski University in Warsaw, Dewajtis 5, 01-815 Warsaw, Poland

Institute of Physical Chemistry, Polish Academy of Sciences, Kasprzaka 44/52, 01-224 Warsaw, Poland
e-mail: wkutner@ichf.edu.pl

List of Abbreviations

AAPH	2,2′-Azobis(2-amidinopropane) hydrochloride
ABCN	1,1-Azobis(cyclohexanecarbonitrile)
a.c.	alternating current
ACN	Acetonitrile
AFM	Atomic force microscopy
AIBN	2,2′-Azobis(2-methylisobutyronitrile)
APTMS	3-(Aminopropyl)trimethoxysilane
BAW	Bulk acoustic wave
cAMP	Cyclic adenosine 3′,5′-monophosphate
cGMP	Cyclic guanosine 3′,5′-monophosphate
ChemFET	Chemical field-effect transistor
CV	Cyclic voltammetry
DCM	Dichloromethane
DDC	N,N'-Didansyl-L-cystine
DDK	N,N'-Didansyl-L-lysine
DMB	Dimethylbenzoate
DMF	N,N-Dimethylformamide
DNA	Deoxyribonucleic acid
DNOC	4,6-Dinitro-o-cresol (2-methyl-4,6-dinitrophenol)
DNT	2,4-Dinitrotoluene
DPV	Differential pulse voltammetry
DVB	Divinylbenzene
DZ	Daminozide
EGDMA	Ethylene glycol dimethacrylate
EIS	Electrochemical impedance spectroscopy
EQCM	Electrochemical quartz crystal microbalance
FIA	Flow injection analysis
FLD	Fluorescence lifetime distribution
FRET	Fluorescence resonance energy transfer

GCE	Glassy carbon electrode
GC–MS	Gas chromatography–mass spectrometry
HEMA	2-Hydroxyethyl methacrylate
HPLC	High performance liquid chromatography
IAA	Indole acetic acid
IIP	Ion-imprinted polymer
ISE	Ion-sensitive electrode
ITO	Indium-tin oxide
IUPAC	International Union of Pure and Applied Chemistry
LOD	Limit of detection
LSV	Linear sweep voltammetry
MAA	Methacrylic acid
MA-Ade	Methacryloylamidoadenine
MES	Methylated salicylate
MIB	Methylisoborneal
MIP	Molecularly imprinted polymer
MIP-CP	Molecularly imprinted polymer-carbon paste
MIPPy	Molecularly imprinted polypyrrole film
MISPE	Molecularly imprinted solid-phase extraction
MPA	Methylphosphonic acid
NAD	Nicotinamide adenine dinucleotide
NADP	Nicotinamide adenine dinucleotide phosphate
N-CBZ-Asp	N-Carbobenzoxy-aspartic acid
NIP	Non-imprinted polymer
NIP-CP	Non-imprinted polymer-carbon paste
OC_1C_{10}-PPV	Poly[2-methoxy-5-($3'$,$7'$-dimethyloctyloxy)]-1,4-phenylene vinylene
PAH	Polycyclic aromatic hydrocarbon
PEDOT	Poly(3,4-ethylenedioxythiophene)
phi-NO_2	O,O-Dimethyl(2,4-dichlorophenoxyacetoxyl)($3'$-nitrophenyl) methinephosphonate
PM	Piezoelectric microgravimetry
PMA	Poly(methacrylic acid)
PMMA	Poly(methylmethacrylate)
PMP	Pinacolyl methylphosphonate
PPV	Poly(1,4-phenylene vinylene)
PVC	Poly(vinyl chloride)
PZ	Piezoelectric
QCM	Quartz crystal microbalance
RAFT	Reversible addition fragmentation chain transfer
RCM	Ring closing metathesis
RDX	Hexahydro-1,3,5-trinitro-1,3,5-triazine
RNA	Ribonucleic acid
RSD	Relative standard deviation
SAM	Self-assembled monolayer
Sarin	Isopropyl methylphosphonofluoridate
SAW	Surface acoustic wave

SCE	Saturated calomel electrode
SDS	Sodium dodecyl sulphate
SECM	Scanning electrochemical microscopy
SEM	Scanning electron microscopy
SERS	Surface enhanced Raman scattering
SH-SAW	Shear-horizontal surface acoustic wave
SLM	Supported liquid membrane
Soman	Pinacolyl methylphosphonofluoridate
SPE	Solid-phase extraction
SPME	Solid-phase micro extraction
SPR	Surface plasmon resonance
ssDNA	Single-stranded deoxyribonucleic acid
ssRNA	Single-stranded ribonucleic acid
STW	Surface transverse wave
SWV	Square wave voltammetry
TCAA	Trichloroacetic acid
TEGDMA	Tri(ethylene glycol)dimethacrylate
TEOS	Tetraethylorthosilane
TMS	Trimethoxysilyl
(TMS)en	N-[3-(Trimethoxysilyl)propyl]ethylenediamine
TNT	2,4,6-Trinitrotoluene
TRIM	Trimethylolpropane trimethacrylate
TSM	Thickness shear mode
T-SPR	Transmission surface plasmon resonance
UV–vis	Ultraviolet-visible
V-65	2,2′-Azobis(2,4-dimethyl)valeronitrile
vb-DMASP	*trans*-4-[1,4-(*N,N*-Dimethylamino)styryl]-*N*-vinyl-benzylpyridinium chloride
VPD	4-Vinylpyridine
VX	*O*-Ethyl-*S*-2-diisopropylaminoethylmethylphosphonothioate
ZnPP	Zinc(II)-protoporphyrin

List of Symbols

α	Separation factor (selectivity)
Δf	Change in the resonant frequency of the quartz resonator
Δm	Change in the mass of the quartz resonator
$\varepsilon, \varepsilon_0$	Electric permittivity of an insulator and free space, respectively
μ_q	Shear modulus of the AT-cut quartz crystal
ρ_q	Density of quartz
A	Acoustically active area of the quartz crystal resonator
A_s	Surface area of the capacitor plate

B_{max}	Density of the imprinted binding sites (molecular cavities)
C	Capacitance
D	Distance between two parallel plates of a capacitor
DS	Degree of substitution
f	Frequency of a.c. voltage
f_0	Fundamental resonant frequency of the unperturbed quartz resonator
f_a	Fraction of easily accessible cavities
I	Fluorescence intensity in the presence of analyte
I_0	Initial fluorescence intensity in the analyte absence
i_{pa}	Current of anodic peak in LSV or CV
k	Retention factor
K_a	Acid dissociation constant
K_d	Complex dissociation constant
K_{SV}^a	Stern–Volmer constant for quenching inside MIP cavities
$K_{NO_3^-, ClO_4^-}^{pot}$, $K_{NO_3^-, I^-}^{pot}$	Potentiometric selectivity coefficients
K_{MIP}	Stability constant of the MIP–analyte complex formation
K_{NIP}	Stability constant of the NIP–analyte complex formation
$K_{selectivity\ Cu^{2+}/Ni^{2+}}$	Ratio of the selectivity coefficients of the imprinted Cu^{2+} and non-imprinted Ni^{2+} polymers
$[Q]$	Analyte concentration
Z_{im}	Imaginary part of impedance

1 General Introduction

1.1 Sensors

A sensor is a device, which responds to a physical or chemical stimulus in order to produce a measurable detection signal or to control another operation [1]. Sensors are encountered in innumerable applications and have become an integral part of our day-to-day life. Examples of everyday use of sensors include a thermocouple, which responds to the change in temperature by an output voltage, or a touch-sensitive sensor of an interactive monitor screen. Basically, a sensor can respond, that is change its signal, to a single factor being sensed, i.e. either to the change of temperature or pressure in the above examples.

Although the construction of sensors for external physical stimuli, such as light, heat or pressure, is relatively simple, it becomes more complicated when the target stimuli come from atoms or molecules. These types of sensors are often referred to as chemical sensors or chemosensors and biochemical sensors or biosensors (see below in Sects. 1.2 and 1.3). For the latter types, a sensing material should be used that can respond to the presence of the target analyte. This response may or may not be obviously true with vague information. Hence, chemo- and biosensors should be selective and able to detect the analyte presence under prevailing complications, like those encountered in real samples. In general, the test solution of the analyte of interest comes along with interfering substances at different concentrations. Apparently the sensors should not only respond to analytes but also be insensitive to interferants. Generally, this issue is being approached in analytical chemistry with separation methods applied before detection and quantification of a particular analyte. Typically, this comes in the form of either chromatographic or electrophoretic separation, or membrane extraction to quantify subsequently each analyte individually. In order to avoid these initial laborious separation steps, it would be advantageous to devise bio- or chemosensors with a high selectivity with respect to their target analytes, i.e. with a selectivity much higher than the selectivity of its possible interferants.

1.2 Chemosensors and Biosensors

In the trend of a major advancement of the field of sensors recognizing atoms and molecules is the growth of chemo- and biosensors. Their apt definitions have been recommended by IUPAC. That is, a chemosensor is a device that transforms chemical information, ranging from the concentration of a specific sample component to total composition analysis, into an analytically useful signal; this information may originate from a chemical reaction of the analyte or from a physical property of the system investigated [2]. A biosensor is an integrated receptor-transducer device, which provides selective quantitative or semi-quantitative analytical information using a biological recognition element [3]. A schematic view of a chemo- or biosensor is shown in Scheme 1.

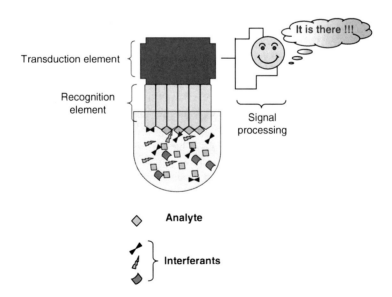

Scheme 1 A representative view of a comprehensive chemo- or biosensor (adapted from [4])

Typically, the recognitive binding of the target analyte by a selective element is transduced into a useful detection signal, like an electrochemical, optical, thermal or gravimetry signal. This signal can be related to concentration or activity of the analyte in the sample solution. In biosensors, the molecular preferential treatment or recognition has been accomplished by integrating transducing elements with either biological systems, such as in vitro tissues, microorganisms, organelles or cells, or biological macromolecules, such as antibodies, enzymes, proteins, nucleic acids, etc. [5]. Classification of biosensors can be based on the transduction mechanism, which predominately involves catalysis or affinity [4].

Biosensor development encountered limitations which have significantly obstructed the progress in the field for the past two decades. The key problems are associated with physical and chemical properties of biological recognition materials, and their stability in particular [1]. Poor tolerance of these biomaterials to environmental effects, such as large changes in solution acidity or temperature, wetting with several organic solvents or exposure to different irradiations, is largely blamed. However, it is the inadequate reproducibility, low availability of biological receptors and high cost of molecular bioreceptors, such as antibodies or enzymes, that have created the demand for the development of inexpensive, renewable and robust alternative synthetic materials capable of mimicking the identification properties of natural recognizing materials towards development of alternative chemosensors.

1.3 Molecularly Imprinted Polymers

Molecular imprinting is a promising way to overcome most of the limitations imposed on biosensors (see Sect. 1.2 above). This imprinting is an inexpensive procedure for preparation of synthetic polymer receptors, i.e. molecularly imprinted polymers (MIPs), with appreciable affinity, selectivity and toughness. MIPs are competent materials for molecular recognition in chemosensors and catalysis as well as in chromatographic and solid-phase microextraction (SPME) separations. Applications of MIP-based materials for the detection and quantification of herbicides [6], pesticides, chemical warfare, environmental pollutants, antibiotics, etc. have extensively been reviewed [1, 7–10]. Inspiration behind MIP elaboration involves striving against designing a polymer or, more generally, a synthetic solid material with desired recognition molecular sites surrounding an individual template or analyte molecule [11]. In other words, it is a template-induced in situ formation of recognition sites in a sensing material. Subsequent removal of the template molecules leaves the material in a moulded shape of sites, which can fit and coincides with the size of the template molecules. In general, the molecules of analytes are bound, as provisioned by balancing forces, with the MIP recognition sites such as, for instance, the lock-and-key model of enzymatic reactions. The forces between the analyte and these sites range from relatively strong covalent bonds and ionic interactions, through mild hydrogen bonds to weak van der Waals interactions [12]. Collectively, they afford relatively strong selective capture of the analyte. Moreover, the MIP materials reveal a certain range of special features as compared to those of biorecognition materials. That is, stability of the MIP receptors under harsh environmental conditions is superior in comparison to that of biological systems or biomolecules [9].

Affinity and selectivity of MIPs are as good as those of natural receptors [13]. Like most polymer materials, MIPs are processable. That is, they are compatible with the fabrication techniques and engineering conditions like micromachining, laser ablation and surface patterning. Therefore, the use of MIPs opens up the possibility for integration of the chemosensors with the electromechanical systems. Evidently, the low manufacturing cost and inexpensive preparation have made MIPs interesting for fabrication of so-called plastic antibodies. In recent years the field of MIPs has attracted significant interest, revealed in several reviews [14–18].

Generally, two different procedures have been adopted for preparation of MIPs. They involve either covalent or non-covalent complex formation of a template and complementary monomers with apt functional groups. [19]. Co-polymerization of this complex with a cross-linking monomer in a porogenic solvent solution, followed by removal of the template, results in formation of the porous polymer material with recognition sites complementary in size and shape to molecules of the target compound that can next be determined as an analyte.

Both of these MIP preparation procedures have their advantages and limitations [20, 21]. For instance, the size of the analyte molecule is not a discriminating criterion in covalent imprinting since the template selectively determines the interaction sites. In contrast, non-covalent imprinting has the advantage of being simpler since an additional synthetic step is not required to introduce the template into the polymer matrix. Moreover, removal of the template via extraction with the suitable solvent solution is simple and mostly complete for the non-covalent imprinting.

In the main, photo- or thermo-induced radical polymerization is engaged in MIP preparation, with electrochemical polymerization being another unique convenient procedure [16, 22–24].

Typically for a chemo- or biosensor, the recognition element, containing artificial or neutral receptor sites, respectively, is assembled over a transducer surface to generate an output detection signal. The same operational principle is also used for the MIP-based sensors. Preferential binding of the analyte by the receptor sites induces changes in the physicochemical properties of the chemosensor system. These changes are detected using proper signal transduction. Several transduction schemes have been combined with MIP-aided sensing, including optical [25–27], PZ (acoustic) [8, 28] and electroanalytical [23, 24] techniques.

The present chapter critically encompasses developments and achievements reached in MIP-based selective sensing combined with optical, piezoelectric (PZ) and electrochemical signal transduction. General procedures of MIP preparation along with methods of MIP immobilization for chemosensor fabrication are presented. Protocols of analyte determination involving measurement complexities, like template presence or absence, have been addressed in detail. Moreover, analytical parameters, such as detectability, sensitivity, selectivity, linear dynamic concentration range, reproducibility and response time, are compared and critically evaluated.

2 Optical Chemosensors

2.1 Introduction to Optical Chemosensors

Optical chemosensing has been a significant research area for several decades now. The reason for the recent rapid expansion of this field is the progress in information technology and the development of optoelectronic devices. The availability of low-cost light sources and detectors, the increasing demand for multi-analyte array-based detection platforms as well as progress in microfluidics technology and imaging techniques are the basic reasons for the development of noteworthy advances in optical sensing systems.

In principle, optical chemosensors make use of optical techniques to provide analytical information. The most extensively exploited techniques in this regard are optical absorption and photoluminescence. Moreover, sensors based on surface plasmon resonance (SPR) and surface enhanced Raman scattering (SERS) have recently been devised.

Generally, the analyte can be determined either directly or indirectly. In the former case, the analyte is determined due to its intrinsic optical features, such as absorption or photoluminescence. In the latter, mostly relevant to optically inactive analytes, chemical or physical interactions of these analytes with optically respon-sive media are involved. These interactions are characteristic for the analyte being determined. Consequently, they affect optical properties of the medium and the resulting changes in these properties can be detected. For efficient transduction of a chemical or physical detection signal into an analytically useful electrical signal, both qualitative and quantitative information must be made available. Advanta-geously, optical techniques are very useful for sensing purposes due to their fast response, reliability and relatively high detectability. In most cases, however, organic dyes, commonly used as the medium, suffer from a lack of selective qualitative analytical response upon interaction with the analyte. That is, optical properties of the dye are frequently altered in the same way by many different analytes. For selectivity enhancement, optical transduction techniques are, there-fore, successfully combined with recognition based on MIPs.

The present section reviews the progress made in the last decade in the field of optical chemosensors based on MIP technology. MIP sensing devices employing optical absorption, photoluminescence and SPR have been critically considered in terms of their binding properties and recognition mechanisms, as well as sensitivity and selectivity against different analytes. Perspectives for the future and potential improvements in the sensing abilities, measurement facilitation and signal trans-duction have also been addressed.

2.2 UV-vis Spectroscopic Chemosensors

UV–vis spectroscopy is a common technique for the detection of organic compounds, especially if the analyte exhibits a high degree of conjugation. However, most analytes are unsuitable for this kind of detection because they absorb in the UV range where interfering compounds also absorb. The use of UV–vis spectroscopy as a detection tool becomes advantageous if it is combined with molecular imprinting that enables greater selectivity. The main advantages of this technique include its reliability and simplicity, which allows for a rapid and facile determination of the samples under investigation.

The fast detection offered by UV–vis spectroscopy is particularly useful in large screening assays. Often, in this type of analysis, the intrinsic absorbance of the template is measured by monitoring the release and rebinding of the species to its MIP. A full use of these features is presented in work involving the synthesis of mini-MIPs, where an attempt to find the best monomer for the imprinting of different chlorotriazines, commonly used as herbicides, has been made [29]. The small scale (~55-mg) polymer batches were sonicated several times and the vials containing MIPs were analysed by high performance liquid chromatography (HPLC) featuring UV detection at 260 nm of the template in the supernatant solution. Rebinding quantification was processed in a similar manner. That is, a 1 mM solution of the template was added to the vial and, after sonication for 1 h, the concentration of the remaining (free) template in the supernatant was determined by reverse-phase HPLC. A series of six monomers was examined to find the best functional monomer for the synthesis of polymer receptors for different chlorotriazines. Eventually, methacrylic acid (MAA) appeared to be the functional monomer with the best binding capacity towards the studied class of compounds. This work can provide a platform for automating MIP preparation, thus accelerating studies that otherwise would be extremely time-consuming.

Combining UV spectroscopy with a solid-phase extraction (SPE) column or cartridge packed with the imprinted polymeric material is one of the most common uses of the UV detection technique. MIP in-tube SPME for selective determination of propranolol in biological fluids has been used successfully [30]. The column fabricated in this case presented remarkable reproducibility and allowed for the UV detection of propranolol at 254 nm within the linear concentration range of 0.5–100 μg mL^{-1} (Table 1). Other excellent results were presented for determination of nicotine in tobacco samples through molecularly imprinted solid-phase extraction (MISPE) [31]. Both nicotine and myosmine were able to bind to the MIP matrix. Then, by using 20-μL volumes of slightly differing eluents, nicotine or myosmine were selectively extracted and then optically detected at 254 nm.

A very important issue in the synthesis of MIPs is the study of the pre-organized complexes formed between the functional monomer(s) and the template. Preferably, this complex should be sufficiently stable to withstand polymerization conditions on the one hand and satisfactorily labile to allow both facile release of a template after polymerization and fully reversible rebinding of an analyte on the other.

Table 1 Structural formulas of analytes and functional monomers, accompanied by characteristic values of merit, used in the UV–vis MIP sensors

Analyte	Functional monomer	Value of merit	References
Terbutylazine	Methacrylic acid	Separation factor, $\alpha = K_{MIP}/K_{NIP}$, i.e. ratio of complex stability constants: $\alpha = 11$	[29]
Propranolol	Methacrylic acid	% Extraction yield: 1.73 LOD: 0.32 μg mL^{-1}	[30]
Nicotine	Methacrylic acid	% Binding: 99.5 LOD: 1.8 μg mL^{-1}	[31]
4-L-Phenylalanylamino-pyridine	Methacrylic acid	Complex stability constant: $K_{MIP} = 9.7 \times 10^4$ M^{-1} Complex stability constant:	[32] [33]

(continued)

Table 1 (continued)

Analyte	Functional monomer	Value of merit	References
Z-Glutamate	1-(4-Styryl)-3-(3-nitrophenyl)urea	$K_{MIP} = (6{,}498 \pm 170)\ M^{-1}$	
Copper(II) ion	1,2-Bis(triethoxysilyl)ethane	Selectivity coefficient: $K_{\text{selectivity}}\ _{Cu^{2+}/Ni^{2+}} = 0.67\text{–}7.4$	[34]
9-Ethyladenine	5,10,15-Tris(4-isopropylphenyl)-20-(4-methacryloyloxyphenyl)porphyrin Zinc(II) complex	Retention factor: $k = 6.0$	[35]

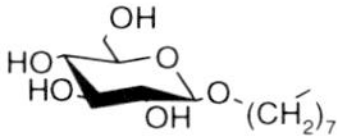

n-Octyl-ß-glucopyranoside
and

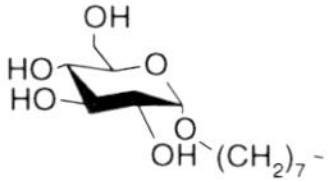

n-Octyl-α-glucopyranoside

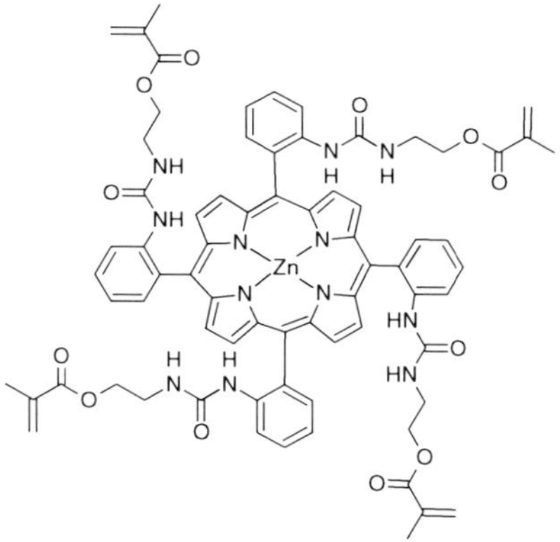

Meso-α,α,α,α-tetrakis(2-(3-o-phenylureido)ethyl methacrylate) porphyrin zinc(II) complex

Complex stability constants:
$K_{MIP} = (6.2 \pm 0.40) \times 10^4$ M^{-1}
n-Octyl-ß-glucopyranoside
and
$K_{MIP} = (13 \pm 2.1) \times 10^4$ M^{-1}
n-Octyl-α-glucopyranoside

[36]

Glucose (1)
Galactose (2)
Mannose (3)
Maltose (4)

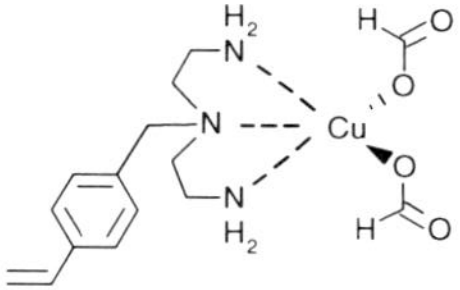

[(4-(N-Vinylbenzyl)-diethylenetriamine)copper(II)] diformate

Complex stability constants
(K_{MIP} in M^{-1}):
$pK_{MIP}(1) = 3.37 \pm 0.31$
$pK_{MIP}(2) = 3.41 \pm 0.43$
$pK_{MIP}(3) = 3.05 \pm 0.41$
$pK_{MIP}(4) = 3.41 \pm 0.26$

[37]

(continued)

Table 1 (continued)

Analyte	Functional monomer	Value of merit	References
Paracetamol	4-[(4-Methacryloyloxy)phenylazo] benzenesulphonic acid	Complex stability constants (K_{MIP} in M^{-1}): log K_{MIP} = 5.12–5.82 Density of imprinted binding sites: B_{MAX} = 0.20–1.73 $\mu M\ g^{-1}$ of hydrogel	[38]

The pre-polymerization solution was analysed by measuring the absorbance of the template and monomers separately. These results were compared to the resulting spectra of a solution containing both the monomer and template. In another example [32] this approach has been used to confirm the interactions between MAA and 4-L-phenylalanylamino-pyridine. A UV–vis spectroscopy study has shown the increase in absorbance of the band corresponding to the template (λ_{max} = 245 nm) upon increasing concentration of the functional monomer. The data were also used as an indication of the formation of a 1:2 template:monomer complex. The final polymer obtained was highly enantioselective.

Moreover, high enantioselectivity has been reported for an MIP that was selective for Z-glutamate [33]. Here, changes in the UV–vis absorbance of the 1-(4-styryl)-3-(3-nitrophenyl)urea monomer were monitored, as opposed to the previous studies where the intrinsic absorbance of the template was measured. The UV–vis spectroscopy titration of the functional monomer with different concentrations of the template in solution led to a bathochromic shift from 349 to 364 nm. The colour change of both the solution and the resulting bulk polymer upon addition of Z-glutamate was visible even to the naked eye, with a 1:1 monomer-to-template ratio demonstrated.

Another recognition mechanism within MIPs was investigated by comparison of UV spectra of the template-monomer complex and the template-loaded MIP [34]. Copper(II) ions were complexed in solution with N-[3-(trimethoxysilyl)propyl] ethylenediamine. Then the absorbance of the two complexes formed, i.e. Cu (TMS)en^{2+} (1:1 stoichiometry, λ_{max} = 651 nm) and Cu(TMS)en$_2^{2+}$ (1:2 stoichiometry, λ_{max} = 547 nm) were measured before and after polymerization. These results were compared with the spectra obtained for the imprinted and non-imprinted bulk polymers using a 0.5 mM solution of copper(II). The authors demonstrated that a 1:2 complex was formed in the case of MIP, since for MIP the value of λ_{max} was 595 nm, being similar for the Cu(TMS)en$_2^{2+}$ complex in solution. For the non-imprinted polymer (NIP), however, the spectra exhibited λ_{max} of 623 nm, a value which is closer to that of Cu(TMS)en^{2+}, suggesting a 1:1 complex stoichiometry.

A very unique approach using a dye molecule, Methyl-Red, for "imprinting" of protons has recently been presented [39]. That is, two different procedures were used to attach the pH sensitive dye to a sol-gel matrix. One involved acidic media while the other – basic media. Due to the presence of two different forms of the dye, i.e. protonated or deprotonated, during formation of the sol-gel, the two resulting sol-gels exhibited two different pK_a values, with a gap of four pH units (2.6 for the deprotonated form and 7.4 for the protonated form) proving that tuning of the pK_a value was possible through "proton-imprinting".

Metalloporphyrins have been used for the development of MIP for detection of a 9-ethyladenine nucleobase derivative [35]. With the increase of the template concentration, the bulk polymer exhibited a red shift in the absorbance spectra. This shift allowed for the quantitative detection of the template, which showed formation of a 1:1 monomer–template complex. Moreover, cinchonidine was imprinted using metalloporphyrin and MAA as the functional monomers [40]. The resulting MIP provided high selectivity against a diastereomer of the cinchonine template. Further, a porphyrin-based MIP approach to recognition of carbohydrates has been proposed [36]. The functional porphyrin monomer was appended with four urea substituents to provide complementary functionality and quality binding sites throughout the polymer matrix. The UV–vis titration of the functional monomer with the n-octyl pyranoside derivatives demonstrated formation of a 1:1 complex. This stoichiometry was determined by fitting the absorbance values of the monomer at 428 nm as a function of carbohydrate concentration. Batch rebinding experiments were conducted using 1,4-nitrophenyl analogues of the sugars with the absorbance detection at 300 nm. The resulting bulk polymers showed high affinities towards carbohydrates and presented high selectivity to the template and structurally related analogues. Another approach regarding determination of carbohydrates has also been presented [37]. In this approach, the binding interactions between several carbohydrates and three different copper(II) complexes were studied to assess the potential of the complexes to act as functional monomers in MIPs. Characterization of the resulting tertiary ligand–copper(II)–carbohydrate complexes via UV spectroscopy allowed for determination of the stability constants and the 1:1 stoichiometry of the monomer–template complex.

A very interesting recent report presented a photoresponsive molecularly imprinted hydrogel that was capable of chemical/drug delivery in aqueous media [38]. A water-soluble functional monomer based on a 4-[(4-methacryloyloxy)phenylazo]benzenesulphonic acid azobenzene dye was cross-linked with bisacrylamide and bismethacrylamide using paracetamol, i.e. N-(4-hydroxyphenyl)acetamide, as the template to produce imprinted polyacrylamide hydrogels. The hydrogel produced with the most commonly used polyacrylamide cross-linker, N,N'-methylenebisacrylamide, did not allow for cis–$trans$ photoisomerization of the monomer because the polymer matrix was too rigid and the cavities containing the monomer were too small due to the high degree of cross-linking. However, application of a longer spacer between two polymerizable moieties in the cross-linker decreased the substrate binding strength, as the monomers became too flexible to form a stable complex with the template. Therefore, the optimum cross-linker appeared to be N,N'-hexylenebisacrylamide. The final imprinted hydrogel was able to bind and release the substrate upon irradiation at 440 and 353 nm, respectively, with extremely high reproducibility and selectivity towards the template.

All structures of both functional monomers and analytes along with characteristic values of merit presented within this section are compiled in Table 1.

2.3   Photoluminescent Chemosensors

Photoluminescence spectroscopy is a well-established technique and a very powerful analytical method, which has proven its advantages in chemical sensing. Due to its high sensitivity and reliability it can provide fast and precise information about recognition by variations in the optical signal. Additionally, progress in materials technology as well as advances in microelectronics and computer science have offered opportunities to develop new analytical methods using photoluminescence spectroscopy. Particularly, photoluminescent sensors based on MIPs represent a class of recently developed materials that utilize optical signalling as the analytically relevant tool. Four major approaches are applied to produce photoluminescent MIPs: (1) the use of the intrinsic photophysical properties of the analyte, (2) the exchange of the non-luminescent analyte with its luminescent analogue that has the same characteristic functional groups, (3) the use of a so-called luminescent reporter molecule within the polymer or (4) a luminescent polymer, which selectively interacts with the analyte. This section is devoted to the MIP sensing systems, in which the photoluminescent signal comes from the reporter molecules incorporated into the polymer materials and then interacts with the analyte [(3) and (4)].

2.3.1   Determination of Ions

A very interesting application of the MIP technology to chemosensor development is the preparation of ion-imprinted polymers (IIPs). This technology allows monitoring of ions, in trace amounts, even in the presence of interfering species. Usually, ions are not straightforwardly imprinted into polymers for the purpose of photoluminescent MIP sensing, as in the case of SPE or electrochemical sensing. Rather they are coordinated with a luminescent host and, subsequently, incorporated together with the host into the polymer. In principle, this sensing is based on changes in photoluminescence of its host as a consequence of the formation or breaking of the coordination bonds between the ion and the host. Although the ions are mainly monitored via electrochemical methods, a few examples for the detection of Al^{3+} employing photoluminescent methods have been described.

One interesting example is the preparation and characterization of an IIP for aluminium ion detection [41]. Al^{3+} is widely used in industry and, therefore, may be present in drinking water and cause numerous illnesses [42]. In this case, morin (3,5,7,2',4'-pentahydroxyflavone) was used as a fluorescent molecule able to coordinate the aluminium ion. In contrast to the author's claim, however, this morin was used as a fluorescent functional molecule rather than the template with Al^{3+} being the only component involved during extraction and rebinding. Morin remained hindered within the polymer and functioned as a host for the Al^{3+}. The polymer was prepared according to the classical method of MIP preparation, i.e. it was bulk polymerized and then ground and sieved. Finally, the MIP microparticles were placed in a mini-reactor and investigated under flow-through conditions. To prevent displacement of the chemosensor particles from the flow-through cell, the MIP bed was kept in place by nylon net. Toluene, dichloroethane and acetonitrile (ACN) were used as solvents; however, water would have been more relevant for the real life applications. When excited at 450 nm, MIP showed an increase of its fluorescence intensity at 535 nm upon binding of Al^{3+} to the morin complexing site. Under well-controlled analytical conditions, the chemosensor linearly responded to the Al^{3+} concentration up to 1 μg mL^{-1} at the limit of detection (LOD) of 0.01 μg mL^{-1}.

This value is comparable to those for the luminescent ion chemosensor systems based on fibre optics or flow-through measurements. In general, the effect of other ions on MIP may be assumed to be negligible, as confirmed by the low MIP fluorescence measured during cross-reactivity tests. However, in toluene in the presence of Mg^{2+} the MIP fluorescence increased significantly (~50% of the total intensity) as compared to that in the presence of Al^{3+}. Moreover, Be^{2+} in ACN yielded almost equal response (90%) to that measured in the presence of Al^{3+}.

Another example of IIPs also considered development, characterization and optimization as the recognition element for Al^{3+} [43]. Here, 8-hydroquinoline-5-sulphonic acid was spatially arranged around the Al^{3+} ion and complexed to produce a supramolecular structure. This structure was subsequently fixed by thermal co-polymerization of 2-hydroxyethyl methacrylate (HEMA) in the presence of ethylene glycol dimethacrylate (EGDMA). After MIP was formed by bulk co-polymerization it was then powdered. In contrast to the previous example, aqueous solutions were used in this study. The same analytical flow-through cell system was used in this study to investigate the MIP properties. The MIP, when excited at 365 nm, showed an increase in its fluorescence emission at 495 nm upon binding of Al^{3+}. The response of the chemosensor was also measured in the presence of interfering cations, such as Cu^{2+}, Ca^{2+}, Mg^{2+}, Pb^{2+}, Co^{2+}, Cd^{2+} and Zn^{2+}. If the ratio of Al^{3+} to the interfering ion was 1:1, there was no significant change in the response. However, these cations interfered at ratios of 1:5 and 1:10 in determination of Zn^{2+}. The responses of other investigated cations were within the tolerable limit. The linear dynamic concentration range after 10 min of reaction was linear up to 1.0×10^4 M and the LOD was 3.62 μM, as shown in Table 2.

The detection of contamination with mercury is a very serious issue due to the extreme toxicity and ease of reaction of this heavy metal. Mercury is commonly used in many everyday applications, such as thermometry, barometry or scientific apparatus. Therefore, it is important to note the development of a fluorescent IIP membrane for detection of mercury [63]. This membrane was based on the combination of two fluorescent functional monomers, namely 4-vinyl pyridine and 9-vinylcarbazole. Binding of Hg^{2+} was proposed to be based on complexation of the metal ion with lone pairs of both pyridine and carbazole. The IIP membrane was capable of recognizing Hg^{2+} in the linear range of concentrations from 5×10^{-7} to 1×10^{-4} M. Selectivity of the chemosensor was studied for a wide group of possible competitive ions and the chemosensor responded to all of them in the presence of a fixed amount of Hg^{2+} at 1×10^{-5} M. Nevertheless, the highest influence on recognition properties was observed in the presence of Cu^{2+} (4.87%) and Pb^{2+} (4.57%), which are still acceptable values for the successful detection of Hg^{2+}. This progress stimulates further development of the MIP detection of ions.

2.3.2 Determination of Biomolecules

Natural recognition species, such as enzymes or antibodies, are specific to target biomolecules but, unfortunately, suffer from low chemical stability. They are

Table 2 Structural formulas of analytes and functional monomers, accompanied by characteristic values of merit, used in the photoluminescent MIP chemosensors

Analyte	Functional monomer	Value of merit	References
Al^{3+}	Morin (3,5,7,2′,4′-pentahydroxyflavone)	LOD: 0.01 μg mL^{-1}	[41]
Al^{3+}	8-Hydroquinoline-5-sulphonic acid	LOD: 3.62 μM	[44]
3′,5′-Cyclic adenosine monophosphate	Trans-4-[1,4-(N,N-dimethylamino)styryl]-N-vinylbenzylpyridinium chloride	$K_{MIP} = (3.5 \pm 1.7) \times 10^5$ M^{-1}	[45, 46]
3′,5′-Cyclic guanosine monophosphate	1,3-Diphenyl-6-vinyl-1H-pyrazolo[3,4-b]quinoline	$K_{MIP} = 10^5$ to 10^6 M^{-1}	[46–48]

(continued)

Table 2 (continued)

Analyte	Functional monomer	Value of merit	References
1-Benzyluracil	2,4-Bis(acrylamido)-6-piperidinopyrimidine	$K_{MIP} = 596 \pm 85 \ M^{-1}$	[49]
Cyclobarbital	2,6-Bis(acrylamido)pyridine	–	[50]
Cyclobarbital	2-Acrylamidoquinoline	–	[51]
Cyclic guanosine 3′,5′-monophosphate	2-Acrylamidopyridine	$K_{MIP} = 3 \times 10^5 \ M^{-1}$	[52]

Histamine	Zinc(II)-[7,12-diethenyl-3,8,13,17-tetramethyl-21H,23H-porphine-2,18-dipropanoato(4−)-κN21,κN22,κN23,κN24]	$K_{MIP} = 4500$ M^{-1}	[53]
D-Fructose	Cyclic 2,3:4,5-bis[[2-[methyl[[10-[[(2-methyl-1-oxo-2-propenyl)oxy]methyl]-9-anthracenyl]methyl]amino]methyl]phenyl]boronate] (9CI)	Linear concentration range: 0–50 mM	[54]
2-Hydroxy-benzoic acid methyl ester	Eu^{3+}	–	[55]

(continued)

Table 2 (continued)

Analyte	Functional monomer	Value of merit	References
9-Ethyladenine		LOD: 0.15 mM	[56]
(−)-Cinchonidine	5,10,15-Tris(4-isopropylphenyl)-20-(4-methacryloyl-oxyphenyl)porphyrine zinc(II) complex	Linear concentration range: 0–2 mM	[57]
Pinacolyl methylphosphonate		Linear concentration range: 0–200 ppb	[58]
Dicrotophos	Tris-(β-diketonate)-dithiobenzoate europium complex	LOD: 7 ppt	[58–60]

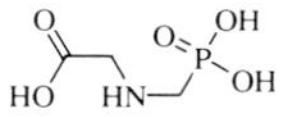

Glyphosate

LOD: 9 ppt

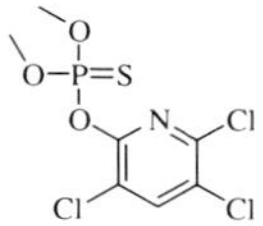

Chloropyrifos-ethyl

Eu^{3+}

LOD: 5 ppt [60]

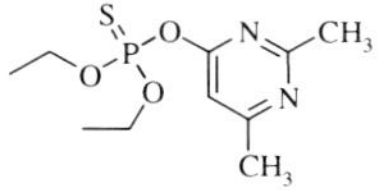

Diazinon

LOD: 7 ppt

(continued)

Table 2 (continued)

Analyte	Functional monomer	Value of merit	References
2,4,6-Trinitrotoluene	Three-dimensionally cross-linked poly(1,4-phenylene vinylene)	$K_{MIP} = 0.28–3.5\ M^{-1}$	[61]

Fluorene

N-(9-Fluorenyl-methoxycarbonyl)-β-phenyl-D-phenylalaninol

Nitrobenzofuran attached to the sol-gel matrix

Linear concentration range: [62]
1–10 ppb

thermally stable in a very limited temperature range and active in a narrow pH range. Additionally, fabrication of these bioreceptors is very complex and often requires experiments on animals (e.g. in antibody production). In contrast, synthetic MIP chemosensors may have numerous advantages over biosensor counterparts because of their high physical and chemical resistance towards various degrading factors. The sensing of biomolecules via MIPs is generally based on mimicking reactions or interactions, which can be encountered in biological systems. MIPs can also work repeatedly and, hence, are highly useful for continuous monitoring, especially in flow injection analysis (FIA). Keeping in mind that photoluminescence techniques are widely used in biology and medicine, one must admit that photoluminescent MIP sensors can have great potential for various applications in diagnosis of biological processes. However, MIP chemosensors also reveal certain limitations. For example, incomplete removal of the template from MIP decreases its sensitivity. Moreover, MIPs often reveal low biocompatibility being important when working in biological media. Further, they may suffer from the presence of non-selective recognition sites, which reduces their selectivity for their analytes. Although many developed photoluminescent MIP chemosensors perform well in sensing, no general protocol of their design and fabrication has so far been established.

A fluorescent MIP chemosensor was prepared or determination of $3',5'$-cyclic adenosine monophosphate (cAMP) [45]. Here, a *trans*-4-[1,4-(N,N-dimethylamino)styryl]-*N*-vinylbenzylpyridinium chloride (vb-DMASP) dye was used as the MIP functional monomer. For this chemosensor, a fluorescent MIP was fabricated for the first time. In this MIP, the dye acted both as recognition and measuring element, in this case, for cAMP determination. The positively charged vb-DMASP dye interacted with the negatively charged cAMP to form a pre-polymerized complex. Next, MIP was prepared in bulk and then ground and sieved. Finally, fluorescence was measured for different concentrations of aqueous cAMP solutions. The fluorescence emission of MIP at ~595 nm decreased upon interaction with the analyte when excited at 469 nm. Importantly, the vb-DMASP emission appeared in a spectral range where autofluorescence of proteins is negligible. The linear dynamic concentration range for this MIP chemosensor was limited from 10 to 100 μM due to saturation of the accessible sites. The complex stability constant was $K_{MIP} = (3.5 \pm 1.7) \times 10^5$ M^{-1} (Table 2). The cAMP imprinted polymer displayed high affinity and selectivity for aqueous cAMP, as confirmed by a cross-reactivity test with the structurally similar $3',5'$-cyclic guanosine monophosphate (cGMP).

The application of vb-DMASP to MIPs was continued in subsequent works [64, 65]. In these investigations, time-resolved fluorescence spectroscopy was applied to study bulk fluorescent MIP. The imprinted polymer fluorescence quenching with increasing concentrations of aqueous cAMP was determined from the fluorescence lifetime parameters. Two components in the fluorescence decays were identified and assigned to two different types of cavities present in the polymer matrix. One was accessible and open to binding, whereas the other was inaccessible, being buried inside the bulk polymer. The fluorescence lifetime decreased due to the increase in the concentration of the initial target analyte. However, the accessible

Fig. 1 The steady-state fluorescence excitation spectrum of (1) PAQ based MIP and (2) its emission spectra in the absence and (3) in the presence of cGMP (adapted from [47])

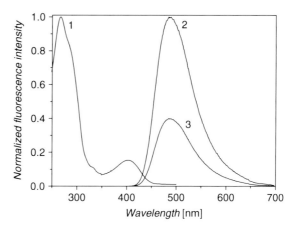

recognition sites situated on the surface of the particles, formed after grinding, seemed to be noticeably damaged. To overcome this problem, a thin-layer MIP sensitive to cAMP was synthesized [46]. Similarly to the bulk MIP, fluorescence of MIP with incorporated vb-DMASP was quenched when it was deposited as a thin film on a quartz substrate. More importantly, quenching caused by interaction with cAMP was much more effective, yielding the complex stability constant in the range of $10^5 \leq K_{MIP} \leq 10^6$ M^{-1}. The difference in the K_{MIP} values was explained in terms of better accessibility of the imprinted sites for cAMP, as compared to the layer with the bulk polymer.

Subsequently, fluorescent MIPs for cGMP were fabricated [46–48, 66, 67]. For that, 1,3-diphenyl-6-vinyl-1H-pyrazolo[3,4-b]quinoline (PAQ) was introduced as the fluorescent indicator to interact with cGMP in a thin-layer fluorescent MIP chemosensor. Both steady-state (Fig. 1) and time-resolved fluorescence spectroscopy were used as two independent analytical techniques for investigation of the chemosensor properties in the presence of cGMP. Steady-state fluorescence spectroscopy is a common technique applied to MIP sensing. Nevertheless, the use of time-resolved fluorescence spectroscopy combined with microscopy was a new approach to MIP sensing.

Fluorescence emission of MIP with the maximum at 470 nm was quenched in the presence of cGMP when excited at 400 nm. Similar to that determined for the vb-DMASP-based MIP, the complex stability constant was in the range $10^5 \leq K_{MIP} \leq 10^6$ M^{-1}. The decays in the fluorescence lifetime were measured from various locations on the surface of the MIP film by using time-resolved fluorescence microscopy. These decays were then analysed by constructing a distribution profile of all fluorescence lifetimes measured [66]. Binding of the analyte to the MIP was analysed by means of changes in the half width and position of these distributions. After extraction of cGMP, the fluorescence lifetime of the MIP increased while that of the corresponding NIP was changed insignificantly. The fluorescence liftime distribution (FLD) profiles are shown in Fig. 2 and their characteristic values are presented in Table 3. Apart from the changes in the

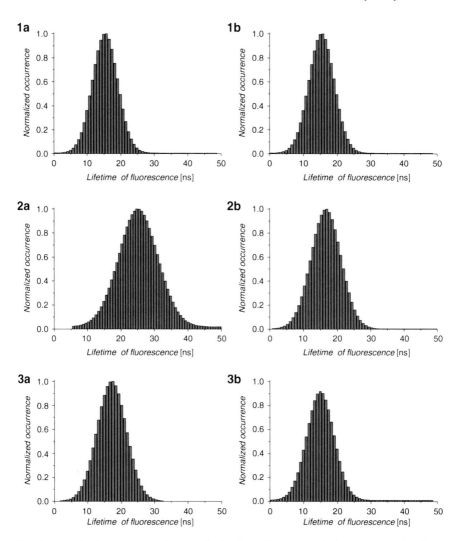

Fig. 2 Normalized fluorescence lifetime distributions (FLDs) for (**a**) the cGMP-imprinted and (**b**) non-imprinted thin-layer polymer film obtained from the fluorescence microscopy measurements (adapted from [47, 66])

Table 3 Characteristic values of the fluorescence lifetime distribution for the cGMP-imprinted and non-imprinted thin-layer polymer film obtained from fluorescence microscopy measurements [47]

Process	Imprinted polymer film		Non-imprinted polymer film	
	Width (ns)	Lifetime (ns)	Width (ns)	Lifetime (ns)
Polymerization	7.6	15.3	7.7	15.2
Extraction	11.7	25.1	9.1	16.2
Re-adsorption	8.5	16.1	8.1	14.9

position of the distribution, the half width became broader. Again, the distribution became narrower and returned to its initial position after rebinding. Similarly, the distribution was broader and the fluorescence intensity changed its position when the analyte was extracted from MIP and, again, was narrower and positioned lower after MIP immersion in the cGMP solution. The NIP showed negligible changes within 5% of intensity and half width. However, the half width and position of the distribution both for the fluorescence lifetime and intensity did not return exactly to their initial values. The rebinding was incomplete and decreased as the extraction-(rebinding) cycle was repeated. The broadening and narrowing of distributions, accompanied by their shift, can be explained in terms of the bound and unbound state of the fluorescent indicator in the presence and absence of the analyte, respectively. The fluorescent indicator in the bound state was non-radiatively relaxed by the energy transfer to the analyte and, consequently, the fluorescence lifetime became shorter.

After extraction, the fluorescent indicator was in the unbound state and gave input to the radiative relaxation. Therefore, the fluorescence lifetime increased and, consequently, the intensity as well. After MIP contacting with the analyte, the non-radiative processes were again efficient compared to the radiative processes and, subsequently, fluorescence was quenched. With steady-state fluorescence spectroscopy the cross-reactivity test towards structurally similar biomolecules was performed that yielded selectivity factors for guanosine, cAMP and cCMP of 1.5, 2.5 and 5.1, respectively.

Two pyrimidines, namely 2,4-bis(acrylamido)-6-piperidinopyrimidine and 2,4-bis(acrylamido)-6-ethoxypyrimidine, were used for sensing of 1-benzyluracil [49]. From the fluorescence emission spectrum of the monomer, information was obtained on both the monomer incorporation in the polymer and the presence of the template. Fluorescence of the bulk MIP was rapidly and selectively quenched upon the 1-benzyluracil addition, with the degree of quenching correlating with the binding affinity and the amount of the template bound to the polymer. The fluorescence signal excited at 270 nm was collected at 370 nm. A 1:1 stoichiometry was assumed of the 2,4-bis(acrylamido)-6-piperidinopyrimidine or 2,4-bis(acrylamido)-6-ethoxypyrimidine complexation with 1-benzyluracil, yielding the complex stability constants $K_{MIP} = 596 \pm 85$ M^{-1} and $K_{MIP} = 561 \pm 37$ M^{-1}, respectively (Table 2). In contrast to this result [49], another work [50] reported enhancement in fluorescence of the bulk MIP containing 2,6-bis(acrylamido)pyridine upon binding cyclobarbital. The interaction principle was the same as that for 1-benzyluracil, namely mimicking the Watson–Crick deoxyribonucleic acid (DNA) base pairing. The enhancement of fluorescence at 380 nm (excitation at 270 nm) was interpreted in terms of increasing rigidity of the monomer residues due to the formation of multiple hydrogen bonds. It was postulated that possible factors contributing to this contradictory result may be related to different UV absorptions of the templates at the excitation wavelength of the dye and intrinsic fluorescence of barbiturates. Moreover, barbiturates are known to undergo chemical transformations upon UV illumination [40]. Interestingly, under certain conditions this transformation led to the formation of a new absorption band with its maximum at 270 nm.

Unfortunately, neither the former [49] nor later [50] research showed absorbance spectra of the templates. Since rather highly energetic illumination wavelengths were used in both works, the excitation of both the fluorescent monomer and the template seems plausible.

Another example of MIP dedicated for determination of cyclobarbital was the application of 2-acrylamidoquinoline as a fluorescent functional monomer [51]. Again, fluorescence of this functional monomer was enhanced upon binding of the template to the bulk MIP. Here, the excitation wavelength was raised to 330 nm in order to decrease the background fluorescence. The fluorescence signal was acquired at 376 nm. The steady-state fluorescence spectroscopy experiments were supported by those using ^{1}H NMR of the quinoline derivative as a function of increasing cyclobarbital concentration. In fact, the study using $CDCl_3$ confirmed the influence of cyclobarbital on free 2-acrylamidoquinoline yielding a fluorescence increase of ~400% after adding a 125-μM sample of cyclobarbital. In case of MIP, the fluorescence enhancement was ~10% after addition of the same amount of cyclobarbital, thus supporting the assumption of the fluorescence quenching caused by non-radiative interactions with the surrounding polymer.

In similar investigations [52], 2-acrylamidopyridine was used as the fluorescent reporter in bulk MIP for cGMP detection. This MIP displayed high affinity and selectivity for aqueous cGMP over structurally similar GMP and cAMP. The polymer was excited at 355 nm and its emission was collected at 460 nm. The concentration-dependent quenching of fluorescence reached a maximum level at ~100 μM yielding a complex stability constant, $K_{MIP} = 3 \times 10^5\,M^{-1}$. These results were in good agreement with those of earlier studies [45].

In another approach [53], zinc(II)-protoporphyrin (ZnPP) was used as a luminescent functional monomer. Similarity of metal-complexed porphyrins to haemoglobin is a direct indication of a possible application of this system for molecular detection in biological media. Recognition of histamine was realized via binding of a Lewis Zn site to the imidazol moiety of histamine through axial coordination. The steady-state fluorescence intensity of the ground bulk polymer decreased upon exposure to histamine. The complex stability constant was $K_{MIP} = 4500\,M^{-1}$, in the histamine concentration range of 0.1–1 mM and $K_{MIP} = 270\,M^{-1}$, in the range of 4–10 mM (Table 2). The linear dynamic concentration range was 0.1–4 mM associated with a 40% change in emission intensity at 583 nm. Unfortunately, the information concerning the excitation wavelength for this MIP was lacking. Additionally, cross-reactivity of MIP towards structurally related compounds was not investigated.

A fluorescent MIP chemosensor for determination of 9-ethyladenine was fabricated [56]. It contained porphyrin as a luminescent functional monomer. The interaction of 9-ethyladenine with the porphyrin quenched the MIP luminescence at 605 nm when excited at 423 nm. The polymer was sensitive to 9-ethyladenine in the range of 0.01–0.1 mM; however, it was already saturated at 0.15 mM. The same researchers used vinyl-substituted zinc(II) porphyrin and methacrylic acid as functional monomers for imprinting of (–)-cinchonidine [57]. The MIP luminescence, when excited at 404 nm, was significantly quenched at 604 nm upon binding of (–)-cinchonidine, even in the low concentration range of 0.01–2 mM.

The luminescence quenching of MIP was negligible when the polymer was exposed to (+)-cinchonine, confirming the presence of binding sites highly complementary to (–)-cinchonidine.

Moreover, bulk MIP was prepared [56] exhibiting diastereoselectivity for cinchona alkaloids. In the presence of MIP in solution, the cinchonidine fluorescence emission was hypsochromically shifted with the increase of the cinchonide concentration. That is, maximum of the fluorescence emission was at 390 nm in the absence of cinchonidine whereas it was at 360 nm at higher concentrations of cinchonidine. When cinchonidine was examined in the absence of MIP or in the presence of NIP, there was no spectral shift or this shift was negligibly small. This shift has been explained on the basis of protonation of the nitrogen atoms present in the cinchonidine structure.

A fluorescent bulk MIP was synthesized to fabricate a fluorescent chemosensor for determination of D-fructose [54]. The recognition involved interaction of the sugar with boronic acid. The fluorescence signal was provided by anthracene coupled with the boronic acid moiety. As a consequence of interaction between D-fructose and the fluorescent monomer, the fluorescence of this MIP at 423 nm was enhanced when excited at 370 nm. This increase was attributed to the formation of a boronate ester with the *cis* diol structure causing an increase in acidity of the boron atom. In turn, this increase affected photoinduced electron transfer. The linear dynamic concentration range extended from 0.01 to 100 mM D-fructose with full saturation at 50 mM (Table 2). The cross-reactivity test of MIP towards other sugars showed moderate selectivity. That is, both D-mannose and D-glucose were interacting with this MIP but the resulting response was ten times weaker than that to D-fructose.

2.3.3 Determination of Warfare Agents and Toxins

Progress in chemical and biological warfare technology and in agricultural industry has stimulated the development of complex, toxic substances. Although chemical and biological weapons have been banned, nerve gases and toxins are still being stocked [68]. Additionally, pesticides commonly used in agriculture are often being released into the environment, causing significant contamination. Chemosensors based on MIP technology are capable of selective binding of the warfare analytes, in contrast to, e.g. traditional nerve gas sensors, which can produce false signals often generated in the presence of organophosphate pesticides [69]. Photoluminescent sensors serve as useful tools in the detection of nerve agents because an increase in the optical signal can easily be detected and transferred, giving immediate information about a possible threat.

MIPs were doped with Eu^{3+} for optical detection of methylated salicylates (MES), a chemical warfare agent simulant [55]. Eu^{3+} absorbs in the near UV region and doped MIP can, therefore, be excited with a commonly available laser diode at 375 nm. MIP doped with Eu^{3+} was prepared as a thin film on a quartz slide substrate. Both the MIP and NIP films were tested towards MES and a structurally similar compound, methylene-3,5-dimethylbenzoate (DMB), in hexane. For MES, the fluorescence intensity of the MIP was significantly stronger than that for its NIP while for DMB the fluorescence signal was stronger for its NIP. The authors proposed that these ion-based MIPs could be used in portable chemical sensors that evaluate differential fluorescence from their respective MIP/NIP pairs.

Prepared by bulk polymerization, an MIP for the detection of dicrotophos based on the Eu^{3+} complex has recently been presented [58]. The authors used reversible addition fragmentation chain transfer (RAFT) polymerization followed by ring closing methathesis (RCM) to obtain the star MIP with arms made out of block copolymer. The star MIP containing Eu^{3+} exhibited strong fluorescence when excited at 338 nm with a very narrow emission peak (half width ~10 nm) at 614 nm. This MIP was sensitive to dicrotophos in the range of 0–200 ppb, but showed saturation above this limit. Cross-reactivity of this MIP was evaluated with respect to structurally similar compounds: dichlorvos, diazinon and dimethyl methylphosphonate. In these tests no optical response of the polymer was detected even at concentrations much higher than the initial concentration of dicrotophos (>1000 ppb).

The same research group evaluated hydrolysis products of soman (pinacolyl methylphosphonofluoridate) in water [58–60]. They presented a luminescent bulk MIP chemosensor for pinacolyl methylphosphonate (PMP) based on Eu^{3+}. The recognition principle of this fibre optics chemosensor was based on coordinative binding to Eu^{3+} in polystyrene. The MIP was excited at 466 nm, resulting in a narrow emission band at 614 nm. The response time of the chemosensor was in the range 15–30 min depending upon the thickness of the active layer and pH of the surrounding media. The LOD for the hydrolysis product was 7 parts per trillion (ppt) in solution with a linear concentration range between 10 ppt to 10 ppm for the 200 μm thick MIP film at pH = 12.

Furthermore, MIPs were prepared for the detection of non-hydrolysed organo-phosphates, including pesticides and insecticides [70]. In contrast to sensors with the pinacolyl methylphosphonate imprinted polymer, these MIP chemosensors showed an increase in the luminescence intensity upon interaction with the analyte. This increase was attributed to the binding of the analyte into the coordination sphere of europium and the concomitant exclusion of water. After excitation of these MIPs at 466 nm, the luminescence of Eu^{3+} increased between 610 and 625 nm. These MIPs were prepared as a 400 μm thick layer coated over the distal end of an optical fibre. The LODs of these MIP chemosensors were less than 10 ppt with a wide dynamic concentration range, and a response time of less than 15 min. Three different compounds, belonging to three different classes, were used as templates, namely methyl-chloropyrifos (pyridine organothiophosphate), diazinon (pyrimidine organophosphate) and glyphosate (planar organophosphate). LOD values for these MIPs were 9, 5 and 7 ppt with linear dynamic concentration ranges up to 100 ppm and the response times of 12, 14 and 15 min at pH = 10.5 for glyphosate, methyl-chloropyrifos and diazinon, respectively (see Table 2). For cross-reactivity, the MIP sensors were tested with structurally similar molecules, such as chloropyrifos-ethyl, coumaphos or dichlorvos, all of which showed no significant increase in the MIP luminescence.

In another approach, a fluorescent conjugated polymer was used as the material for the preparation of a chemosensor to detect 2,4,6-trinitrotoluene (TNT) and its related nitroaromatic compounds. To this end, microparticles, made of three-dimensionally cross-linked poly(1,4-phenylene vinylene) (PPV) via emulsion polymerization, were synthesized [61]. This material was chosen due to its high fluorescence intensity and sensitivity to changes in its microenvironment. The chemosensor was exposed to vapour containing different amounts of TNT and quenching of the polymer luminescence at 560 nm was observed after excitation at 430 nm. The dependence of the fluorescence signal in response to the analyte was described by a modified Stern–Volmer equation that assumes the existence of two different cavity types. The authors proposed the modified Stern–Volmer equation as follows:

$$\frac{I_0}{I_0 - I} = \frac{1}{f_a} + \frac{1}{f_a \, K_{SV}^a \, [Q]} \tag{1}$$

where I_0 is the initial fluorescence intensity in the absence of analyte, I is the fluorescence intensity in the presence of analyte, f_a is the fraction of easily accessible cavities, $[Q]$ is analyte concentration and K_{SV}^a is the Stern–Volmer constant for quenching inside the cavities. The fit of experimental data to this equation was a good one, yielding an f_a equal to 0.9. The MIP chemosensor was exposed to TNT vapour for 1 h, which led to a plateau in the response after approximately 20 min, indicating the filling of most cavities. The cross-reactivity was tested with respect to eight structurally related compounds. Unexpectedly, the MIP microparticles showed stronger fluorescence quenching for 2,4-dinitrotoluene (DNT) than that for TNT itself. This phenomenon was explained in terms of the spatial arrangement of DNT within the TNT imprint, which was assumed to be more likely to occur than that for TNT. The Stern–Volmer constants calculated on the basis of (1) were in the range 0.28–3.5 depending on the quencher used as presented in Table 2.

Moreover, the use of MIP microparticles with quantum dots (QDs) as signal transducers for the detection of nitroaromatic explosives has been very recently presented [71]. LOD for aqueous solutions was 30.1 μM and 40.7 μM for DNT and TNT, respectively. Although the LODs of the presented system are 100 times lower than those for other already developed TNT sensitive systems, this example presents a new interesting approach in the MIPs technology. If the colloidal stability and size distribution of the microparticles were improved, this example would present a reasonable approach to MIP chemosensor preparation.

The study of QDs and MIPs in the development of the detection scheme for pyrethroids has very recently been reported [72]. Pyrethroids are pesticides commonly used against insects being widely used in agriculture, medicine, industry and the household. The QDs were embedded in silica nanospheres having lambda-cahalothrin imprinted sites on their surfaces. As also presented in the previous example, the principle of recognition was based on a decrease of fluorescence with an increase of analyte concentration. The lambda-cahalothrin imprinted nanospheres showed good selectivity (approximately three times) over other pyrethroids, namely cypermethrin, permethrin and deltamethrin. The limit of lambda-cahalothrin detection was found to be as low as 3.6 mg L^{-1} at a recovery of ~96%.

All the previously described approaches employed QD as signal transducers embedded within a colloidal system such as microparticles or nanoparticles. Nevertheless, another method considering formation of a polymeric shell around a QD is worth noticing as well. This approach is advantageous over the entrapment in particles method due to a larger surface being available for an analyte to bind and stronger influence on QD luminescence due to shorter distance between a recognition site and a QD. The main drawback of this system can be associated with the colloidal stability of QDs during rebinding. An example of MIP utilizing such an approach has been presented for imprinting of pentachlorophenol and its detection in water [73]. Pentachlorophenol belongs to a class of organochlorine compounds used as herbicides, insecticides or in wood preservation. Short-term exposure of pentachlorophenol can be harmful for important organs such as liver, kidneys or lungs. The Mn-doped ZnS QD based MIP optosensors working based on luminescence changes were used to determine pentachlorophenol in water samples. The MIP was capable of recognizing the analyte with the limit of detection in the nanomolar range of concentrations. Interestingly, in the presence of other organochlorines to a certain limit, NIP was also responding to the analyte.

A highly selective MIP employing halogenated bisphenol as a recognition element has been presented for the recognition of carbaryl (1-naphthyl methylcarbamate) in water [74]. Carbaryl belongs to the carbamate class of compounds. It is mainly used as an insecticide for home gardens, commercial agriculture and to protect forests. The polymers were prepared with and without a halogen, either bromine or iodine, by bulk polymerization. Introducing heavy halogen atoms makes it possible to detect the recognition phenomenon on the basis of heavy-atom-induced room temperature phosphorescence. This approach overcomes common drawbacks associated with classical fluorescence detection such as short lifetime and scattering, but on the other hand makes the analytical signal temperature sensitive. In the case of halogenated MIP the response to carbaryl was 4–10 times higher than for its non-imprinted counterpart.

Among other herbicides, maleic hydrazide, being a regulator for the growth of buds in vegetables during storage, was used as an analyte in design and development of a fluorescent MIP chemosensor [75]. The recognition was based on the production of chemiluminescence in a flow injection system. Upon reaction of maleic hydrazide with the alkaline luminol-potassium periodate the latter produces strong post-chemiluminescence which can easily be detected by a photomultiplier. After the reaction, the absorbed analyte was removed by the flow solution and left cavities on the MIP ready for the next adsorption assay. The dynamic linear range of MIP to the concentration change of maleic hydrazide was in the range of 3.5×10^{-4} to 5.0×10^{-2} mg mL^{-1} with LOD of 6.0×10^{-5} mg mL^{-1}, which is claimed to be lower than those offered by conventional methods of this herbicide determination.

A fluorescent sol-gel MIP for fluorene by introduction of fluorene functionalized silane has been presented [62]. Subsequent cleavage of the carbamate link led to removal of fluorene, leaving a secondary amino group behind. A polarity sensitive molecule, 7-nitrobenz-2-oxa-1,3-diazole, was then attached to this amino group and served as a fluorescent indicator. The recognition was based on differences in polarity caused by the interaction of fluorene with the indicator. The thin MIP films were deposited on the glass slide substrates and excited at 477 nm, which resulted in the maximum fluorescence emission at 540 nm. The dynamic concentration range covered fluorene concentration from 1 to 10 ppb. Such a small range can indicate a rather small number of imprinted sites that are already saturated at low concentrations of fluorene. Therefore, no more changes in the fluorescence signal can be detected at higher concentrations. Structurally similar compounds, such as fluorene-2-carboxyaldehyde, anthracene, fluoranthene and naphthalene, were also examined using the fluorene-templated MIP. In a methanol solution, the fluorene MIP showed three to five times higher sensitivity towards fluorene than to its structural analogues. In the methanol–water solution, however, the response of this MIP to the analogues was almost the same as that for fluorene. This phenomenon was explained in terms of the polarity changes. Because 7-nitrobenz-2-oxa-1,3-diazole is a polarity sensitive probe, the polarity would naturally change after the addition of water. As a consequence, non-specific binding ensures that non-specific fluorescence signals are produced. The same group has recently presented an MIP for the parts per trillion detection of TNT based on similar semi-covalent imprinting approach [76]. In this case the analytically useful signal was generated by waveguide interferometry.

All structural formulas of both functional monomers and analytes along with characteristic values of merit presented within this section are gathered in Table 2.

2.4   Surface Plasmon Resonance Chemosensors

SPR spectroscopy has been gaining relevance in the past few years for analysing MIP recognition processes. Theoretical principles of this analytical technique consist in the measurement of a difference in the refractive index of two transparent media separated by a thin metal film. When polarized light at a certain angle incites on the higher refractive index medium, at which total reflection is observed, an evanescent wave is generated. The resonance energy transfer between the evanescent waves, resulting in the surface plasmons, produces a sharp signal of the surface plasmon resonance. The angle of the incident light causing the resonance is dependent on the characteristics of the thin metal layer. Therefore, if a certain analyte is adsorbed on this layer, the concentration of this analyte can be correlated with the change in the resonance angle. The adsorption of an MIP layer on a silver film can, for example, be used to study phenomena such as binding and release of the template to and from the MIP film.

One of the main disadvantages of combination of the MIP recognition with the SPR transduction is the low sensitivity of the latter for detection of small molecules. The evanescent field generated at the sensing surface is often insufficiently sensitive to detect small changes in the refractive index upon rebinding of the analyte by the MIP layer. Nevertheless, some very promising work combining the two techniques has already been demonstrated [57].

An SPR chemosensor chip using an MIP layer have been devised for determination of sialic acid [57]. The MIP layer was synthesized by photoinduced polymerization of two functional monomers, namely HEMA and N,N,N-trimethylaminoethyl methacrylate, directly on the surface of the chip. In the end, selectivity was confirmed against galacturonic acid. The linear concentration range was 0.1–1.0 mg mL^{-1} sialic acid (Table 4). These values illustrate limitations of the SPR detection towards small molecules. Another successful approach has been demonstrated by using an oligopeptide derivative as the imprinting material for selective recognition of adenosine [77]. Binding sites were generated using 9-ethyladenine as the template and cross-reactivity studies with guanosine and adenine proved selectivity of the material towards the latter.

A method for determination of the NAD(P)$^+$ and NAD(P)H cofactors was developed using gold surfaces coated with MIP films [78]. These films were prepared using acrylamide and 3-acrylamidophenylboronic acid as functional monomers. The SPR determination was possible since binding of the reduced or oxidized substrates resulted in swelling of the polymer films through the uptake of water. The recognition sites were selective with respect to different imprinted cofactors, allowing for analysis within the linear concentration range of 1 μM to 1 mM. In order to illustrate the method, the kinetics of the biocatalysed oxidation of lactate to pyruvate in the presence of NAD$^+$ and lactate dehydrogenase was studied using the NADH-imprinted polymer film.

In another approach to the use of MIP layers grafted on a gold chip surface, the MIP film templated with domoic acid was synthesized by direct photografting on the chip surface [79]. The analyte was then determined by SPR. The grafting consisted of (1) surface functionalization with a self-assembled monolayer (SAM) of 2-mercaptoethanol, followed by (2) covalent attachment of the photoinitiator, 4,4′-azobis(cyanovaleric acid), and finally (3) polymerization using 2-(diethylamino)ethyl methacrylate as the functional monomer and EGDMA as the cross-linker. This sequential procedure allowed for the formation of a 40 nm thick homogeneous MIP film. LOD for domoic acid was 5 μg L^{-1}. This value is comparable to that of the biosensor using a monoclonal antibody, equal to 1.8 μg L^{-1}, but with higher stability than that of the latter (Table 4). In a very similar approach, thin homogeneous MIP films were synthesized to determine dansylated amino acids [47].

Recent research involved embedding gold nanoparticles into an MIP layer [82]. This procedure enormously increased sensitivity of the method and, therefore, made it suitable for determination of a low molecular weight analyte, such as dopamine. The method was based on the swelling of the MIP gel upon template binding. This swelling resulted in the increase of the distance between the

Table 4 Structural formulas of analytes and functional monomers, accompanied by characteristic values of merit, used in the MIP surface plasmon resonance chemosensors

Analyte	Functional monomer(s)	Value of merit	References
	N,N,N-Trimethylaminoethyl methacrylate		
	2-Hydroxyethyl methacrylate		
	p-Vinylbenzeneboronic acid[a]		
Ganglioside		Linear concentration range: 0.1–1.0 mg mL^{-1}	[57]
Adenosine	Aminomethylated polysulphone derivatized with the N-carboxyanhydride of γ-benzyl-L-glutamate (DS = X + Y = 0.9)	Complex stability constant: K_{MIP} = 1.30– 1.60 × 10^4 M^{-1}	[77]

(continued)

Table 4 (continued)

Analyte	Functional monomer(s)	Value of merit	References
	Methacrylamide		
NAD⁺			
NAD(P)⁺ and their respective reduced forms	Acrylamidophenylboronic acid	LOD: 1 μM	[78]
Domoic acid	2-(Diethylamino) ethyl methacrylate	LOD: 5 μg L^{-1}	[79]

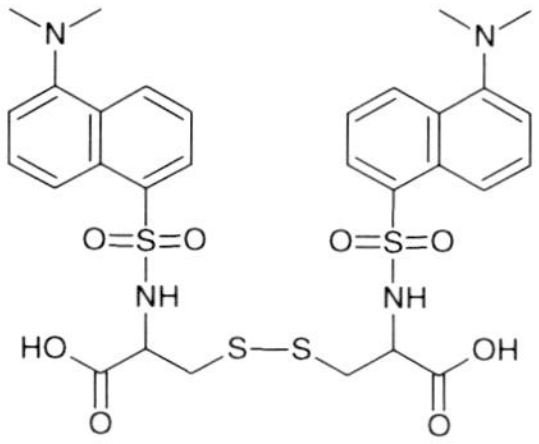

N,N'-Didansyl-L-cystine (DDC)

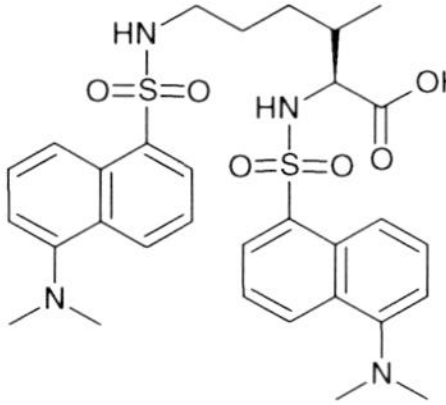

Didansyl-L-lysine (DDK)

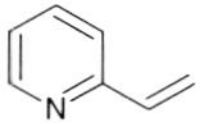

Zearalenone

2-Vinylpyridine

Polypyrrole (PPy)

Complex stability constants: $K_{DDC} = 6.38 \times 10^4\ \mu M^{-1}$ and $K_{DDK} = 60.98 \times 10^4\ \mu M^{-1}$

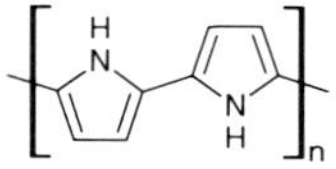

[80]

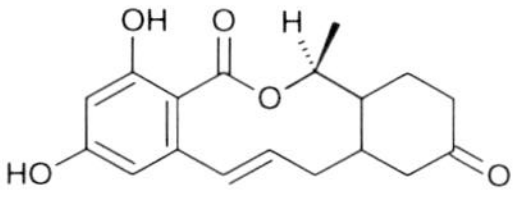

LOD: 0.3 ng g^{-1}

[81]

(continued)

Table 4 (continued)

Analyte	Functional monomer(s)	Value of merit	References
Dopamine	Acrylic acid N-Isopropylacrylamide	Range of studied concentrations: 1 nM to 1 mM	[82]
Atrazine	Methacrylic acid	Concentration range: 5 pM to 5 nM	[83]
Cholesterol	Poly(2-vinylpyridine)	–	[84]
Hexahydro-1,3,5-trinitro-1,3,5-triazine (RDX)	Bisaniline–cross-linked gold nanoparticles composite	Complex stability constant $K_{MIP} = 3.4 \times 10^7$ M^{-1} LOD: 4 nM	[43]

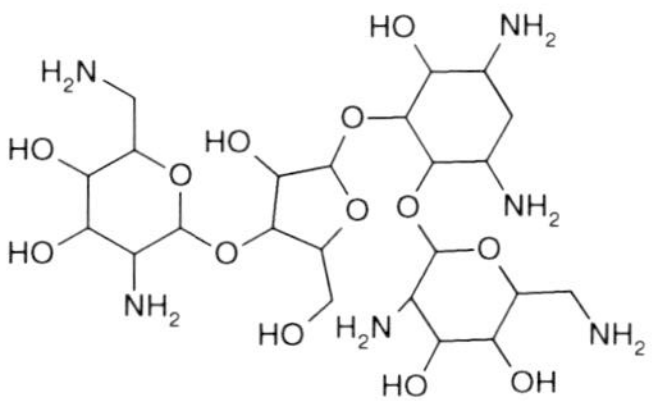

LODs:
NE: (2.00 ± 0.21) pM
KA: (1.00 ± 0.10) pM
ST: (200 ± 30) fM

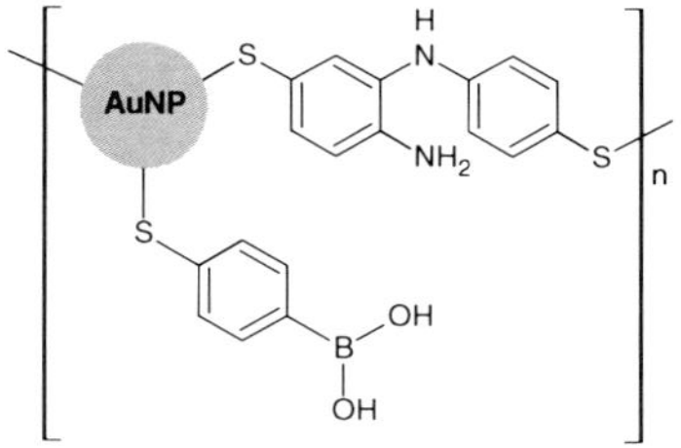

Bisaniline–cross-linked gold nanoparticles
functionalized with thiophenylboronic acid

Neomycin (NE)

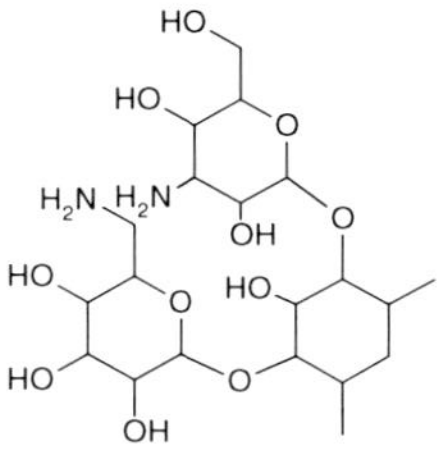

Kanamycin (KA)

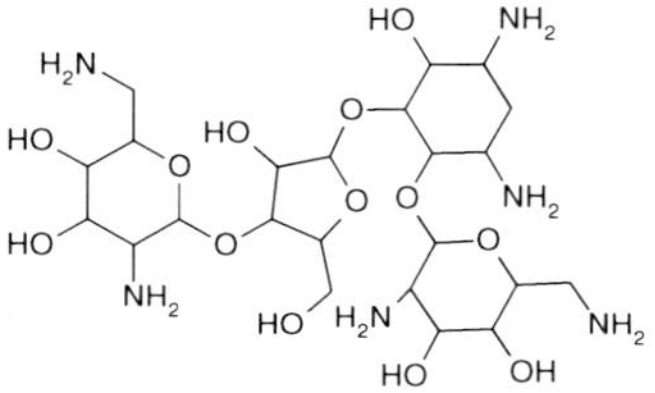

Streptomycin (ST)

[85]

(continued)

Table 4 (continued)

Analyte	Functional monomer(s)	Value of merit	References
Theophylline	N-Propylacrylamide	LOD: 10^{-6} M	[86]
S-Propranolol	Methacrylic acid	LOD: 1×10^{-7} M	[87]
S-Propranolol	Methacrylic acid	Complex dissociation constants: $K_d(S\text{-Prop.}) = 59 \pm 7.7$ µM LOD: 1 µM	[87]
S-Atenolol			

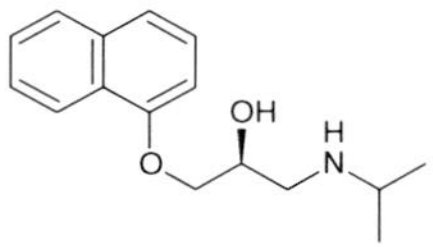

S-Propranolol

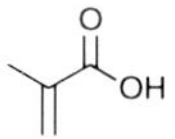

Methacrylic acid

[88]

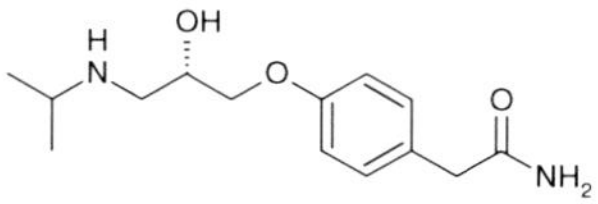

—

S-Propranolol

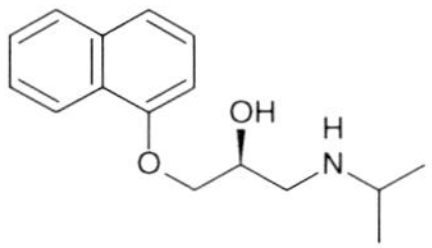

Methacrylic acid

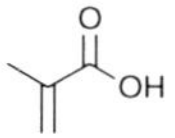

R-Propranolol

K_d (S-Prop.) = 0.4 mM
K_d (R-Prop.) = 1 mM

[89]

[a]This monomer was introduced in the polymer, being covalently attached to the template and remaining in the cavities upon removal of the template

nanoparticles and the SPR substrate leading to a shift in the SPR angle. Comparison between MIPs with and without gold nanoparticles proved the effectiveness of the method with respect to the sensitivity enhancement. This research opens up new perspectives for the use of SPR in combination with MIPs since it overcomes one of the major limitations of minute signal changes in case of sensing small analytes.

An ultrathin MIP chemosensor system (31.5 ± 4.1 nm) has been devised for determination of cholesterol through transmission surface plasmon resonance (T-SPR) [84]. The MIP film was synthesized in five consecutive steps, with special emphasis put on the last one, i.e. on the adsorption of gold nanoparticles on top of the MIP film. The rebinding of cholesterol onto the MIP film resulted in a 56-nm shift in the absorption maximum of the SPR band, i.e. from 654 to 598 nm. The selectivity of the sensing system towards cholesterol was compared to that towards structurally related stigmasterol, digitoxigenin and progesterone. Although these interferants were also recognized with the absorption maximum shifts of 14, 26 and 30 nm, respectively, the response to cholesterol was at least twice as high.

Beside the use of gold nanoparticles, the swelling capacity of some types of polymer nanospheres is another useful feature that can be exploited for the signal enhancement of SPR-based sensing [86]. Illustratively, theophylline was detected using slightly cross-linked imprinted poly[N-(N-propyl)acrylamide] that was spin-coated onto a gold surface. The refractive index of the polymer nanospheres changed significantly after binding of the template in aqueous solutions. LOD was of the order of 1 μM. High selectivity was proven by comparison of determination of the analyte to that of caffeine, which did not cause any response at concentrations up to 10^{-2} M.

All structural formulas of both functional monomers and analytes along with characteristic values of merit presented within this section are summarized in Table 4.

2.5   Conclusions and Perspectives of Optical Chemosensors

The present section shows how significant progress has been made during the last decade in the area of absorbance, photoluminescence and surface plasmon resonance MIP chemosensors. It also reveals that some papers are rather vague in terms of their practical application and some issues have not been studied in sufficient detail. Frequently, important sensing properties of the fabricated materials are defined in different ways, making it difficult to assess the quality of the results (Tables 1, 2, and 4). The use of typical parameters, and most importantly the complex stability constant, LOD, selectivity and the response time, should be mandatory in order to compare the newly obtained nanomaterials with previously published systems. Most approaches to optical sensing are based on measurement of the fluorescence intensity at a single wavelength, characteristic for the indicator dye. Interactions between the indicator and the analyte are then assessed by changes of the intensity. However, the measurement of the fluorescence lifetime as well as the fluorescence polarization would be advisable to provide insight into mechanisms of molecular interactions. The SPR-based MIP chemosensors may be competitive with those exploiting fluorescence but reproducibility of the former still remains challenging. Many indicators used within the presented studies require UV excitation or display sensing properties in the UV range. This property may be useful in laboratory research but excitation in the UV range is inappropriate for real applications, e.g. sensing or probing of chemical species in a biological matrix. With excitation below 400 nm, the intrinsic fluorescence of the sample is so strong that it blurs the signal of the indicator. Additionally, the filtering effect of a biological matter can be intense and, consequently, no practical signals may be obtained. One of the alternatives to overcome these problems is the use of dyes utilizing the fluorescence resonance energy transfer (FRET) effect, which is quite well operative in the visible and near-infrared region. Alternatively, the longer wave absorbing and emitting indicators featuring specific groups for the analyte determination could be synthesized. However, development of such indicators may be related to a risk of low solubility of intermediates formed during the MIP preparation as a result of their large molecular size. Consequently, well optimized indicators are highly desired to be applied in MIPs. A comprehensive analysis of all crucial aspects, i.e. chemical and photochemical stability, sensitivity towards the analyte and appropriate spectral properties, make the indicator an excellent candidate to be applied in MIPs. Overall, the use of MIPs for sensing of analytes has given insight into many recognition processes at a molecular level. Future perspectives are driven by the development of new more specific indicators, better defined MIP structures and application of the MIP technology on the nanoscale. The use of new indicators to nanoscale imprinting would provide a great opportunity to enhance selectivity of molecular sites. Since the surface of a nanoparticle is significantly reduced, the diffusion of the analyte into and out of the polymeric nanoparticle can be more efficient compared to bulk or thin-layer MIPs. Furthermore, using core-shell architectures, one can develop a nanoparticle in which only its shell is imprinted. By adjusting the thickness of the imprinted shell, which is usually in the range of few nanometres, the response time of the chemosensor can be significantly reduced. With the application of more sensitive and more selective indicators incorporated in the shell of nanoparticles, continuous sensing of various analytes using the flow-through systems or biochip array technology may become feasible. It would lead to a platform of practical applications that extends from medical research in the laboratory to process analysis in industrial facilities.

3 Piezoelectric Chemosensors

3.1 Introduction to Piezoelectric Chemosensors

In recent decades, PZ materials have been used at frontiers of scientific research and technology. Owing to their sensitivity, acoustic wave PZ devices have been widely used in electronics. These devices, such as wave or bandpass filters, have been employed for fabrication of telecommunication components, etc. One of the most important applications of the PZ materials is fabrication of acoustic wave PZ sensors. Several sensors applied in the automotive industry (e.g. torque and tyre pressure sensors), environmental monitoring (e.g. vapour, humidity and temperature sensors) and medicine (e.g. chemosensors and biosensors) rely on acoustic wave PZ materials. To this end, mass-sensitive PZ devices that enable subnanogram weighing have emerged as attractive tools for chemical sensing [90]. Mass is the universal property of matter. Therefore, piezoelectric microgravimetry (PM) can be applied for a wide range of sensing. All the acoustic wave devices make use of PZ materials for the generation of acoustic waves. Any change in the resonator mass disturbs the wave characteristics like velocity or amplitude, which is crucial in solving the sensing mechanism. For instance, velocity changes in the wave motion can be inferred from the resonant frequency change, which can be correlated with the physical quantity being measured.

Discovery of piezoelectricity by the Pierre and Paul-Jacques Curie brothers brought about an abrupt change in electronics and sensors in particular. Piezoelectricity consists in generating electric charges, upon application of mechanical stress, at opposite sides of a PZ crystal. In contrast, the converse PZ effect is caused by applying alternating current (a.c.) voltage to the PZ crystal [90]. This voltage generates an electric field which produces an electromechanical standing wave, so that the crystal resonates in time with its deformation [91–93]. The latter effect is being exploited in PZ sensors. Quartz is the PZ material most widely used for fabrication of these sensors [94]. By virtue of its intrinsic property, the temperature dependence of the quartz resonant frequency can be altered by selecting a suitable crystal cut angle and the wave propagation direction [95]. The acoustic wave PZ sensors can be classified according to the way of propagation of these waves. That is, these waves can be transmitted either through the bulk or on the surface of the PZ crystal [96]. In the latter case, the device is known as a surface acoustic wave (SAW) chemosensor [97]. Among these devices, the most common is a shear-horizontal surface acoustic wave (SH-SAW) chemosensor and a surface-transverse-wave (STW) chemosensor. If the wave propagates through the bulk of the PZ substrate then the chemosensor is called a thickness shear mode (TSM) bulk acoustic wave (BAW) chemosensor, and most commonly a quartz crystal microbalance (QCM). Thickness and surface shape of the resonator as well as angle of the crystal cut determine the resonant frequency as well as the mode of oscillation, respectively [91]. The resonant frequency is very sensitive to changes in the mass amount of the substance loaded onto or removed from its surface. An efficient transducer of a chemical or physical signal into a convenient electric signal must provide both qualitative and quantitative information. A PZ resonator is very well suited as a transducer due to its exceptional mass detectability reaching down to a subnanogram level. However, it does not provide qualitative analytical information as deposition of any species may alter its frequency. Selectivity here is the major criterion for successful transduction and, hence, sensing applications. In this direction, selectivity can be incurred by successfully combining molecular imprinting and PZ transduction.

The present section critically compiles the progress made in the last decade in the development of PZ chemosensors in combination with the MIP recognition. As a result of this combination, efficiency of the MIP-PZ sensors has been collectively enhanced in terms of binding, recognition and sensitivity. Herein, both covalent and non-covalent MIP-PZ chemosensors have been highlighted to signify the achievements. Besides, improvement in the sensing capabilities, strategies adopted and measurement complexity have been addressed.

3.2 Preparation and Immobilization of MIPs on Piezoelectric Transducers

MIPs gained great attention in the field of chemosensors owing to their high selectivity, appreciable stability as well as simple and inexpensive methods of preparation [98, 99]. Both sensitivity and selectivity of the MIP-PZ chemosensors is an attractive area of improvement [100].

The MIP-PZ chemosensors offer appreciably low LOD with the possibility of highly economical miniaturization and automation. In the past decade, progress in the field of these chemosensors has been widely reviewed [101, 102].

For consistent and reproducible operation it is important to control chemosensor features like thickness of the MIP coatings. Generally, MIPs were prepared and immobilized on PZ resonators by using two distinct procedures, namely in situ assembling on the resonator surface [103] or physical entrapment of the pre-prepared MIP particles in an inert polymer matrix attached to this surface [21]. Typically, thickness of the polymer is 20 nm to 5 μm. In situ immobilization of MIPs (see below in Sects. 3.2.1–3.2.4) can be accomplished by surface grafting, sandwich casting, electropolymerization, physical entrapment or chemical coupling [102].

3.2.1 MIP Immobilization by Surface Grafting

A very effective way of MIP film immobilization on a PZ transducer involves self assembling of an azo initiator monolayer by surface grafting [104, 105] (Scheme 2). Then this transducer is exposed to a solution of a functional monomer, cross-linker and template. Subsequently, polymerization is initiated by shining UV light. Finally, the template is extracted by rinsing the resulting MIP film with a suitable solvent solution. Typically, SAMs of the carboxy-group-terminated alkylthiols are used for surface modification of the resonator surface. 2-Ethyl-5-phenylisoxazo-lium-3-sulphonate or 2,2-azobis(2-amidinopropane) hydrochloride is mostly used to activate carboxy groups of these SAMs. This procedure improves the adherence of the MIP film to the PZ transducer surface [106]. Further, alkylthiol SAM allows

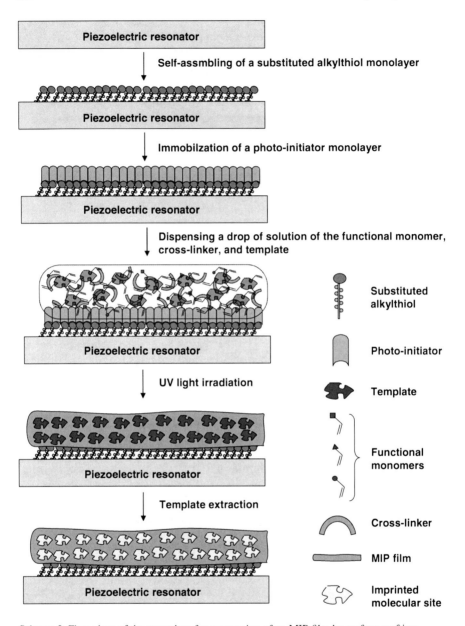

Scheme 2 Flow chart of the procedure for preparation of an MIP film by surface grafting

immobilizing of the MIP film and makes it compatible for simultaneous transmitive, reflective and other optical sensing. For instance, S-propranolol, a β-adrenergic blocker, has been determined using the MIP-PZ chemosensor with the MIP recognizing film immobilized by surface grafting [107].

3.2.2 MIP Immobilization by Sandwich Casting

Sandwich casting permits one to prepare an MIP film with uniform thickness [28, 106, 108, 109]. In this procedure, a drop of the solution containing a monomer, cross-linker, template and initiator is dispensed on the surface of a PZ transducer and covered with a microscope quartz slide. Then this assembly is exposed to UV light in order to initiate polymerization that results in a thin MIP film. The polymerization can be performed either on the activated immobilized initiator PZ transducer surface or on the bare transducer surface. For example, sialic acid has been determined with an MIP film immobilized on a platinum-film electrode of the quartz resonator using the former procedure [57]. That is, 1-butanethiol has been used for modification of the Pt surface. An indole-3-acetic acid plant hormone served as the template. An MIP-PZ chemosensor prepared that way operated reproducibly. That is, the coefficient of variation of the chemosensor performance was 9% for three different sensors.

3.2.3 MIP Immobilization by Electrochemical Polymerization

Electrochemical polymerization is a versatile procedure adopted for preparation of well-defined MIP films with consistent properties, like thickness, porosity and visco-elasticity [110, 111]. Electropolymerization is simple. It allows for formation of robust MIP films adhering well to the surface of the metal film electrode of the resonator. There is no need for the use of any polymerization initiator, UV light or heat. Instead, a polymer film can be deposited directly on the electrode under potentiodynamic, potentiostatic or galvanostatic conditions. Moreover, thickness of the MIP film can easily be controlled with the amount of charge transferred for the electropolymerization [111]. In addition, suitable selection of a solvent and supporting electrolyte can readily tune porosity, surface morphology and visco-elastic properties of the MIP film. For instance, sorbitol, slowly metabolized sugar alcohol, has been determined using a 1,2-phenylenediamine MIP film deposited onto a quartz resonator of QCM by electropolymerization [112]. Linear concentration range and LOD was 1–15 mM and 1 mM sorbitol, respectively.

3.2.4 MIP Immobilization by Physical Entrapment or Chemical Coupling

Physical entrapment or chemical coupling is a well-established procedure for MIP preparation. First, a complex is formed between a functional monomer and template in an appropriate solvent solution. Then the complex is immobilized by polymerization in excess of a cross-linker. Predominantly, free-radical polymerization thermally launched with a 2,2-azobis(isobutyronitrile) (AIBN) initiator, is performed. In the case of photo-radical polymerization, a benzophenone or acetophonone derivative is also used as the initiator [101]. Next, the template is extracted by rinsing the resulting MIP block with a suitably selected solvent solution. The bulk polymer prepared that way is dried, then thoroughly powdered and sieved to form MIP beads of a uniform size. These beads can subsequently be immobilized in a matrix of an inert polymer, like poly(vinyl chloride), PVC, on the resonator surface. Alternatively, the beads can be dissolved in an organic solvent to form a homogeneous solution. Drop- or spin-coating of this solution, if dispensed on the surface of the PZ substrate, can result in a thin film. Typically, 4-vinylpyridine (VPD), MAA and acrylamide derivatives are used as functional monomers.

In order to minimize swelling, MIP is often synthesized in a polar solvent, such as dimethylformamide (DMF) or ACN [113]. Most importantly, a (cross-linker)-to-(functional monomer) mole ratio affects the chemosensor performance. That is, a high (5:1) ratio enhances mechanical strength of an MIP film and favours formation of binding sites. But a low (1:5) ratio leads to a flexible MIP film of improved template memory and stronger adherence to the transducer surface [114]. Therefore, this ratio must be compromised for preparation of MIP of superior performance [114].

3.3   Analytical Applications of MIP Piezoelectric Chemosensors

In mass-sensitive devices, such as QCM, a quartz crystal resonator is used as the PZ transducer. This resonator is agitated for vibrations with a.c. voltage applied to conducting coatings, i.e. electrodes, typically deposited on opposite sides of the crystal wafer. In most of these devices only one of the electrodes is wetted by the test solution. Besides inducing oscillations of the resonator, this wetted electrode serves as the substrate for analyte adsorption, absorption, or either chemical or electrochemical reaction [90] affecting the electrode mass. The change in this mass determines the change in the resonant frequency, Δf. This frequency is opposite to the mass change, Δm, of a mass-loaded resonator, as the Sauerbrey Equation [94] predicts:

$$\Delta f_m = \frac{-2\,\Delta m\,f_0^2}{A\sqrt{\rho_q\,\mu_q}} \tag{2}$$

where f_0 is the fundamental resonant frequency of the unperturbed quartz resonator, μ_q is the shear modulus of the quartz crystal, ρ_q is the density of quartz and A is the acoustically active area of the crystal resonator. Typically for QCM resonators, the electrode area is made to match the acoustically active central crystal area.

Although QCM has originally been devised as a mass sensor for operation in vacuum or gas, it also appeared to be suitable for measurements of the mass and visco-elastic changes at a solid–liquid interface. That was possible due to elaboration of the dedicated oscillator circuitry [115]. The Sauerbrey Equation (2) was derived for resonator oscillations in vacuum. However, it also holds for solution measurements provided that (1) the deposited film is rigid and (2) it is evenly distributed over the entire acoustically active area as well as (3) the resonant frequency change due to the mass loading does not exceed 5% of the resonant frequency of the unperturbed resonator.

Numerous examples have been reported on the efficacy of the QCM studies to measure the changes in mass, density, and viscosity at the solid–liquid interface [116]. Moreover, QCM has been applied to study of adsorption or desorption of analytes, such as proteins, antibiotics, antibodies, nucleotides and even entire cells [117–123]. Detailed description of these studies extends beyond the scope of this section and, therefore, only representative examples of PZ transduction combined with MIP recognition are discussed below. To this end, an MIP-PZ chemosensor with a SAW resonator has been used for biomimetic recognition of biorelevant compounds [100]. Determination of analytes in the gas and liquid media can be distinguished among analytical applications of MIP-PZ sensors. In both these determination types, either steady-state (batch analysis) or dynamic (flow injection analysis, FIA) mode of measurement can be adopted. The latter mode is generally preferred because of lower sample consumption, shorter analysis time that is based on a transient signal measurement and online methods to carry out sometimes sophisticated operations of separation and preconcentration. Briefly, the sample is injected in FIA into a flowing carrier stream of reagents in gaseous or liquid phase [105]. As the injected zone moves downstream, the desired physical parameter, such as the change in spectrophotometric absorbance, fluorescence, current, potential, capacity, mass, etc. can be measured. Scheme 3 illustrates the QCM operation under FIA conditions.

3.3.1 Gas-Phase MIP Piezoelectric Chemosensors

For gas-phase sensors, both remarkable selectivity and very low LOD are important. Sensors featuring MIP recognition combined with SAW transduction can meet these requirements. The MIP-PZ chemosensors operating in gases are devised for two main applications, namely for indoor gas inspection and online monitoring of volatile organic compounds. The latter is essential to protect humans from threats of environmental atmospheric pollutants.

For instance, aromatic solvent vapours were determined with polyurethane MIPs combined with SAW transducers [124]. That is, first, the hydrophilic quartz surface of SAW was hydrophobized with N,N-dimethylaminotrimethylsilane. Then a solution for polymerization was prepared by mixing functional monomers, such as 4,4′-dihydroxydiphenyldimethylmethane, 4,4′-diisocyanatodiphenylmethane and 30% 2,4,4′-triisocyanatodiphenylmethane, with the 1,3,5-trihydroxybenzene cross-linker in the ethyl acetate or ethanol template used also as the solvent for polymerization. Subsequently, the hydrophobized resonator surface was spin-coated with an aliquot of this solution. Finally, the free-radical polymerization has been initiated thermally to form a polyurethane MIP film. The desired vapour concentration and relative humidity of the analyte were achieved by mixing dry air and saturated steam with solvent vapours generated with thermoregulated bubblers.

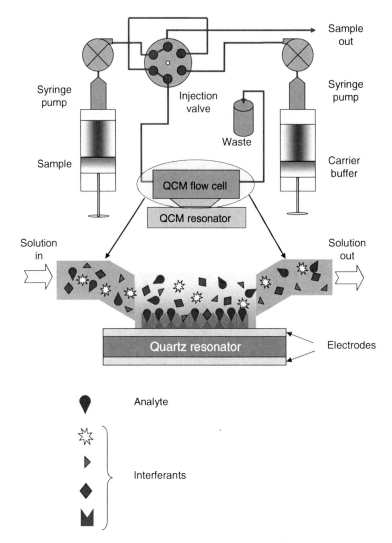

Scheme 3 Diagram of a quartz crystal microbalance (QCM) setup operating under flow injection analysis (FIA) conditions

The chemosensor response to the solvent, which was used for imprinting, was higher than that to other solvents. For instance, the MIP film, prepared using ethyl acetate as both the polymerization solvent and template, more strongly bound ethyl acetate than the ethanol interferant. The amount of the cross-linker played a decisive role in discrimination of this interferant. The sensitivity and selectivity of the 433-MHz MIP-SAW chemosensor and 10-MHz MIP-QCM chemosensor were compared. It appeared that affinity for 1,2-xylene of the SAW chemosensor using the 1,2-xylene imprinted MIP film was pronounced. That is,

LOD for 1,2-xylene at the SAW chemosensor was as low as 0.1 μL L^{-1} at the appreciable resonant frequency change of 6 Hz. In contrast, LOD for 1,2-xylene at the QCM chemosensor, coated with a similar film, was rather high and equal to 4 μL L^{-1} at the resonant frequency change of merely 0.7 Hz. Apparently, 40-fold lower LOD for the SAW chemosensor indicates its prospective application for mass-sensitive gas sensing.

A gas-phase chemosensor using the MIP film and QCM transducer has been devised for determination of a 2-methylisoborneal (MIB), an odorant produced by algae in natural waters [125]. For that purpose, a MAA functional monomer, EGDMA cross-linker, 2,2′-azobis(2,4-dimethyl)valeronitrile (V-65) initiator and MIB template were dissolved in hexane and the resulting solution used for the synthesis of an MIP film. Accordingly, a drop of this solution was dispensed on a gold electrode of the quartz resonator and thermo-radical polymerization initiated at 40 °C. Finally, the MIB template was removed by rinsing the film with ethanol. Nitrogen was used as the carrier gas at a flow rate of 0.2 L min^{-1}. The MIP film selectively recognized MIB with the response, which was 1.1–1.3 times higher than those of the NIPs at an LOD of 200 μg L^{-1} MIB. Affinity of this chemosensor towards other odorants, such as hexane, geosmin, terpineol, citronellol and β-ionone, was substantially lower. Moreover, MIB produced by microorganisms was determined with this chemosensor in real time. However, the difference between the responses of the frequency change for the MIP- and NIP-film coated resonator was insignificant. Subsequently, detectability of this chemosensor has been improved by coating the resonator surface with a nylon film [126]. On top of this film, an MIP film with binding sites for MIB has been deposited to boost the selectivity of the chemosensor towards MIB. At this configuration, LOD was 10 μg L^{-1} MIB, being 20-fold lower than that at the former configuration [125].

In other studies, MIP-PZ chemosensors were used for determination of volatile organic solvents, such as toluene and 1,2-xylene [113], using QCM transducers. Preparation of receptive MIPs involved co-polymerization of MAA and divinyl-benzene (DVB) in toluene or 1,2-xylene serving both as the solvent and template, in the presence of the benzoyl peroxide initiator. The thermo-radical polymerization was then performed at 80 °C. The MIP block obtained was ground to a fine powder and dried at 80 °C to remove the template. Next, the MIP powder was mixed in a 5% poly(methylmethacrylate) (PMMA) acetone solution. With the resulting suspension, a polymer film was then spin-coated on the silver electrode of the quartz resonator. The toluene or 1,2-xylene vapour was determined using dry nitrogen as the carrier gas at a flow rate of 50 mL min^{-1}. Predominantly the chemosensor sorbed either toluene or 1,2-xylene, if prepared from the respective solvent, and this sorption was reversible (Fig. 3). However, the response time of 60 min was remarkably long. Presumably, it was due to the presence of the PMMA matrix that might impede the analyte diffusion to the recognition sites in MIPs.

A TNT explosive was determined in the gas phase using an MIP-QCM chemo-sensor [127]. The MIP preparation procedure involved dissolution of an appropriate amount of the TNT template, MAA monomer, EGDMA cross-linker and Irgacure$^{®}$ 369 [2-benzyl-2-dimethylamino-1-(4-morpholinophenyl)-butanone-1] photoinitiator

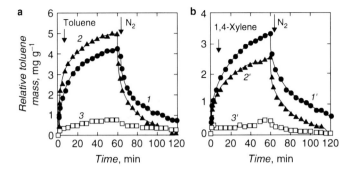

Fig. 3 Relative mass change with time of the QCM sensors featuring MIP–PMMA and NIP–PMMA blend films at 30 °C for sorbed vapours of (**a**) toluene and (**b**) 1,4-xylene. At zero time the toluene or 1,4-xylene vapour is passed and at the time of 60 min nitrogen gas is passed; (*1, 1′*) 1,4-xylene-MIP–PMMA, (*2, 2′*) toluene-MIP–PMMA, (*3, 3′*) NIP–PMMA (adapted from [113])

of Ciba® in chloroform, in the presence of the poly(vinyl alcohol) emulsifier. A drop of this solution was dispensed on the quartz resonator and photo-radical polymerization initiated by shining UV light. The template was then removed by consecutive rinsing of the resulting MIP film with water, chloroform, methanol, and toluene. For the TNT determination under flow conditions, dry nitrogen was used as the carrier gas at the flow rate of 10–20 mL min^{-1}. Affinity towards gaseous TNT of this chemosensor was pronounced, manifesting itself with the TNT uptake of 150 pg μg^{-1} h^{-1}. Advantageously, there was no cross-reactivity of dinitrotoluenes (DNTs). Moreover, the MIP films for TNT determination were prepared using five other functional monomers and three different solvents. The sensing results appeared to be the best for the poly(acrylic amide) MIP film synthesized in chloroform. Furthermore, an MIP film for selective recognition of DNT has been prepared.

3.3.2 Liquid-Phase MIP Piezoelectric Chemosensors

Most SAW transducers are inadequate for operation in liquids because most types of surface acoustic waves are completely damped in this viscous environment. Therefore, QCM transducers are used instead. For successful operation of an MIP-QCM chemosensor in liquid, the MIP film should be sufficiently stable with respect to dissolution and peeling off from the resonator surface. Moreover, this film should be relatively rigid, neither shrinking nor swelling in the test solution.

An MIP film attached to a quartz resonator of the QCM transducer has been used successfully to devise a chemosensor for determination of disinfection by-products of haloacetic acids in drinking water [128]. An ACN solution of trichloroacetic acid (TCAA), VPD, EGDMA and AIBN, used as the template, functional monomer, cross-linker and initiator, respectively, has been polymerized thermo-radically. Next, the resulting MIP was spin-coated on a quartz resonator to form a thin film. The TCAA template was then extracted by rinsing the film with water. This TCAA chemosensor selectively discriminated other haloacetic acids, such as dichloro-, monochloro-, tribromo-, dibromo- and monobromoacetic acid. Unfortunately, its response time of ~10 min was long and its LOD of 20 μg L^{-1} TCAA was high compared to those of MIP electrochemical sensors for the haloacetic acids (see below in Sect. 4.4.1).

Hexachlorobenzene has been determined selectively using an MIP-QCM chemosensor [129]. A ~20-μL sample of a dichloromethane (DCM) solution of the 1,4-diacryolyloxybenzene functional monomer, benzyl methacrylate cross-linker, 1,5-bis(2-acetylaminoacryloyloxy)pentane initiator and hexachlorobenzene template was spin-coated on the resonator surface. Then the film was exposed to UV light to initiate photo-radical polymerization. Next, the template was extracted from the resulting film with DCM. The chemosensor thus prepared was capable of detecting hexachlorobenzene at a picomolar concentration level with the appreciably short response time of 10 s. Affinity of this chemosensor towards TNT, anisole, chlorobenzene, benzene and cyclohexane has been examined. A relative affinity factor, Δf_{max}(hexachlorobenzene)/Δf_{max}(anisole), was high, equalling 5.1. The frequency change of both the MIP and NIP film for hexachlorobenzene was 38-fold higher than that for the other compounds studied.

An MIP-QCM chemosensor for determination of carbamate pesticides, such as carbaryl, has been devised [130]. The chemosensor featured a thin film of PVC, containing carbaryl-imprinted polymer microspheres, which was deposited on top of the gold-sputtered quartz crystal transducer. The microspheres were prepared by thermo-induced co-polymerization in ACN of MAA and EGDMA, used as the functional monomer and cross-linker, respectively, in the presence of carbaryl and AIBN serving as the template and initiator, respectively. The chemosensor performance was evaluated for determination of carbaryl exhibiting the linear concentration range of 10–1000 ng mL^{-1} in the Britton–Robinson buffer of pH = 8.0. This chemosensor was highly stable. It selectively discriminated carbaryl from its structural counterparts, such as carbofuran and aldicarb, with LOD of 1.25 ng mL^{-1} carbaryl.

Biogenic amines, such as histamine [131], adenine [132], dopamine [133] and melamine [134], have been determined using chemosensors combining MIP recognition and PM transduction at QCM. Electronically conducting MIPs have been used in these chemosensors as recognition materials. Initially, functional electroactive bis(bithiophene)methane monomers, substituted either with the benzo-18-crown-6 or 3,4-dihydroxyphenyl, or dioxaborinane moiety, were allowed to form complexes, in ACN solutions, with these amines as templates. Subsequently, these complexes were oxidatively electropolymerized under potentiodynamic conditions. The resulting MIP films deposited onto electrodes of quartz resonators were washed with aqueous base solutions to extract the templates.

The histamine [131], adenine [132] and dopamine [133] amines are electroactive in the positive potential range, in which the thiophene is electropolymerized. Therefore, these amines could be oxidized at the electrode surface in the course of deposition of the MIP film. That way, products of these oxidations might be available in the electrode vicinity for imprinting rather than the desired pristine amines themselves. Moreover, these products might foul the metal electrode surface by adsorption [135] thus most likely deteriorating adherence of the MIP film itself. Therefore, a conducting barrier layer of poly(bisthiophene) was first deposited by electropolymerization onto the electrode surface prior to the MIP film deposition. Advantageously, this layer conducted electrons well, enabling MIP film electropolymerization on its top on the one hand and effectively preventing electro-oxidation of the amines on the other.

Thickness of the barrier layer, optimized at ~220 nm [133], played a crucial role with respect to the chemosensor sensitivity, selectivity and LOD. So, eventually, the chemosensor architecture comprised a gold-film electrode, sputtered onto a 10-MHz resonator, coated with the poly(bithiophene) barrier layer, which was then overlaid with the MIP film. This architecture enabled selective determination of the amine at the nanomole concentration level. LOD for histamine was 5 nM and the determined stability constant of the MIP–histamine complex, $K_{MIP} = 57.0$ M^{-1} [131], compared well with the values obtained with other methods [53, 136, 137]. Moreover, due to the adopted architecture, the dopamine chemosensor could determine this amine with the stability constant for the MIP–dopamine complex, $K_{MIP} = (44.6 \pm 4.0) \times 10^6$ M^{-1} and LOD of 5 nM [133], which is as low as that reached by electroanalytical techniques [138]. The MIP-QCM chemosensor for adenine [132] also featured low, namely 5 nM, LOD and the stability constant determined for the MIP–adenine complex, $K_{MIP} = (18 \pm 2.4) \times 10^4$ M^{-1}, was as high as that of the MIP–adenine complex prepared by thermo-induced co-polymerization [139]. The linear concentration range for determination of these amines extended to at least 100 mM.

Melamine is electroinactive in the potential range of the thiophene electropolymerization [134]. Therefore, the MIP film was directly deposited on the electrode of the quartz resonator in the absence of any barrier layer by electropolymerization of the preformed melamine complex with the bis(bithiophene)methane derivative bearing the benazo-18-crown-6 substituent. The presence of bithianaphthene cross-linking monomer and a porogenic ionic liquid in the solution used for this electropolymerization greatly improved rigidity and enhanced porosity of the MIP film. The template-extracted MIP film was characterized by XPS, UV–visible spectroscopy and scanning electrochemical microscopy (SECM) to evaluate rebinding of melamine used as the analyte. The chemosensor performance, examined with PM under FIA conditions, showed a linear concentration range of 5 nM to 1 mM with low, namely 5 nM melamine, LOD (Fig. 4). This value is appreciably low compared to that reached by the MIP electrochemical sensor [140].

All the MIP-QCM chemosensors for biogenic amines above described were quite selective with respect to functionally and structurally similar compounds (inset to Fig. 4).

An MIP-QCM chemosensor has been devised for determination of thymine [141]. A new functional monomer, vis. methacryloylamidoadenine (MA-Ade), as well as the EDGMA cross-linker and AIBN initiator have been used for preparation of the thymine-imprinted polymer film by photo-induced polymerization. This film was deposited on the QCM resonator, featuring homogeneous binding sites for

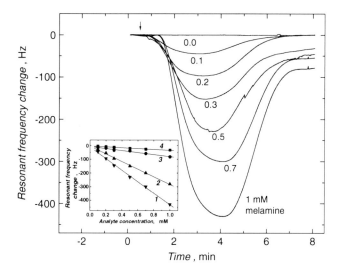

Fig. 4 Resonant frequency changes with time due to repetitive FIA melamine injections, for the MIP-QCM chemosensor. Melamine concentration is indicated with number at each curve. *Inset* shows FIA calibration plots for (1) melamine and its interfering compounds, such as (2) ammeline, (3) cyanuric acid, and (4) cyromazine. Volume of the injected sample solution was 100 μL. The flow rate of the 1 mM HCl carrier solution was 35 μL min^{-1}. The MIP film was prepared by electropolymerization of 0.3 mM bis(2,2'-bithienyl)-benzo-[18-crown-6]methane functional monomer and 0.3 mM 3,3'-bis[2,2'-bis(2,2'-bithiophene-5-yl)]thianaphthene cross-linking monomer, in the presence of 0.1 mM melamine, in the trihexyl(tetradecyl)phosphonium tris(pentafluoroethyl)-trifluorophosphate ionic liquid:ACN (1:1 v/v) solution, which was 0.9 mM in trifluoroacetic acid (pH = 3.0). The melamine template was extracted from the MIP film with 0.01 M NaOH before the determinations (adapted from [134])

thymine ($K_{MIP} = 1.0 \times 10^5$ M^{-1}) and heterogeneous binding sites for uracil. This chemosensor was selective towards uracil, ssDNA and ssRNA.

An array of MIP-QCM chemosensors, integrated to form an artificial tongue, have been fabricated for FIA determination of bitter and sweet taste-causing compounds, such as quinine and saccharine, respectively, in bitter drinks, like commercial tonic water [142]. The recognition MIP particles were prepared by thermally induced co-polymerization of MAA, EGDMA and AIBN in the presence of the quinine or saccharine template. The resulting chemosensor exhibited high sensitivity of 2.04 and 32.8 mg L^{-1} as well as the appreciably wide linear dynamic concentration range of 10–1080 mg L^{-1} and 51–3420 mg L^{-1} for quinine and saccharine, respectively. Most substances commonly found in drinks, except sucrose, interfered insignificantly in the determination of these analytes. This chemosensor selectively discriminated quinine and saccharine from likely interferants, such as caffeine, vanillin and citric acid, with the sample throughput of 17–19 determinations per hour of high repeatability.

Caffeine has been determined extensively with the MIP-QCM chemosensors [39, 143–146]. MIP preparation involved, first, complexation of the caffeine template by the MAA functional monomer in chloroform. Next, this complex was thermo-radically polymerized with the excess of the EGDMA cross-linker in the presence of the AIBN initiator. A solid polymer block obtained that way was thoroughly powdered and sieved to yield fine beads. These beads were immobilized in an inert PVC or poly(cyanoacrylate ester) film on a quartz resonator. The caffeine template was removed by rinsing the MIP film with the methanol–acetic acid (9:1 v/v) mixed solvent solution. The template-free MIP beads were capable of rebinding caffeine with high affinity and selectivity. Moreover, LOD of 5 nM for caffeine was significantly low. Analytical parameters of other MIP-QCM chemosensors in application to caffeine determination in real samples are summarized in Table 5.

A molecularly imprinted polypyrrole film coating a quartz resonator of a QCM transducer was used for determination of sodium dodecyl sulphate (SDS) [147]. Preparation of this film involved galvanostatic polymerization of pyrrole, in the presence of SDS, on the platinum-film-sputtered electrode of a quartz resonator. Typically, a 1-mA current was passed for 1 min through the solution, which was 0.1 mM in pyrrole, 1 mM in SDS and 0.1 M in the TRIS buffer (pH = 9.0). A carbon rod and the Pt-film electrode was used as the cathode and anode, respectively. The SDS template was then removed by rinsing the MIP-film coated Pt electrode with water. The chemosensor response was measured in a differential flow mode, at a flow rate of 1.2 mL min^{-1}, with the TRIS buffer (pH = 9.0) as the reference solution. This response was affected by electropolymerization parameters, such as solution pH, electropolymerization time and monomer concentration. Apparently, electropolymerization of pyrrole at pH = 9.0 resulted in an MIP film featuring high sensitivity of 283.78 Hz per log(conc.) and a very wide linear concentration range of 10 μM to 0.1 mM SDS.

L-Aspartic acid, pervasive in biosynthesis, has been recognized enantioselectively with an MIP-QCM chemosensor [148]. The MIP film was prepared by the galvanostatic L-aspartic acid templated electropolymerization of pyrrole on the gold-film electrode of the quartz resonator. First, a current of constant density of 10 μA cm^{-2} was passed through aqueous electrolyte solutions of different concentrations of L-aspartic acid and pyrrole. Next, the MIP films obtained were overoxidized to remove the L-asparate anionic template by applying a constant current density of 25 μA cm^{-2} at pH = 7.0. The overoxidized template-free polypyrrole films served as recognition elements. The resulting chemosensor was much more sensitive to L- than to D-aspartic acid. That is, surface mass density of the MIP film increased by 1.0 μg cm^{-2} when L-aspartic acid was injected to the FIA carrier KCl–HCl solution (pH = 1.6) instead. However, density of the same film increased merely by 0.05 μg cm^{-2} when the same amount of D-aspartic acid was injected (Fig. 5). Apparently, this enantioselective MIP-QCM chemosensor was 20-fold more selective with respect to the templating L-aspartic acid than to its non-templating D-aspartic acid counterpart. Electropolymerization parameters, such as pH and composition of the electrolyte solution, strongly influenced the enantioselectivity.

An MIP film was deposited on a quartz resonator of the QCM transducer for determination of pyrimethamine, a medicine used to cure protozoal infections [149]. MIP particles for this chemosensor were prepared by thermo-radical polymerization, in the ACN solution, of MAA, EGDMA, AIBN and pyrimethamine used as

Table 5 Comparison of performance of quartz crystal microbalance chemosensors using molecularly imprinted polymer for determination of caffeine

Application	Linear concentration range	LOD	Recovery (%)	RSD (%)	References
Coffee, tea	10–1000 ng mL^{-1}	5.5 ng mL^{-1}	90–104[a]; 94–109[b]	5	[143]
Not applied	10^{-9} to 1 mg mL^{-1}	5.9 × 10^{-11} mg mL^{-1}	Not determined	9	[144]
Not applied	10^{-7} to 10^{-3} mg mL^{-1}	3.76 × 10^{-11} mg mL^{-1}	52–122[a]	17	[145]
Human serum or urine	5 nM to 100 μM	5 nM	96–106[b]	1.4–2.1	[39]

[a]Synthetic samples
[b]Real samples

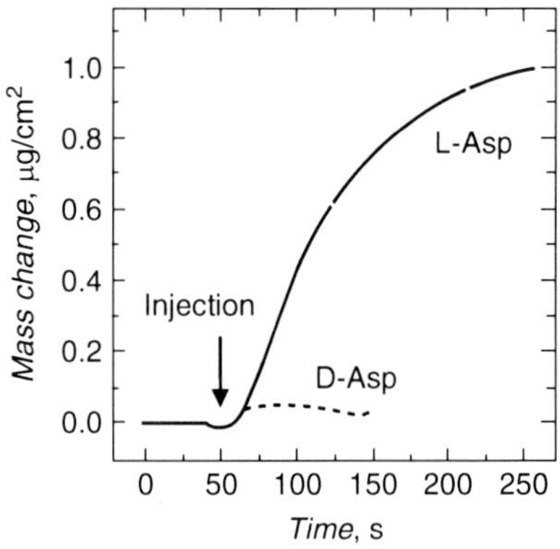

Fig. 5 Mass change, under FIA conditions, of the MIP film (imprinted with L-aspartic acid) due to injection of a 10-mM sample of L-aspartic acid or D-aspartic acid into the KCl–HCl carrier solution (pH = 1.6). The MIP film was deposited on a gold-coated quartz crystal resonator held at a constant potential of −0.4 V (adapted from [148])

Fig. 6 Dependence of the resonant frequency change of the quartz crystal microbalance modified with the molecularly imprinted polymer film, MIP-QCM, chemosensor on the pyrimethamine concentration. A quartz resonator was coated with the film of (1) MIP or (2) NIP (adapted from [149])

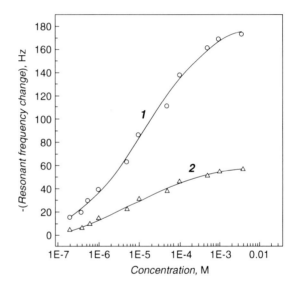

the cross-linker, functional monomer, initiator and template, respectively. The solid MIP block obtained in that way was powdered thoroughly and then sieved to yield fine particles. Next, these particles were rinsed with a 10% acetic acid solution of benzyl alcohol to remove the pyrimethamine template. Finally, the particles were immobilized on the resonator surface in a PVC film. The chemosensor was remarkably stable under harsh environmental conditions, such as high temperature, presence of organic solvents, strong bases or acids, etc. The measured resonant frequency change decreased linearly with increase of pyrimethamine concentration in the range of 0.6–100 μM (Fig. 6). LOD for pyrimethamine was 0.2 μM.

A daminozide (DZ) pesticide, a plausible carcinogen, has been determined using an MIP-QCM chemosensor [150]. Preparation of MIP particles used for fabrication of this chemosensor involved thermo-radical co-polymerization, in methanol solution, of the MAA monomer and EGDMA cross-linker in the presence of the DZ template, and AIBN initiator. The MIP block obtained was ground scrupulously and then sieved to yield fine beads. Next, these beads were suspended and then refluxed in 10% NaOH for 12 h to remove the DZ template. Subsequently, the beads were immobilized on a quartz resonator in a PVC film. The size distribution of the beads was determined by scanning electron microscopy (SEM) and atomic force microscopy (AFM) imaging. The resonant frequency decrease was linear with respect to the increase of the DZ concentration in the range of 1–100 mg mL^{-1} at LOD of 5 ng mL^{-1} DZ.

Small peptide mammalian hormones, such as oxytoxin and vasopressin, have been discriminated using an MIP-QCM chemosensor [83]. For this discrimination, a new cross-linking monomer, namely (N-Acr-L-Cys-NHBn)$_2$, has been synthesized. The surface of a gold-film electrode of the quartz resonator was first modified with a film of this monomer. Next, the acrylic acid and acrylamide functional monomers, N-benzylacrylamide initiator, and the oxytoxin or vasopressin template were dissolved in an ACN–water mixed solvent solution. A drop of this solution was dispensed on the modified Au electrode of the resonator. Subsequently, the resonator was UV light illuminated for 6 h for photo-radical polymerization. The resulting MIP film was rinsed with the phosphate buffer (pH = 3.0–4.0) to remove the template. This film was able to differentiate oxytocin from vasopressin at a nanomole concentration level. Values of the stability constant of the MIP complex of oxytocin and vasopressin were $K_{MIP-oxytocin} = 9.0 \times 10^7$ M^{-1} and $K_{MIP-vasopressin} = 5.0 \times 10^7$ M^{-1}, respectively. LOD for the templating peptides was 10–100 times lower, in terms of the K_{MIP} values, compared to those for similar non-templating peptides.

Determination of α-bilirubin, a lypophilic bile pigment, is clinically important because disorders in the α-bilirubin metabolism may lead to certain diseases, like hyperbilirubinaemia. MIP films with binding sites templated by α-bilirubin have been prepared by photo-grafted polymerization [105]. To that end, an allyl thiol monolayer was first self-assembled from a methanol solution on a gold-film electrode of the quartz resonator of the QCM transducer. Next, a thin film of the benzophenone initiator was drop coated as a graft on this monolayer. On top of this film, an MIP film was then drop coated using a solution of the α-bilirubin template, VPD functional monomer and DVB cross-linker. Subsequently, this film was exposed to UV light to initiate photo-radical polymerization. Finally, the MIP-film-coated resonator was immersed in 10 mM NaOH to remove the template. α-Bilirubin was determined under FIA conditions. A mixed ACN–water carrier solution was flown at the rate of 1.0 mL min^{-1}. The resonant frequency change of the chemosensor linearly decreased with the increase of the α-bilirubin concentration in the range of 0.46–11.0 mg dL^{-1}. The chemosensor response was stable for 7 months. Its sensitivity with respect to α-bilirubin was 31.2-fold higher than to its biliverdin analogue. Relatively, however, LOD for α-bilirubin of 0.46 mg dL^{-1} was high and the response time of 41 min was long.

An MIP-QCM chemosensor was devised for determination of a pirimicarb pesticide, a likely mutagen [151]. Three different types of MIP particles have for that purpose been prepared by bulk polymerization in chloroform or precipitation polymerization in chloroform or ACN. A solution of pirimicarb, MAA, EGDMA and AIBN serving as the template, functional monomer, cross-linker and initiator, respectively, was thermo-radically polymerized at 60 °C. The resulting MIP particles were rinsed with a methanol–(acetic acid) (9:1 v/v) mixed solvent solution to remove the template and then immobilized in a PVC film on a gold-film electrode of a quartz resonator. Performance of this MIP film was characterized by cyclic voltammetry (CV) and electrochemical impedance spectroscopy (EIS). The linear concentration range was 5 μM to 4.7 mM and LOD was 0.5 μM pirimicarb. Moreover, pirimicarb was determined onsite in contaminated vegetables.

An indole-3-acetic acid (IAA) plant hormone has been determined using an MIP-QCM chemosensor [152]. An MIP recognition film for this chemosensor was prepared by sandwich casting followed by photo-radical polymerization. For that purpose, a chloroform solution of the IAA template, EGDMA cross-linker, N,N-dimethylaminoethyl methacrylate functional monomer and AIBN initiator was dispensed onto a Pt-film electrode of the quartz resonator and covered with a trimethylchlorosilane-modified quartz slide. Next, UV light was shown for 5 min to initiate photo-radical polymerization. Affinity to IAA of the fabricated MIP-QCM chemosensor was 12.5–25 times higher than that to structurally similar compounds, such as indole-3-butyric acid, indole and indole-3-ethanol. The resonant frequency change of the chemosensor linearly decreased with the IAA concentration increase in the range of 10–200 nM.

An MIP-QCM chemosensor was devised for determination of sialic acid [153]. For this chemosensor, the MIP film was prepared by sandwich casting followed by photo-radical polymerization. For that, 1,4-vinylbenzeneboronic acid and/or N,N,N-trimethylaminoethyl methacrylate were used as functional monomers along with EGDMA and 2,2'-azobis(2.4-dimethyl valeronitrile) (V-65) as cross-linker and initiator, respectively. The dynamic concentration range was linear between 20 and 250 nM sialic acid. The chemosensor was selective to this analyte over interfering similar compounds, like galacturonic acid, with the relative response to galacturonic acid equal to 30% of that to sialic acid.

MIP films, applied to a QCM transducer, have been employed for chiral recognition of the R- and S-propranolol enantiomers [107]. MIP films were prepared for that purpose by surface grafted photo-radical polymerization. First, a monolayer of 11-mercaptoundecanoic acid was self-assembled on a gold electrode of the quartz resonator. Then, a 2,2'-azobis(2-amidinopropane) hydrochloride initiator (AAPH), was attached to this monolayer. Subsequently, this surface-modified resonator was immersed in an ACN solution containing the MAA functional monomer, enantiomer template and trimethylolpropane trimethacrylate (TRIM) cross-linker. Next, the solution was irradiated with UV light for photo-polymerization. The resulting MIP-coated resonator was used for enantioselective determination of the propranolol enantiomers under the batch [107] conditions and the FIA [107] conditions with an aqueous-ACN mixed solvent solution as the carrier. The MIP-QCM chemosensor was enantioselective to S-propranolol at concentrations exceeding 0.38 mM [107].

A biomimetic MIP recognition material was devised for determination of a paracetamol (also known as acetominphen) pain reliever using a QCM transducer [109]. To this end, MIP was prepared with two different functional monomers, namely VPD and MAA. An EGDMA cross-linker and AIBN initiator were also present in the solution. The thermo-radical polymerization was performed at 60 °C. The resulting MIP powder was dispersed in a PVC matrix and then drop coated onto a quartz resonator. A linear decrease of the resonant frequency change with the increase of the paracetamol concentration ranged from 50 nM to 10 mM at LOD of 50 nM paracetamol. The chemosensor was successfully applied for the paracetamol determination in real samples, such as human serum and urine.

3.4 Analytical Characteristics of MIP Piezoelectric Chemosensors

Typically for a biosensor, a biological molecular recognition material is immobilized on a transducer surface to generate a detection signal. The same principle is adopted in an MIP-PZ chemosensor except that the biological receptor is replaced by an MIP artificial receptor operating according to the biomimetic mechanism. In MIP-PZ chemosensors, the change in the resonant frequency signals the analyte concentration. The analytical characteristics, such as sensitivity, selectivity and LOD, are affected by the properties of MIPs as well as the sensing capabilities of PZ transducers. Importantly, MIPs in application to PZ sensing should be sufficiently rigid and capable of resisting solvent swelling and shrinking. Only under those conditions can the frequency changes be ascribed unambiguously to the mass changes resulting from the analyte ingress or egress rather than, for instance, to the change in the MIP elasticity or solution viscosity.

3.4.1 Sensitivity and Detectability of MIP Piezoelectric Chemosensors

In principle, sensitivity of an MIP-PZ chemosensor can be improved in two ways, i.e. either by using a resonator of higher resonant frequency or by increasing the number of recognition sites in MIP.

Although SAW devices can operate in a gigahertz frequency range in gases revealing impressive sensitivity, their oscillations, for most kinds of surface acoustic waves, are completely damped in liquids. Therefore, the use of higher oscillation harmonics (overtones) of TSM-BAW resonators, capable of oscillation in liquids, would be an appealing alternative for sensitivity improvement. However, it is difficult to obtain a sufficiently stable detection signal of frequency at higher harmonics because the maximum mass load that can be measured under these conditions is very low.

Bulk rather than film imprinting can be adopted towards increasing the number of the recognition sites in MIP. In the former imprinting, the template molecules determine both the diffusion pathways and interaction sites. Application of a bulk

MIP material is, therefore, advantageous in terms of higher analyte accumulation and, hence, higher sensitivity provided that the analyte diffusional transport within the material is not hindered.

Detectability of analytes of small molecules and low molecular mass can be increased by preconcentration of the analyte in the test solution. Then, a reasonable amount of the analyte should bind to the receptor sites of MIP showing a pronounced decrease in the measured resonant frequency change.

Interestingly, the analytes of smaller sizes than that of the functional monomer used for MIP preparation were determined based on the MIP-PZ sensing. For instance, acetaldehyde was determined by using the methacrylate functional monomer for MIP preparation [154]. The fabricated MIP-PZ chemosensor was 11-fold more selective for acetaldehyde than for acetone.

Application of MIP chemosensors imprinted with two different templates can also improve detectability. For instance, polycyclic aromatic hydrocarbons (PAHs) have been determined using polymers molecularly imprinted with two different PAH templates [155]. Compared with MIPs of a single-template imprint, the detection signal of the resonant frequency change was enhanced by a factor of five and LOD for pyrene was as low as 60 $\mu g \ L^{-1}$. This two-template largely improved detectability can be attributed to easier accessibility of the recognition sites through different diffusion pathways.

3.4.2 Selectivity and Response Time of MIP Piezoelectric Chemosensors

Affinity of MIP towards the target analyte should be examined prior to fabrication of the chemosensor. Batch binding assays are used to test selectivity of suitable MIPs. Especially, affinity of MIP to compounds, which are structurally related to the target analyte, should be tested. If MIP binds similarly with these compounds as the template, then cross-reactivity is manifested [156]. This effect was exploited for determination of adenine and its derivatives with the use of MIP templated with 9-ethyladenine. Nevertheless, the cross-reactivity, if undesired, can be avoided by suitable sample pretreatment, e.g. by interferant extraction with a supported liquid membrane (SLM) coupled to the MIP-PZ chemosensor. The Fluoropore™ membrane filter of submicrometre porosity can serve that purpose. That way, this membrane holds interferants, thus eliminating the matrix effect. The SLM-involving determination procedure is cheaper than traditional laborious sample pretreatment used to remove the interfering substances. For instance, caffeine [143] and vanillin [157] in food samples have been determined using this procedure.

Selectivity of the MIP-PZ sensors can be improved by separately optimizing the binding and determination medium. MIPs combined with PZ transducers are unique in selectivity with respect to enantiomers. The proper choice of functional monomers used for imprinting can improve this selectivity at a very low LOD. For instance, paracetamol has been determined with the MIP-QCM chemosensor using VPD and MAA as the MIP functional monomers [109]. Affinity of this chemosensor to paracetamol was increased when VPD and MAA were used in a mixture rather than separately. Interference of the phenacetin and antifebrin analogues was negligible at their concentrations below 10 μM.

Discrimination of similar compounds with less pronounced functionality most profoundly substantiates the selectivity of MIP-PZ chemosensors. For example, aromatic solvents, such as benzene, toluene and 1,2-xylene, have been determined successfully in a mixture with the use of the polystyrene MIP beads [158]. When imprinted with toluene, these beads bound this solvent most favourably by showing a very low affinity towards 1,2-xylene and benzene. Apparently, the second methyl group in the *ortho* position made the 1,2-xylene analyte unfavourable to enter the recognition sites imprinted with toluene. Notably, the MIP-PZ chemosensor was capable of selective discrimination of analytes that merely differed by a single methyl substituent. Similarly, nandralone, an anabolic steroid occurring naturally at low concentration in humans but used for illegal doping in sports, has been discriminated selectively from its testerone and episterone analogues [118]. This result is crucial in demonstrating the selectivity of MIP since nandralone differs from these analogues by just one methyl substituent.

Moreover, both selectivity and response time can be improved by employing an array of MIP-PZ sensors. For instance, six resonators of a QCM transducer, each coated with different MIP film, comprised a chemosensor array, which was used for the online monitoring of the water, ethyl acetate, limonene and 1-propanol lead analytes in a composting process [103]. The resonant frequency changes with time showed a distinct pattern. That is, the frequency for 1-propanol initially decreased with the increase of the concentration of ethyl acetate while limonene was detected in the later stages of the composting. The compost degradation was independently monitored with the (gas chromatography)-(mass spectrometry) (GC–MS) reference measurements. Both the MIP-PZ and GC–MS measurements showed the same time profiles of relative detection signals for all four analytes.

PZ transducers coupled to MIP recognition elements are very effective in improving LOD. Determination of atrazine can serve as an illustrative example [159, 160]. For the MIP-PZ chemosensor, LOD for atrazine was as low as 2 μM, being six times lower than that for an electrochemical chemosensor (see Sect. 4.4 below).

3.5 Conclusions and Perspectives of MIP Piezoelectric Chemosensors

The state of the art of MIPs has affected the progress in MIP-PZ chemosensors in the last decade. Generally, chemosensor selectivity has collectively been enhanced by combining MIP recognition and PZ transduction.

Although PZ gas chemosensors are susceptible to artefacts, like temperature drift as well as humidity and pressure changes, their essential virtues, consisting in low cost, ease of measurements and manufacturing, make them prospective for application in integrated portable devices.

It appeared, however, that the progress in development of MIP-PZ chemosensors is considerably slower than that of optical and electrochemical MIP chemosensors [101]. Partially, this relatively slow progress may be due to difficulties in immobilization of MIPs on PZ transducers and much more effort-demanding transduction. Hopefully, recent remarkable progress in acoustic sensing technology is promising for fabrication of devices that can advantageously be integrated with the MIP recognition elements to provide a new generation of MIP-PZ chemosensors for determination of biologically and environmentally important compounds both inside and outside the laboratory.

4 Electrochemical Sensors

4.1 *Introduction to Electrochemical Sensors*

Electrochemical sensors comprise one of the most exciting areas of electroanalysis. Basically, these sensors use either solid polarizable electrodes or complete galvanic cells featuring permselective membranes as transducers for transferring of an electric signal. This signal is controlled by diffusional transport, adsorption, electrochemical reaction or membrane equilibria of the analyte. Moreover, these sensors can be regarded as chemosensors or biosensors wherein chemically or biologically modified electrodes [161], respectively, are used.

Electrochemical sensors operate efficiently if (1) a suitable biological or chemical recognition element is selected, (2) immobilized onto an electrode surface or in a membrane and (3) an appropriate electroanalytical transduction scheme is applied.

Electrochemical sensors can easily be fabricated and miniaturized in a most economical way. Their appealing features, such as selectivity, sensitivity and reproducibility, have drawn attention for the past couple of decades. In this direction, immobilization of biomolecules on the electrode surface for molecular recognition is a reasonable choice, thus gaining importance in the field of electrochemical sensors. A large number of materials, ranging from biocomposites [162] to sol-gel matrices [25], prepared by using different immobilization procedures, have been used for electrochemical sensing. Carbon composites incorporating both the recognition and transduction elements in the same materials have been widely used since surfaces of these sensors can easily be renewed, for instance, by abrasion [163].

The use of synthetic materials that imitate recognition characteristics of biological materials has been explored. Particularly, MIPs can be thought of as viable alternates to replace natural receptors. Due to easier methods of in situ preparation as films on electrode surfaces or in membranes and, hence, easier fabrication, the field of chemosensors featuring artificial receptors has received broad attention showing a pronounced progress.

4.2 Procedures of MIP Preparation for Electrochemical Sensor Application

Now, bulk polymerization in the presence of a template is just one among many frequently used procedures for the fabrication of MIPs [19]. First, the preformed complex of the functional monomer and template is co-polymerized with excess of the cross-linking monomer in a porogenic solvent. This step results in a solid matrix of the highly cross-linked polymer. Upon removal of the template molecules, molecular cavities featuring recognition sites are formed. These cavities are suitable for accommodation of the templating compound used as the analyte in this step. The resulting insoluble solid polymer material is often ground to yield a fine powder for preparation of the polymer-film coated electrodes [164] and membranes [165].

In situ polymerization, and electrochemical polymerization in particular [22], is an elegant procedure to form an ultra thin MIP film directly on the transducer surface. Electrochemical polymerization involves redox monomers that can be polymerized under galvanostatic, potentiostatic or potentiodynamic conditions that allow control of the properties of the MIP film being prepared. That is, the polymer thickness and its porosity can easily be adjusted with the amount of charge transferred as well as by selection of solvent and counter ions of suitable sizes, respectively. Except for template removal, this polymerization does not require any further film treatment and, in fact, the film can be applied directly. Formation of an ultrathin film of MIP is one of the attractive ways of chemosensor fabrication that avoids introduction of an excessive diffusion barrier for the analyte, thus improving chemosensor performance. This type of MIP is used to fabricate not only electrochemical [114] but also optical [59] and PZ [28] chemosensors.

MIP beads or microspheres are also widely used for sensing purposes [166]. They are prepared by precipitation polymerization and then they are embedded in a dedicated matrix, which is immobilized on the transducer surface. Moreover, the MIP beads are used to serve as stationary phases in HPLC [167] and for catalytic purposes. Other systems, such as self-assembled monolayers, SAMs [168], sol-gel matrices [169] and preformed polymers [170] have also been utilized for fabrication of MIP constructs.

4.3 Procedures of MIP Immobilization for Electrochemical Sensor Fabrication

Integration of MIPs with the transducer surface (immobilization) is a foremost and arduous step of chemosensor fabrication. The polymer material must perfectly adhere to this surface, neither dispersing away nor peeling off in aggressive solvent solutions, under extreme pH or ionic strength conditions, or under analytical flow conditions. A range of methods is being used for immobilization of imprinted polymer materials. The most effective involves electrochemical polymerization [171, 172]. This polymerization is an in situ process that combines polymerization and immobilization. That is, the redox monomer molecules polymerize in solution in close proximity to the electrode at a certain potential or current waveform applied. When grown so large that they cannot stay in solution, they precipitate onto the electrode surface. Moreover, the transducer surface can be coated with an MIP film by drop coating, spray coating, spin-coating, etc. [7]. For those purposes, a solution of the polymer material in a suitably selected solvent or, alternatively, a solution for polymerization is prepared and applied to the transducer surface. These coating procedures result in formation of thin MIP films of controlled thickness. Besides, the MIP microparticles, i.e. microspheres or beads, are incorporated in an inert polymer or gel film. For instance, sol-gel matrices have been used to hold the MIP particles for sensing applications [25]. Moreover, a film of an inert soluble polymer, like PVC, has been used as a matrix for dispersion of the MIP particles to fabricate chemosensors [149].

Preparation of an MIP-based composite is a valuable method for immobilization of MIP on the transducer surface. For instance, a carbon-based material, such as graphite, is mixed in this procedure with MIP particles to form a composite [173]. This composite and transducer surface (graphite) is then brought in contact to enhance binding of the sensing element and the conducting substrate. The most important aspect of this procedure is that abrasive polishing can readily renew the surface of the chemosensor.

Electrochemical MIP-based chemosensors have gained particular attention owing to robustness of the material used for their fabrication, appreciable detectability, sensitivity and promising possibility of miniaturization.

The present section is devoted to MIP-aided electrochemical sensors. Here, versatility in using electroanalytical techniques for determination of the analytes and their advantage over other detection techniques have been emphasized. A range of recognition MIP materials has been described. Several review papers, related to the electrochemical sensing based on MIPs, have already been available [16, 22–24]. The present chapter focuses on recent achievements in this research area.

4.4   Electrochemical Transduction Schemes

Electroanalytical techniques, such as conductometry [174], potentiometry [22], voltammetry [6], chronoamperometry [25] and EIS [175], have been used extensively for transduction of the detection signal in the MIP-based chemosensors. The chemosensor response may be due to different interfacial phenomena occurring at the electrode–electrolyte interface [16], which will be discussed below in the respective sections. The electrochemical transduction scheme can be devised for accurate measurements tailored to the analytes exhibiting either faradic or nonfaradic electrode behaviour. In many instances, the detection medium is an inert buffer solution [24]. In order to enhance the chemosensor response, some of the electrochemical transductions require achievement of the steady-state transport regime with respect to the analyte concentration at the electrode surface. Hence, the electrolyte solution should be agitated at constant speed and the recognition element, i.e. MIP, should be sufficiently robust to resist damaging conditions of solution convection.

4.4.1 Conductometry

Conductometry involves the measurement of migration of ions of opposite signs in solution at the a.c. voltage applied between two electrodes contacting this solution. In conductometry, the detection signal is due to the change in the conductivity of the ions permeating through the MIP recognition material upon binding the analyte. This binding relies on recognition properties of the MIP membrane. The membrane is prepared, for instance, by placing a drop of the radical-polymerizing solution between two quartz slides and exposing it to UV light to form a thick membrane [114]. This membrane is mounted in the centre of a glass cell, thus separating it into two compartments. If filled with the electrolyte solution and furnished with two platinum electrodes, each on either side of the membrane, the cell serves the conductivity measurements [165, 176]. In principle, such a recognition system based on conductometric transduction is a complete MIP conductometric detector rather than an MIP chemosensor. The conductivity of the solution, which is often buffered, is measured as a function of time in the presence and absence of the analyte in solution. In other words, the change in the resistance of the MIP membrane due to the interaction of the analyte with the MIP recognition sites is measured. Binding of the analyte by these sites results in MIP conformational reorganization. Eventually, this reorganization affects the rate of diffusion of co- and counter ions through the membrane. Therefore, the membrane changes its resistance, which depends upon the analyte concentration. Hence, changes in the membrane resistance or conductivity of the electrolyte solution can be used to measure the concentration of the analyte. Moreover, this transduction scheme, combined with the membrane permselectivity, can be used to study permeability of biorelevant analytes.

For instance, conductometric transduction has been used for determination of an atrazine herbicide in the concentration range 4.6–231.8 mM [165]. A sintered glass frit was used as the support for the MIP film prepared by thermo-radical polymerization. However, both the response time of 30 min and the chemosensor recovery time of 12 h were long. For better performance, the MIP film was prepared by photo-radical co-polymerization of a chloroform solution, which contained tri (ethylene glycol)dimethacrylate (TEGDMA), MAA, and oligourethane acrylate, sandwiched between two quartz slides (Table 6).

Application of this procedure improved LOD down to 5 nM atrazine [114]. There are several other examples of conductometric determination of atrazine based on MIPs [114, 165, 176]. These determinations discriminated atrazine from analogous herbicide interferants, such as trazine, simazine and prometryn [114].

Table 6 Performance of representative examples of MIP electrochemical sensors

Analyte	Functional monomer	Cross-linking monomer	Polymerization initiator	Polymerization solution	MIP preparation procedure	Transduction scheme	Linear concentration range	LOD	References
Atrazine	MAA	Oligourethane acrylate	AIBN	Chloroform	Photo-radical polymerization	Conductometric	5–100 nM	5 nM	[114]
TCAA	VPD	EGDMA	AIBN	Acetonitrile	Thermo-radical polymerization	Conductometric	0.5–5 μg L^{-1}	0.5 μg L^{-1}	[128, 177]

Template	Structure 1	Structure 2	Structure 3	Solvent	Polymerization	Sensor type	Concentration range	Detection limit	Reference
L-nicotine	MAA	EGDMA	AIBN	Chloroform	Thermo-radical polymerization	Capacitive	1–10 nM	5 nM	[178]
L-Nicotine	Dopamine	—	—	Phosphate buffer (pH = 7.4)	Electrochemical polymerization	Capacitive	1–25 µM	0.5 µM	[106]
Histidine	Tetraethyl orthosilicate	Phenyltrimethoxysilane	—	Ethanol	Sol-gel film preparation	Capacitive	50 nM to 1 mM	25 nM	[179]

(continued)

Table 6 (continued)

Analyte	Functional monomer	Cross-linking monomer	Polymerization initiator	Polymerization solution	MIP preparation procedure	Transduction scheme	Linear concentration range	LOD	References
	Methyltrimethoxysilane								
Serotonin	MAA	EGDMA	2,4-Dimethylvaleronitrile	Dibutyl phthalate	Thermo-radical polymerization	Potentiometric	100 pM to 1 µM	100 pM	[180]
MPA	Octadecyltrichlorosilane	—	—	Chloroform and carbon tetrachloride	Sol-gel method	Potentiometric	50 µM to 0.62 M	50 µM	[181]
Morphine	3,4-Ethylenedioxythiophene	—	—	Acetonitrile	Electrochemical polymerization	Chronoamperometric (Pt electrode)	0.01–0.2 mM	10 µM	[182]

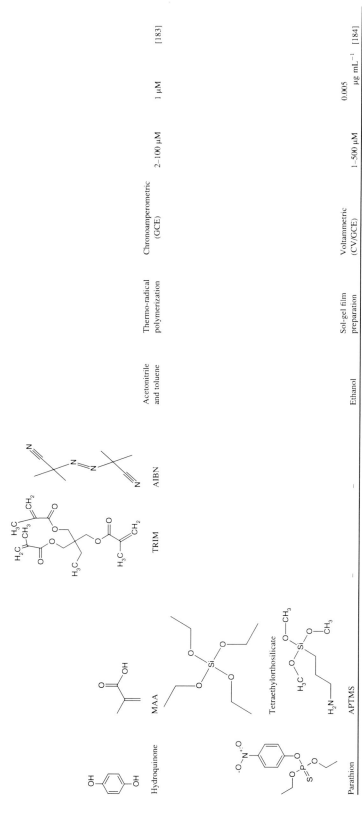

Hydroquinone	MAA	Tetraethylorthosilicate	TRIM	AIBN	Acetonitrile and toluene	Thermo-radical polymerization	Chronoamperometric (GCE)	2–100 µM	1 µM [183]
Parathion	APTMS		–	–	Ethanol	Sol-gel film preparation	Voltammetric (CV/GCE)	1–500 µM	0.005 µg mL⁻¹ [184]

(continued)

Table 6 (continued)

Analyte	Functional monomer	Cross-linking monomer	Polymerization initiator	Polymerization solution	MIP preparation procedure	Transduction scheme	Linear concentration range	LOD	References
Albuterol	Benzyl methacrylate / MAA	2-Hydroxyl ethyl methacrylate	AIBN	Propylene glycol monomethyl ether acetate	Thermo-radical polymerization	Voltammetric (DPV/ Pt electrode)	1–200 μM	1 μM	[185]
Theophylline	Methylene green	–	–	98.9 mM NaNO₃, 9.96 mM NaB₄O₇	Electrochemical polymerization	Voltammetric (CV/GCE)	0.3 μM to 1.00 mM	0.3 μM	[186]

Moreover, other polymer imprinting procedures have been developed for recognition of atrazine, L-phenylalanine, 6-amino-1-polyuracil and sialic acid [176]. In these procedures, EGDMA was used as a cross-linker along with different functional monomers in DMF to generate recognition sites for the templates. For atrazine, the linear concentration range of the conductometric determinations using these MIP materials was 1–50 μM at LOD of 1 μM. Notably, this LOD is lower than that of the spectroscopic atrazine MIP chemosensors, equal to 6.7 μM (1.7 ppm) [187].

Haloacetic acid carcinogens have been determined in drinking water with chemosensors using MIP recognition and conductometric transduction (Table 6) [128, 177]. In these determinations, a lab-on-chip device was used. It was incorporated with a miniaturized liquid handling system. For MIP preparation, TCAA, VPD and EGDMA have been used as the template, functional monomer and cross-linker, respectively, in ACN. The ingress and egress of the target or analyte molecule into the imprinted cavity changed the MIP conductivity. This property has been used to devise a chemosensor. A temperature change affected the resistance signal. This signal decreased by ~1.45% per kelvin with the temperature increase at the applied a.c. voltage of 3 kHz. Therefore, the temperature was kept constant during the determination. The linear concentration range and LOD was 0.5–5 and 0.5 μg L^{-1} TCAA, respectively (Fig. 7). The affinity of the MIP films towards TCAA as well as to other haloacetic acids, such as dichloro-, monochloro-, tribromo-, dibromo- and monobromoacetic acid, was selective.

Disadvantage of the conductometric signal transduction originates from its additivity. For instance, interferants adsorption adds to the conductivity measured as the detection signal. Hence, conductivity is not so widely used for transduction as other electroanalytical techniques. Moreover, any similar difference in the limiting equivalent conductance is insufficient for discrimination between the analyte and its interferants.

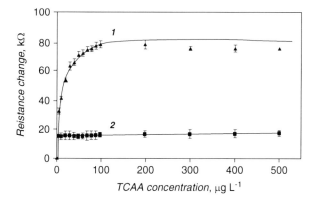

Fig. 7 Dependence of the membrane resistance on concentration of trichloroacetic acid (TCAA) in distilled water at 3 kHz, at room temperature, for (1) MIP chemosensor and (2) reference NIP chemosensor. Each point represents the average of three independent measurements (adapted from [128])

4.4.2 Capacitance and Impedance Measurements

The capacitive or impedimetric transduction scheme can be elucidated from the concept of the electric double layer at the electrode–(electrolyte solution) interface. Capacitance is the amount of charge stored by a capacitor at a given electric potential. The calculation of capacitance requires information on dielectric property of the insulating film, which separates two parallel plate conductors, with definite thickness. Under electrochemical conditions, changes in this capacitance correspond to the changes in the stored charge of the double layer due to its charging and discharging. This capacitance, C, can be described as

$$C = \frac{\varepsilon \, \varepsilon_0 \, A_s}{d} \qquad (3)$$

where ε and ε_0 is the electric permittivity of the insulator and free space, respectively, A_s is the surface area of the capacitor plate and d is the distance between the plates. In (3), all parameters, other than d and ε, are constant. Hence, variation in the thickness or electric permittivity of the insulating film results in change of capacitance. Under these conditions, the double layer capacitance can be expressed as the series combination [188]

$$\frac{1}{C_t} = \frac{1}{C_u} + \frac{1}{C_b} \qquad (4)$$

where C_t is the total capacitance, C_u is the capacitance in the absence of any change in the electric permittivity of the insulating film and its thickness and C_b is the capacitance due to the analyte sorption by the insulator or changes in thickness of the film. In particular, the latter capacitance can be related to the analyte concentration and, therefore, exploited for the detection purposes.

The EIS technique is very convenient for characterizing different electrode–electrolyte interfacial phenomena [189]. It measures the effects that govern physical and chemical properties of a substance present at the interface using a broad range of frequencies at a selected constant potential applied [190, 191]. The capacitance and impedance of the interface can be measured with EIS based on the following relation [117]:

$$C = \frac{1}{2 \pi f \, Z_{im}} \qquad (5)$$

where Z_{im} is the imaginary part of impedance and f is the frequency of the a.c. voltage applied.

For the measurement of impedance or capacitance of the MIP material in sensing applications, this material is treated as a capacitor dielectric, which is placed between the electrode and electrolyte pseudo-plates. Any sorption or binding of the analyte disturbs the dielectric characteristics of the interface, which is manifested by the change in the capacitance. The EIS technique has been widely used for transduction in chemical sensors. The most important criterion for the successful use of this transduction in the MIP-based chemosensors is formation of the ultrathin defect- and pore-free MIP film with predetermined thickness. A surfactant layer can be deposited onto the MIP film for filling its pores and levelling its surface defects [168]. Owing to the small size and opportunity of integrating with microelectronic devices, an ion-sensitive field-effect capacitor [192] has been devised as a tool suitable for constructing chemosensors.

Capacitive detection using an MIP film was first reported for determination of phenylalanine [168]. For that purpose, first, phenol was electropolymerized in NaH_2PO_4 solution (pH = 7.5) in the presence of the phenylalanine template to yield a suitable MIP film. This template was subsequently removed by rinsing the electrode, coated with the resulting MIP film, with the phosphate buffer (pH = 7.5). EIS was then used to determine variations in the dielectric properties of the recognition MIP film. The capacitance decreased upon phenylalanine addition to the test solution. In the presence of other amino acids or phenol, however, the capacitance decrease was very low, indicating a pronounced MIP selectivity towards phenylalanine (Fig. 8). Linear concentration range extended from 0.5 to 8 mg mL^{-1} phenylalanine. However, the response time of 15 min was long. Unfortunately, this chemosensor has not been optimized with respect to temporal stability and reversibility.

A histamine selective MIP chemosensor, based on impedimetric transduction, has been devised [136]. Its preparation involved immobilization of the histamine imprinted MIP particles in a poly[2-methoxy-5-(3,7-dimethyloctyloxy)-1-4-pheny-lene vinylene] (OC_1C_{10}-PPV) film deposited on aluminium electrodes. Preparation of these particles comprised thermally induced co-polymerization of MAA (functional monomer), EGDMA (cross-linker) and AIBN (initiator) in the presence of histamine. This film efficiently rebound histamine in the presence of histidine and

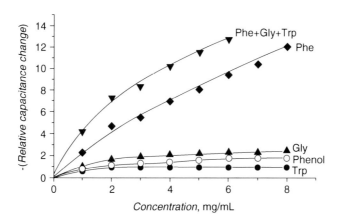

Fig. 8 Change of the MIP chemosensor relative capacitance with the concentration of phenylalanine (Phe), glycine (Gly), tryptophan (Trp) and phenol as well as of the equimolar mixture of phenylalanine, glycine and tryptophan (Phe, Gly and Trp). Phenylalanine was used as the template for the MIP preparation (adapted from [168])

other interfering compounds. The impedance of this chemosensor was linearly decreased by 40%, due to rebinding of histamine, in the concentration range of 1–12 nM. This range is comparable to that of the MIP-PZ chemosensor for histamine determination [131].

Molecular imprinting has been used to devise a chemosensor for L-nicotine (Table 6) [178]. For that, poly(methacrylic acid) (PMA) beads, imprinted with the L-nicotine template in chloroform, were incorporated in a film of the conjugated polymer, OC_1C_{10}-PPV. EIS has then been utilized for the L-nicotine determination in the 1–10 nM concentration range. This MIP chemosensor showed predominant affinity towards L-nicotine over a structurally related L-nicotine metabolite, L-cotinine. Similarly, the polydopamine-imprinted film prepared by electropolymerization in the phosphate buffer (pH = 7.4) has been used to devise a chemosensor for L-nicotine with LOD of 0.5 μM (Table 6) [106]. This LOD is still much higher than that reported for other L-nicotine determination methods based on MIPs, such as SPE combined with differential pulsed elution, which was 6 nM [31].

Herbicides, like demestryn, have been determined using a chemosensor based on the MIP film recognition and capacitive transduction [193, 194]. In this determination, photografting polymerization has been demonstrated as an efficient procedure for fabrication of a capacitive chemosensor. This procedure involved immobilization of an initiator on the electrode. In this case, first, an alkanethiol monolayer was self-assembled on the gold electrode. This monolayer was perfectly dielectric. Then, an MIP film was deposited on top of this monolayer by photo-radical polymerization in the acetone solution with benzophenone, 2-acrylamido-2-methyl-1-propane sulphonic acid, N,N'-methylenediacrylamide and demestryn used as the initiator, cross-linker, functional monomer and template, respectively. Subsequently, the template was extracted with methanol. The capacitance decreased by ~20% upon binding the demestryn analyte by the MIP film. Similarly, a creatine chemosensor was constructed [194].

The EIS measurements, performed with the use of QCM, have been employed for stereospecific recognition of histidine (Table 6) [179]. For that purpose, a thin film of MIP was fabricated by sol-gel imprinting, in ethanol, of the L-histidine template. This chemosensor was appreciably sensitive to L-histidine at LOD of 25 nM with the linear concentration range extending from 50 nM to 1 mM.

Glutathione, a tripeptide vital for redox metabolic chains in living organisms, plays the essential role in drug metabolism and synthesis of several proteins, nucleic acids, etc. A glutathione abnormal concentration in tissue can indicate certain cell disorder and, therefore, seeks accurate determination. Accordingly, an MIP capacitive chemosensor for determination of glutathione in human serum has been devised [195]. Preparation of the appropriate MIP film included, first, electropolymerization of 1,2-phenylenediamine in the presence of the glutathione template, in the phosphate buffer (pH = 6.98). Then, a hydrophobic monolayer of 2-mercaptoethanesulphonate and 1-dodecanethiol was self-assembled to fill pores and defects in the MIP film. The capacitive transduction employed allowed for determination of glutathione with the linear concentration range of 0.025–0.30 mM at LOD of 1.25 μM.

Electropolymerization of 1,2-phenylenediamine in the presence of a glucose template in the acetate buffer (pH = 5.18) on a gold electrode resulted in an MIP film remarkably selective for the glucose analyte [175]. The capacitance of the MIP film recognition element, measured under voltammetric conditions, markedly decreased due to the presence of glucose. But the capacitance was decreased by only 7% for a structurally related compound, such as fructose. For glucose, the linear concentration range and LOD was 0.1–20 mM, and 0.05 mM, respectively. This LOD was appreciably low as compared to that of a QCM glucose chemosensor, equal to 0.07 mM [104].

Tegafur, a widely used chemotherapeutic drug, has been determined with an MIP chemosensor using capacitive transduction [111]. In this chemosensor, the MIP film was prepared by electropolymerization of 1,3-aminophenol, with tegafur as the template, in the phosphate buffer (pH = 6.98), on the gold electrode. The EIS technique using an electrochemical quartz crystal microbalance (EQCM) has been exploited for this determination. A tegafur detection signal was manifested as a pronounced capacitance decrease, as compared to that for a similar compound, such as uridine, 5-flurouracil, uracil, 3-aminophenol, ascorbic acid or mercaptoethylamine. The linear concentration range was 10 μM to 1 mM tegafur and LOD was 10 μM. The chemosensor showed predominant affinity to tegafur with the appreciably high stability constant of the MIP–tegafur complex, $K_{MIP} = 1.06 \times 10^9$ M^{-1}.

Cytidine is a nucleoside component of DNA. It is used as a linker in genetic RNA engineering. This nucleoside has been determined by sol-gel imprinting coupled with the EIS detection performed at EQCM [196]. First, a 1,6-hexanethiol monolayer was self-assembled on the gold electrode. Next, the Au nanoparticles were adsorbed via covalent bond formation with this dithiol monolayer. Then the electrode was coated with a film of 3-(aminopropyl)trimethoxysilane (APTMS), which was assembled layer-by-layer in ethanol. Subsequently, APTMS was electropolymerized under potentiodynamic conditions at negative potentials in the presence of the cytidine template to form an MIP film. This electropolymerization involved hydrolysis and condensation of APTMS. Afterwards, the cytidine template was extracted with an ethanol–water solution. The capacitance was measured as the cytidine detection signal. At pH = 8.0, the apparent stability constant of the MIP–cytidine complex ($K_{MIP} = 1.19 \times 10^7$ M^{-1}) was markedly higher than that for the NIP–cytidine complex ($K_{NIP} = 9.25 \times 10^5$ M^{-1}). These constants could be determined for nanomolar concentrations of the analyte, thereby substantiating the chemosensor detectability with respect to cytidine.

4.4.3   Potentiometry

Potentiometry is one of the versatile electroanalytical techniques widely applied for sensing. A classical example of its application is potentiometric determination of pH, or activity of other inorganic and organic ions by using a pH electrode or ion-selective electrodes (ISEs), respectively. Generally, potential difference across a permeable ion-selective membrane is determined in the measurement involving a potentiometric ISE chemosensor. This difference, measured under no- or low-constant-current-flow conditions, is due to the difference in activity of ions in the test solution and in the membrane. In an MIP-based potentiometric chemosensor, a recognition MIP film plays the role of an ion-selective membrane. In principle, this membrane is immersed in a test solution and the potential difference between the internal and external reference electrode is measured. The MIP membrane selectively binds the target ionic analyte and the resulting membrane potential is measured with respect to that of the external reference electrode, usually Ag|AgCl. This potential is directly related to the activity of the analyte in solution. The analyte quantification is accomplished with respect to a standard analyte concentration. Conducting polymers with imprinted recognition sites and incorporated counter ions have broadly been used in combination with potentiometric detection [19]. Although the potentiometric detection signal is not related to the diffusion of the analyte through the membrane, the establishment of a constant membrane potential requires permanent binding of the analyte to the MIP recognition sites. Importantly, the imprinted template is not removed from MIP after polymerization. Moreover, any template leakage from MIP may lead to indecisive changes in the measured potential values. This effect imposes working conditions onto potentiometric transduction as a trouble-free sensing method [197].

A nitrate-selective potentiometric MIP chemosensor has been devised [197, 198]. For preparation of this chemosensor, a polypyrrole film was deposited by pyrrole electropolymerization on a glassy carbon electrode (GCE) in aqueous solution of the nitrate template. Potentiostatic conditions of electropolymerization used were optimized for enhanced affinity of the resulting MIP film towards this template. In effect, selectivity of the chemosensor towards nitrate was much higher than that to the interfering perchlorate ($K_{NO_3^-, ClO_4^-}^{pot} = 5.7 \times 10^{-2}$) or iodide ($K_{NO_3^-, I^-}^{pot} = 5.1 \times 10^{-2}$) anion. Moreover, with the use of this MIP chemosensor the selectivity of the nitrate detection has been improved, as compared to those of commercial ISEs, by four orders of magnitude at the linear concentration range of 50 μM to 0.5 M and LOD for nitrate of (20 ± 10) μM [197].

Enantiomers of certain chiral amino acids have been discriminated by using a dedicated potentiometric MIP chemosensor [198]. A sol-gel technique using octadecyltrichlorosilane in ethanol has been applied for the fabrication of the relevant MIP film. This chemosensor selectively responded to one of the enantiomers of the racemic (N-carbobenzoxy)-(aspartic acid), N-CBZ-Asp, mixture (Fig. 9). The enantiomeric selectivity coefficient was 9×10^{-3} and 4×10^{-3} for the L- and D-enantiomer, respectively.

The MIP-aided potentiometric sensing has successfully recognized chloropropanol, a contaminant of food products and possible carcinogen. For this recognition, 3-chloro-1,2-propanediol in chloroform has been imprinted with the use of vinylphenylboronic acid as the functional monomer [199]. The MIP film polymerization was thermally initiated. A high concentration of recognition sites imprinted that way, i.e. 123.3 ± 3 μmol g^{-1}, advantageously affected LOD, which reached down to ~200 μM chloropropanol.

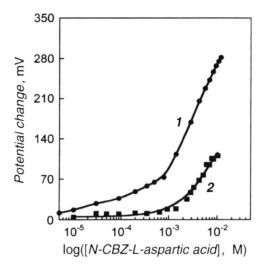

Fig. 9 Potentiometric responses of N-CBZ-L-Asp on the N-CBZ-L-Aspartic acid imprinted sol-gel film, coated on indium-tin oxide (ITO) electrodes (1) with and (2) without the template. No electrolyte other than L-aspartic acid was used. [N-CBZ-L-Asp] stands for the initial concentration of L-aspartic acid. Co-adsorption time was 3 min. Concentration of octadecyltrichlorosilane and N-CBZ-L-aspartic acid in the chloroform-(carbon tetrachloride) solution was 0.8 and 37 mM, respectively (adapted from [198])

A multi-microsensor array of potentiometric MIP chemosensors has been devised for determination of a serotonin neurotransmitter [180]. In the toluene porogenic solvent solution, the MAA functional monomer and the EGDMA cross-linker were polymerized in the presence of the serotonin hydrochloride template (Table 6). Subsequently, the resulting MIPs were immobilized on a plasma polymer layer by swelling and polymerization. Plasma polymerization was performed using styrene or ethylbenzene as the monomer. The chemosensor fabricated that way was appreciably responsive to serotonin while selectivity to serotonin analogues, like acetaminophen ($\log K^{pot}_{serotonine,acetaminophen} = -3.43$), tryptamine ($\log K^{pot}_{serotonine,tryptamine} = -3.88$) and procainamide ($\log K^{pot}_{serotonine,procainamide} = -4.39$), was much lower. The response time was ~12 s with LOD for serotonin of 100 pM.

Methylphosphonic acid (MPA), a degradation product of gas chemical warfare agents, such as sarin (isopropyl methylphosphonofluoridate), soman or VX (*O*-ethyl-*S*-2-diisopropylaminoethylmethylphosphonothioate), has been recognized selectively by an MIP chemosensor using potentiometric transduction (Table 6) [181]. The MIP preparation involved co-adsorption, in ethanol, of the methylphosphonic acid (MPA) template and octadecyltrichlorosilane, followed by silanization on the indium-tin oxide (ITO) electrode surface in the chloroform–carbon tetrachloride solution at 0 °C. Subsequently, the electrode was rinsed with chloroform to remove the template. A potential shift due to the presence of MPA was significant as compared to that due to interferants like methyl parathion, dimethoate, phosdrin, malathion, etc. The linear concentration range varied from 50 μM to 0.62 M MPA at LOD as low as 50 μM and an appreciably short response time of ~50 s.

The promethazine imprinted MIP particles were used as the recognition element to fabricate a relevant potentiometric chemosensor [200]. These particles, prepared with the vinylbenzene functional monomer and the divinylbenzene cross-linker, effectively bound promethazine as compared to the MIP particles prepared with the MAA or vinylbenzene monomer and the EGDMA cross-linker. The MIP particles were embedded in a PVC membrane attached to one end of a glass tube, which was filled with 1 mM promethazine. The chemosensor was able to detect promethazine potentiometrically with LOD as low as 1.0×10^{-7} M and the dynamic linear concentration range from 5.0×10^{-7} to 1.0×10^{-1} M. With this chemosensor, promethazine concentration can be determined in syrup samples and biological fluids with a response time of 50 s.

Melamine has been determined potentiometrically using an MIP film as an ion-selective membrane [140]. To this end, first, melamine was imprinted by thermo-induced co-polymerization of MAA, EGDMA and AIBN in the presence of melamine. After removing this template, the MIP particles were mixed with PVC and 2-nitrophenyloctyl ether in THF for film casting. The film was then mounted to one end of a PVC tube, filed with 0.1 mM melamine, in order to operate as a melamine selective electrode. The electrode potential difference of the ISE setup prepared that way varied linearly with the melamine concentration in the range of 5.0×10^{-6} to 1.0×10^{-2} M. The MIP membrane was doped with different cross-linking agents and the effect of these agents on the performance of the resulting chemosensors was evaluated. This ISE was able to detect melamine with the response time of 16 s at a very low LOD of 1.6 mM. This LOD, however, is rather high compared to that of the melamine chemosensor based on the PZ transduction [134].

An inert polymer matrix, such as PVC, embedding MIP particles has been used for devising a potentiometric chemosensor for MPA [150]. These particles were prepared by thermo-radical polymerization using MAA, EGDMA and AIBN as the functional monomer, cross-linker and initiator, respectively, in a chloroform–(carbon tetrachloride) solution. For immobilization of the MIP particles on the electrode surface, dioctyl phthalate or 2-nitrophenyloctylether was used as the PVC film plasticizer in order to add tensile strength to the matrix. LOD of the chemosensor fabricated that way was appreciably lower than those of the other MIP chemosensors for MPA [181] and equal to 0.05 μM MPA [150].

MIPs with recognition sites for an atrazine herbicide have been used for devising potentiometric sensors [201]. The active analyte was protonated atrazine and, therefore, the measurements were carried out at pH values below pK_a of atrazine, i.e. at pH < 1.8. An MIP membrane was prepared by thermo-radical polymerization using MAA as the functional monomer, AIBN as the initiator and EGDMA as both the cross-linker and solvent. During polymerization this membrane was attached to one end of Teflon[®] tubing. Then this tubing was filled with the internal reference solution of 37 mM atrazine, 30 mM HCl (pH = 1.5) and saturated KCl. An Ag|AgCl reference electrode was immersed in this solution for transduction of the potentiometric detection signal. The test solution of HCl (pH = 1.5) and 20 mM KCl contained atrazine. The cell configuration was Ag|AgCl||AgCl$_{sat}$, KCl$_{sat}$|| 0.34 M KNO$_3$||test solution|membrane|internal reference solution|AgCl|Ag.

A semi-logarithmic plot of the MIP membrane potential vs atrazine concentration increased linearly over the range of 3 μM to 1 mM atrazine. The slope of this plot (i.e. sensitivity) was 25 mV per decade. The chemosensor response time of 10 s was remarkably short, and its stability extended over 2 months.

Benzothiazole derivatives of plausible biological activity have been determined using MIP-based potentiometric sensors. For instance, levamisole hydrochloride, a widely used antitumor agent, has been determined that way [202]. In the levamisole chemosensor, MIP beads were incorporated in an inert PVC matrix, which was cast as a thin membrane. Then, this membrane was attached to one end of Pyrex glass tubing and used for sensing. LOD for levamisole was 1 μM and the response time, significantly, was shorter than 15 s. Moreover, the chemosensor interacted much more strongly with levamisole than with similar compounds, such as thiabendazole, (2-methylthio)benzothiazole or (2-amino)benzothiazole. Appreciably, stability of this chemosensor extended over 4 months.

Successful operation of potentiometric chemosensors opened up the possibility for the fabrication of chemical field-effect transistors (chemFETs) and ion-selective field-effect transistors (ISFETs). A sensing element in these devices, i.e. the MIP film loaded with the molecular, neutral or ionic, respectively, imprinted substance is used to modify surface of the transistor gate area. Apparently, any change in the potential of the film due to its interactions with the analyte alters the current flowing between the source and drain.

4.4.4 Chronoamperometry

In MIP chronoamperometric sensors, current due to diffusion of the analyte towards the electrode surface, coated with the recognition MIP material, is measured at a constant applied potential. The MIP film used for that purpose must be sufficiently permeable in order to allow for unhindered diffusion of the analyte. The measured current is directly proportional to the concentration of the analyte and depends upon the rate of the analyte diffusion towards and, in some cases, away from the electrode surface. The analyte determination involves a preliminary step of removal of the template from the MIP film.

Chronoamperometric transduction can be applied to electroinactive analytes as well as electroactive, which are sorbed by the MIP film and then undergo an electrochemical reaction [25]. In the latter case, the analyte should be able to diffuse freely both towards and away from the electrode surface for the current to flow. The primary requirement of chronoamperometric sensing is a linear relationship between the current measured at the constant potential and the concentration of the analyte. Moreover, the electrochemically generated species should readily diffuse away from the electrode surface coated by the sensing film. By way of example, a few representative chronoamperometric sensors based on MIPs are presented below.

An MIP preparation procedure has been adopted to fabricate a chronoamperometric chemosensor for determination of dopamine [203]. First, quercetin mediator was attached to thioglycolic acid SAM deposited on a gold electrode surface. Then, MAA was photopolymerized, on top of this assembly, in the presence of dopamine. Upon removing the dopamine template, this MIP film was used for selective determination of dopamine at a constant potential of 0.45 V. Chronoamperometry was chosen to determine dopamine. This chemosensor discriminated dopamine from similar compounds that might interfere, such as ascorbic acid, methyldopa and isoproterenol, with sensitivity of 0.445 mA cm^{-2} M^{-1} and LOD as low as 5 μM.

An alkaloid pain reliever, morphine, is an often abused drug. Chronoamperometric MIP chemosensors have been devised for its determination [204]. In these chemosensors, a poly(3,4-ethylenedioxythiophene) (PEDOT) film was deposited by electropolymerization in ACN onto an ITO electrode in the presence of the morphine template to serve as the sensing element [204]. Electrocatalytic current of morphine oxidation has been measured at 0.75 V vs Ag|AgCl|KCl$_{sat}$ (pH = 5.0) as the detection signal. A linear dependence of the measured steady-state current on the morphine concentration extended over the range of 0.1–1 mM with LOD for morphine of 0.2 mM. The chemosensor successfully discriminated morphine and its codeine analogue. Furthermore, a microfluidic MIP system combined with the chronoamperometric transduction has been devised for the determination of morphine [182] with appreciable LOD for morphine of 0.01 mM at a flow rate of 92.3 μL min^{-1} (Table 6).

Moreover, 0.5-μm diameter MIP beads have been prepared for chronoamperometric determination of morphine [204]. These beads were synthesized by thermo-radical precipitation polymerization of the MAA functional monomer, TRIM cross-linker, AIBN initiator and morphine template in the ACN solution. Then the beads were immobilized in a film of the PEDOT conducting polymer, electropolymerized onto the ITO electrode. The morphine detection with the use of the resulting chemosensor was much more sensitive to morphine (41.63 μA cm^{-2} mM for the morphine concentration range of 0.1–2 mM) than to morphine analogues. LOD for morphine was 0.3 mM.

Approximately 0.7-μm diameter MIP beads with hydroquinone recognition sites have been prepared using similar one-step precipitation polymerization (Table 6) [183]. These beads were immobilized in an agarose gel film deposited on GCE. The resulting MIP-modified electrode has been used as a chronoamperometric chemosensor for determination of hydroquinone. The current response of the chemosensor linearly increased with the hydroquinone concentration in the range of 2–100 μM and LOD for hydroquinone was 1 μM.

4.4.5 Voltammetry

Voltammetry is an electroanalytical technique most widely used for transduction as the anodic and cathodic peak potential is characteristic for a particular target redox analyte. In this transduction, current is measured vs potential swept. Similarly to chronoamperometric detection, the analyte diffuses through the MIP film towards and, in some cases, away from the electrode surface. Linear sweep voltammetry

(LSV), CV, differential pulse voltammetry (DPV), square wave voltammetry (SWV), etc. are voltammetric techniques of different potential waveforms. The potential varies either linearly, as in LSV and CV, or linearly with a few-mV superimposed rectangular pulses or square oscillations in DPV and SWV, respectively. Particularly, DPV and SWV offer higher detectability and signal-to-noise ratio than LSV and CV. This is because of the application of programmed current sampling that eliminates a capacity component of the total current measured. Both LSV and CV additionally enable determining of surface concentration of the target electroactive analyte just by coulometric integration of the measured cathodic or anodic current-potential peak. As with chronoamperometric transduction, voltammetric transduction can be applied for determination of both the electroactive and electroinactive analytes. A variety of analytes have been determined using MIP recognition combined with voltammetric transduction. The scope of this chapter is inadequate to discuss all these reports and, henceforth, only a few representative results have been presented below.

For instance, an MIP voltammetric chemosensor for organophosphate pesticides has been fabricated (Table 6) [184]. Both gas- and liquid-phase binding of the pesticides were investigated by CV and PM at QCM. A sol-gel technique has been utilized for preparation of the film featuring sites that recognized a parathion pesticide. Typically, preparation of the MIP film involved silanisation with the use of the ethanol solution containing tetraethyl orthosilicate, phenyltrimethoxysilane, aminopropyltriethoxysilane and concentrated HCl. For imprinting, an aliquot of this sol was mixed with a sample of ethanolic solution of parathion. GCE spin-coated with this sol, after air-drying, has been used for determination of parathion in the phosphate buffer (pH = 7.0). Sensitivity with respect to parathion was higher than that to other organophosphate pesticides (Fig. 10). However, selectivity for detection in the gas phase was much lower than that in the liquid phase.

Recently, a parathion-selective voltammetric MIP chemosensor was designed [205]. The parathion-imprinted polymer particles were prepared by polymerization of an MAA functional monomer, EGDMA cross-linker and AIBN initiator. Subsequently, the powdered MIP particles were blended with a graphite powder, in the presence of n-icosane, to form an (MIP)–(carbon paste) (MIP–CP) electrode. After removal of the template, MIP–CP much more selectively rebound parathion than the (non-imprinted imprinted polymer)–(carbon paste) carbon paste (NIP–CP) electrode. Recognition ability of the MIP–CP electrode was very high compared to that of the NIP–CP electrode. The chemosensor response was calibrated in the linear range of 1.7–900 nM parathion and LOD was 0.5 nM [205]. This chemosensor selectively determined parathion in the presence of its structural and functional counterparts, such as paraxon, in real samples.

A multi-MIP array has been fabricated photolithographically for determination of an albuterol broncholidator (Table 6) [185]. 20-μm diameter acrylic MIP beads have been synthesized by co-polymerization of the benzyl methacrylate functional monomer, MAA functional monomer and HEMA cross-linker in the propylene glycol monomethyl ether acetate porogenic solvent. Thermo-radical polymerization on a Pt electrode was initiated by AIBN. Albuterol was recognized in the

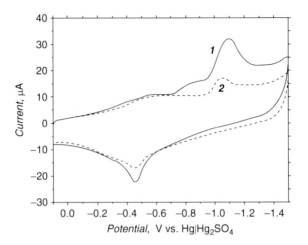

Fig. 10 Cyclic voltammograms for a glassy carbon electrode, coated with the parathion-imprinted MIP film, after 10-min preconcentration in (1) 0.1 mM parathion or (2) 0.1 mM paraoxon. Supporting electrolyte, 0.1 M phosphate buffer (pH = 7.0). Potential scan rate, 100 mV s^{-1} (adapted from [184])

presence of the clenbuterol and terbutaline interferants. The DPV transduction was relatively highly sensitive with the linear concentration range of 1–200 μM and LOD of 1 μM albuterol.

A voltammetric chemosensor for selective determination of 1,4-nitrophenol [206] has been designed with the template-removed imprinted polymer particles as the recognition element. The 1,4-nitrophenol imprinted MIP particles were blended with a graphite powder to yield a composite MIP-CP electrode. With this electrode, 1,4-nitrophenol was voltammetrically determined in the linear dynamic concentration range of 8×10^{-9} to 5×10^{-6} M with LOD of 3 nM 1,4-nitrophenol.

A voltammetric MIP chemosensor has been devised for theophyline, a drug used in therapy for respiratory diseases (Table 6) [186]. For the chemosensor preparation, methylene green azine was electropolymerized in the presence of a theophyline template to form an MIP film on GCE. Followed by extraction of the template, this film was used as a recognition element for determination of the theophyline analyte. Selectivity with respect to theophyline was remarkably high as compared to similar compounds, such as caffeine, with LOD for theophyline of 3 nM. In another study, an MIP film for recognition of theophyline has been prepared on a silanized surface of ITO [207]. That is, the ITO or silicon surface was first silanized with a 10 vol.% solution of 3-(trimethoxysilyl)propyl methacrylate in toluene. The MIP film was then photo-radically polymerized on top of the resulting layer using MAA, EGDMA, AIBN and theophyline as the functional monomer, cross-linker, initiator and template, respectively. In the presence of theophyline, the CV cathodic and anodic peak currents for this MIP were 9.5-fold higher than those for corresponding NIP.

An imprinted poly[tetra(*o*-aminophenyl)porphyrin] film, deposited on a carbon fibre microelectrode by electropolymerization, was used for selective determination of dopamine [208] in the potential range of -0.15 to 1.0 V. This chemosensor has been used successfully for dopamine determination in brain tissue samples. The dopamine linear concentration range extended from 10^{-6} to 10^{-4} M with LOD of 0.3 μM. However, this LOD value is very high compared to that of the dopamine voltammetric detection using polyaminophenol MIPs prepared by electropolymerization [209]. Dopamine was determined by CV and DPV at concentrations ranging from 2×10^{-8} to 0.25×10^{-6} M with LOD of 1.98 nM. This LOD value is lower than that of PM dopamine detection [133].

Recently, dopamine and catechol were detected using voltammetric MIP chemosensors [138]. Here, again, the chemosensors could determine dopamine and catechol in the presence of excess of interfering compounds, such as phenol, resorcinol, hydroquinone, serotonine, ascorbic acid, etc. For catechol, LOD was 228 nM and the linear concentration range was linear from 228 nM to 144 μM.

Caffeine-templated MIP particles were mixed with a graphite powder to devise a voltammetric chemosensor for caffeine [210]. Preparation of the MIP particles involved thermo-induced polymerization of MAA, EGDMA, initiated by AIBN, in the presence of caffeine. The MIP–CP electrode was fabricated by mixing the MIP particles with a graphite powder in *n*-icosane. Subsequently, the template-removed MIP–CP electrodes selectively rebound caffeine. DPV was employed to determine caffeine in the concentration range, which was linear from 6×10^{-8} to 2.5×10^{-5} M. The LOD for caffeine (100 nM) was, however, higher than that of the MIP-PZ chemosensor [143].

An MIP brush grafted onto a sol-gel film was used to fabricate an MIP chemosensor for selective determination of barbituric acid in blood plasma and urine samples [211]. To that end, MIP particles were prepared by co-polymerization of melamine and chloranil, in DMF, in the presence of barbituric acid. Then a TEOS sol-gel slurry, prepared by mixing TEOS and ethanol in 0.1 M HCl, was mixed with the MIP particles and a graphite powder to form an MIP adduct-sol-gel composite. This composite was drop coated onto the graphite electrode to serve as the recognition element. Binding characteristics of this MIP sol-gel modified graphite electrode was compared with that of the NIP sol-gel modified electrode using cathodic stripping DPV. The MIP sol-gel chemosensor assembly was able to determine barbituric acid with the linear concentration range of 4.95–100.00 μg mL^{-1} and LOD of 1.6 μg mL^{-1}.

A TNT chemical explosive was determined using a chemosensor that combined MIP recognition and CV transduction [212]. Initially, the TNT-templated MIP particles were produced and then blended with a graphite powder to fabricate an MIP-CP composite. Subsequently, the TNT template was removed and the composite selectively recognized TNT in the presence of nitrobenzene, 1,4-nitrophenol and benzoic acid. The MIP chemosensor devised that way was used for the SWV determination of TNT. With this transduction, TNT was determined within the dynamic linear concentration range of 5×10^{-9} to 1×10^{-6} M with a relatively low LOD of 1.5 nM. However, this LOD value was still higher than that reached with the MIP–PZ chemosensor for determination of TNT in the gas phase [127].

An MIP film was applied in a chemosensor for determination of a 2-methyl-4, 6-dinitrophenol (4,6-dinitro-*o*-cresol, DNOC) insecticide using SWV transduction [213]. The imprinting was accomplished by electrochemical co-polymerization of aniline with 1,2-phenylenediamine, in the presence of DNOC, which led to deposition of the resulting MIP film onto a carbon fibre microelectrode. After removing the DNOC template, the DNOC analyte was selectively determined, with the linear concentration range of 8×10^{-7} to 10^{-4} M and LOD of 0.2 mM, in the presence of interfering dinitrophenolic pesticides such as binapacryl, and dinobuton.

A voltammetric MIP chemosensor for determination of an ephedrine stimulant has been fabricated using an immobilized MIP film overlaid with a polypyrrole film [214]. That is, first, ephedrine-templated MIP was prepared by thermo-radical polymerization in a chloroform solution of ephedrine, hydroxyethylmethacrylate, EGDMA, and 1,1-azobis(cyclohexanecarbonitrile) (ABCN) used as the template, functional monomer, cross-linker, and initiator, respectively, [14]. Next, after extracting the template, GCE was drop coated with this MIP film. Then a polypyrrole film was grown potentiostatically by electro-oxidative polymerization, on top of this MIP film to form an MIP-modified electrode. The determination procedure involved preconcentration of ephedrine in the 0.5–3 mM concentration range in the Britton–Robinson buffer (pH = 7.0) followed by ephedrine determination by measuring the height of the anodic CV peak at 0.90 V vs SCE. The current of this peak was proportional to the analyte concentration. Affinity of the MIP film towards interferants, such as ascorbic acid, uric acid and glucose, was lower than that towards the analyte and the interference ratio, i.e. i_{pa}(interferant)/i_{pa}(ephedrine) was 0.43.

A thin MIP film of titanium(IV) dioxide (TiO$_2$) with recognition sites for *O,O*-dimethyl-(2,4-dichlorophenoxyacetoxyl)(3'-nitrophenyl)methinephosphonate (phi-NO$_2$) was deposited on GCE [120]. It was fabricated via liquid deposition by dipping GCE in an aqueous solution of ammoniumhexafluorotitanate [(NH$_4$)$_2$TiF$_6$], TiO$_2$, boric acid and phi-NO$_2$. The linear concentration range and LOD were 0.1–5 μM and 0.04 μM phi-NO$_2$, respectively.

4.5   Determination Protocols Using Electrochemical Sensors

General strategies developed for electrochemical detection using MIP chemosensors are discussed herein. In principle, a procedure of the measurement leading to successful determination of the analyte should facilitate maximally interaction between the template and the polymer. It means that interactions between MIP and the supporting electrolyte or solvent should be as weak as possible. For all sensing purposes, inert supporting electrolytes with low ionic strengths are typically preferred. A target analyte should access the MIP film without hindrance. Therefore, the solvent, in which the analyte is dissolved, should be miscible with or extensively wet the film.

Commonly, the transduction mechanisms characteristic for electrochemical MIP chemosensors can fall into two categories, as shown in Scheme 4. For some transductions, like conductometric, impedimetric or potentiometric, sole presence of the target analyte in the MIP film is sufficient to produce an appreciable detection signal. However, this presence is insufficient for application of other techniques like chronoamperometry or voltammetry. As mentioned above in Sects. 4.4.4 and 4.4.5, proper electrode reaction is necessary in the latter techniques to generate the detection signal. Moreover, in the case of chronoamperometry or voltammetry, electrogenerated products may often foul the electrode surface. That is, these

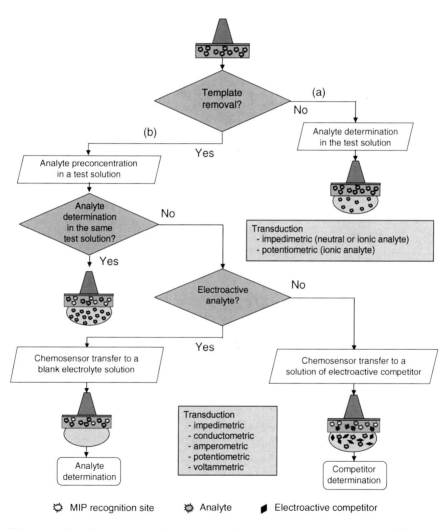

Scheme 4 Flow charts of general analyte determination procedures developed for MIP-based electrochemical sensors (adapted from [24])

products may interact irreversibly with the MIP sensing element and, subsequently, affect the measurement. This critical issue has recently been addressed by fabrication of both surface-renewable [173] and single-use, i.e. disposable [9], chemosensors. For the former, the MIP beads are incorporated in a solid binding matrix, such as a graphite powder. Abrasion after each measurement can renew the chemosensor surface. Removing the template from MIP before use allows for analyte determination with high sensitivity and selectivity. Advantageously, these chemosensors can even be used for electroinactive analytes, as mentioned above in Sect. 4.4.5.

4.5.1   Procedures of Determination without Template Removal

An MIP film is immobilized on the electrode surface and this electrode is maintained in the test solution (Case *a* in Scheme 4). Next, the MIP surface is thoroughly rinsed with abundant solvent to remove a physisorbed analyte and, then, directly used for analyte determination. Herein, impedimetric or potentiometric transduction can be used successfully for determination of either ionic and neutral, or ionic analytes, respectively.

4.5.2   Procedures of Determination with Template Removal

This procedure can be applied irrespectively of whether the analyte is electroactive or electroinactive (Case *b* in Scheme 4). Here, the template is removed from the MIP film with a suitable solvent solution. Next, the chemosensor with template-free MIP is immersed in the test solution for analyte preconcentration. Analyte determination can be carried out with the chemosensor in the same test solution. Alternately, the determination can be performed in another solution. For an electroactive analyte, the chemosensor is removed from the test solution (after preconcentration) and transferred to a blank electrolyte solution followed by the analyte determination. In the case of an electroinactive analyte, the chemosensor is transferred to solution of an electroactive competitor for displacement of this analyte from the MIP film. All the electrochemical transductions, i.e. conductometric, impedimetric, potentiometric, chronoamperometric and voltammetric, are operative under this scheme. In particular, voltammetry and chronoamperometry can be used for determination of both electroactive and electroinactive analytes. This approach as well as the relevant experimental design will be elaborately discussed below in the same section.

Scheme 5 summarizes in detail procedures developed for determination of electroactive analytes after template removal from MIP. In principle, these procedures are based on the competitive binding between an electroactive analyte and electroinactive competitor with the use of the CV, DPV or chronoamperometric transduction.

Case *a* in Scheme 5. First, a template-free MIP chemosensor is equilibrated in a blank electrolyte solution, and then background current is recorded. Next, an electroactive analyte is added. It binds to the recognition sites of the MIP film.

Equilibrium of the analyte partitioning is reached and manifested, for instance, by the constant current flow under chronoamperometric conditions. Subsequently, an electroinactive competitor is added and a new equilibrium of the analyte displacement is attained as the competitor binding results in selective release of the analyte molecules occupying the MIP recognition sites. The current decreases as the displaced analyte diffuses away from the electrode-confined MIP film to the bulk solution. This current decrease can be related to the concentration change of the displaced analyte. Unfortunately, accomplishing this procedure may be time-consuming because diffusion through the MIP film of either the analyte or competitor may be slow. This slow diffusion results in slow filling or emptying recognition sites, and, hence, time-consuming analyte determination. Besides, an interferant that may be present in the bulk analyte solution, may adsorb onto the surface of the MIP film. This surface contamination may deteriorate response of the chemosensor.

Determination of morphine using chronoamperometric transduction may serve as an example of the competitive procedure presented above [25]. The linear dynamic concentration range was 0.1–10 µg mL^{-1} morphine. Codeine played the role of an electroinactive competitor. Microorganism contamination was removed from the MIP film prior to morphine determination using an autoclave. Performance of this chemosensor was much improved with respect to those of traditional biosensors in terms of stability under extremely harsh chemical conditions and elevated temperature.

Case *b* in Scheme 5. This procedure is similar to that described in Case *a* except that a separate electrolyte solution is used in each step. Another requirement is that MIP should be incorporated in a solid binding matrix to form a composite film of a modified electrode, which is capable of mechanical renewal of its surface in order to improve sensing properties. The first step is the same as that described in Case *a* in Scheme 5. That is, it consists in selective binding in the test solution of an electroactive analyte allowing for its preconcentration in the template-free MIP film. Then, after partitioning equilibrium is reached, the electrode is removed from this solution, rinsed with solvent or a solvent solution and transferred to a blank supporting electrolyte solution to record a steady-state background current under chronoamperometric conditions. In the next step, the electroinactive competitor is added to this solution and the current measured again. This competitor selectively displaces the immobilized analyte from the MIP film. A decrease in the current measured is proportional to the concentration of the analyte released from the MIP film. The immobilized analyte in MIP exchanges charge with the electrode substrate and the analyte diffusion within MIP is the key factor of signal generation.

Determination of clenbuterol, a drug for asthma treatment, with isoxsuprine as an electroinactive competitor can be invoked as an operative example of this procedure [173]. DPV was employed for signal transduction and the chemosensor surface was renewed by abrasion after each measurement. LOD was 20 nM clenbuterol.

Detail procedures for determination of electroinactive analytes after template removal from MIPs are presented in Scheme 6. These procedures are complementary to those presented above in Scheme 5 where an electroinactive competitor was used to study binding of an electroactive analyte.

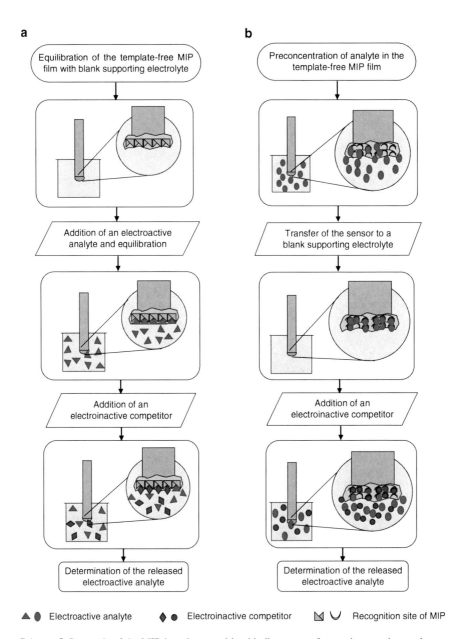

Scheme 5 Protocols of the MIP-based competitive binding assays for an electroactive analyte

Case *a* in Scheme 6. MIP particles are mixed with both the analyte and electroactive competitor, and allowed to reach equilibration. Subsequently, the mixture is centrifuged. The unbound electroactive competitor in the supernatant solution is detected using voltammetry or chronoamperometry with a template-free MIP chemosensor.

For instance, a 2,4-dichlorophenoxyacetic acid systemic herbicide has been determined by adopting this procedure. 2,4-Dichlorophenol and 2,5-dihydroxyphenylacetic acid (also known as homogentisic acid) were used as electroactive competitors and DPV for signal transduction [6]. Likewise, 2-chloro-4-hydroxyphenoxyacetic acid was applied as the electroactive competitor for detection of the

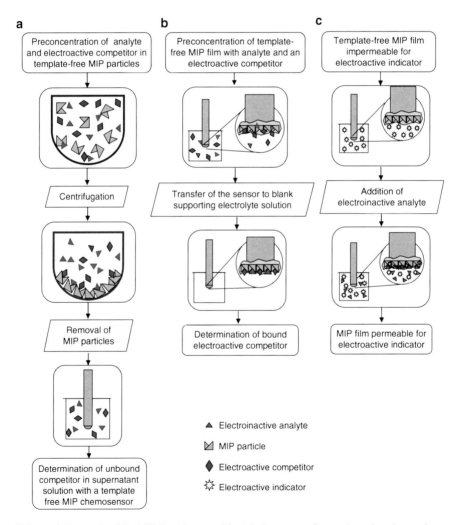

Scheme 6 Protocols of the MIP-based competitive binding assays for an electroinactive analyte

same analyte with the LSV transduction. Disposable screen-printed electrodes served as substrates for MIP films in both these model determinations.

Case *b* in Scheme 6. Here, the MIP-film-coated electrode is equilibrated in solution containing both the electroinactive analyte and electroactive competitor. Then the electrode is rinsed with a solvent and transferred to a blank supporting electrolyte solution. Next, the concentration of the MIP-bound electroactive competitor is determined. Procedures for determination of a range of herbicides have been developed based on this approach.

Case *c* in Scheme 6. This approach discriminates the analyte in the presence and absence of an electroactive indicator in the test electrolyte solution. Importantly, such an electroactive indicator is selected whose diffusion through the MIP film is enhanced in the presence of the electroinactive analyte being determined. In this approach, first, an MIP film is prepared on the electrode surface. Then, after template extraction, the blocking property of the template-free MIP film is examined by CV with the electroactive indicator. Next, electroinactive analyte is added. Molecules of this analyte tend to occupy the MIP receptor sites. In effect, the resulting CV current due to the electrode reaction of the electroactive indicator is significantly enhanced. This enhancement implies an increase in the permeability of the indicator through the MIP film and points to the gate effect. For instance, theophyline was determined using this procedure [215]. That is, the MIP-film-coated ITO electrodes and the $Fe(CN_6)^{4-}$ electroactive indicator have been used for this detection. The theophyline presence resulted in threefold enhancement of the indicator permeability through the MIP film.

4.6   Analytical Characterisation

4.6.1   Detectability of Electrochemical Sensors

LOD is a direct measure of detectability of the MIP chemosensor at a customarily adopted signal-to-noise ratio of at least 3. Desirably, the chemosensor should be able to detect as small concentration amount of the analyte in the test solution as possible. This requirement seeks both very sensitive recognition and transduction. The latter should be able to correlate minute changes in the properties of the recognition element with the binding or releasing of the analyte or competitor, or permeation of the indicator. In terms of transduction, generally, the highest detectability among all the electrochemical transductions is attained with chronoamperometry, SWV and DPV.

Moreover, the higher the concentration of the imprinted sites in the MIP film the lower, advantageously, is LOD. For instance, LOD for the MIP films fabricated from acrylic polymers is, unfortunately, very high. This is because their sensing capabilities are hindered due to a relatively low concentration of recognition sites. Therefore, other different routes of MIP fabrication, using other than acrylic monomers, have been developed to increase this concentration.

4.6.2 Sensitivity of Electrochemical Sensors

Chemosensor sensitivity is the slope of the linear segment of its calibration plot. The absorption of interferences from atmosphere or solution, used either for electropolymerization or template removal, affects the affinity of MIP towards an analyte, thus decreasing the sensitivity of the chemosensor. Frequently, determination is carried out under batch binding conditions and the significance of the interferences is ignored. In potentiometric or impedimetric transduction the same solution (often a buffered electrolyte) is used both for the analyte binding and determination. Voltammetry is a useful technique for evaluation of chemosensor sensitivity. In the determination procedure using voltammetric transduction the binding and the determination are separately examined. That is, the sensitivity of an MIP film is evaluated in the presence of interferants. Moreover, formal redox potential is a distinct analyte feature, which often enables discrimination of structurally related compounds from the analyte.

4.6.3 Stability, Reproducibility and Response Time of Electrochemical Sensors

In general, the MIP-based recognition materials fabricated for biomimetic sensing are stable with respect to dissolution in different highly solubilizing organic solvents. In particular, MIPs prepared from vinylic and acrylic monomers are appreciably stable with over 4-month storage lifetimes under mild conditions [201, 202]. For MIP beads, the response time (defined as the time required achieving 95% of the magnitude of the maximal signal) is generally longer than that for the MIP thin films. Presumably, this time lag is due to the higher complexity and longer path for the analyte to reach the recognition site by diffusion. Hence, ultrathin MIP films are generally preferred for applications that promote a short response time of sensing. Electrochemical polymerization offers flexible and convenient approaches to accomplish this task.

4.7 Conclusions and Perspectives of Electrochemical Sensors

In the last decade, the field of chemosensors with MIPs as effective alternatives to natural recognition materials has grown tremendously. Disadvantages of using biological and biomolecular receptors in traditional biosensors, such as a high cost and susceptibility to matrix effects, have mostly been eliminated with MIP technology. However, there are some practical difficulties in using MIPs as recognition elements. These involve limitations, such as lack of generalized procedure for preparation of MIP and complexity in immobilization on a transducer surface, that still remain to be addressed. Application of the MIP approach in chemosensor development should greatly expand with the advancement of new functional monomers devised for selective immobilization of template molecules featuring unique binding sites. Combination of MIPs with electrochemical transduction has redefined application of chemosensors in a facile manner. Most importantly, the robust nature of MIP qualifies its usefulness as a recognition element of the electrochemical sensor by immobilizing MIP in a composite or as a thin film on an electrode surface.

Electropolymerization has several advantages over other preparation procedures of MIP films. In particular, film thickness can conveniently be controlled with the electrochemical parameters of the deposition. Moreover, the use of suitable solvents and supporting electrolytes can tune visco-elastic properties and porosity of the MIP film. Further, a conducting polymer matrix of MIP prepared by electropolymerization facilitates the charge transport between the electrode substrate and the analyte occupying the MIP molecular cavities. Charge conduction is largely increased if conducting nanofibres, and carbon nanotubes in particular, are present in the MIP matrix, thus forming a highly conducting composite.

Clearly, MIPs are becoming increasingly useful as promising candidates both for inspiration and fabrication of plastic antibodies and enzymes. This outcome ultimately strengthens the importance of MIPs for various envisioned chemosensor and biosensor applications.

Acknowledgments S.S. and P.J.C. equally contributed to this chapter. W.K. thanks the European Regional Development Fund (ERDF, POIG.01.01.02-00-008/08 2007-2013) for financial support. S.S. is grateful to the European Commission for financial support through the *Nanomaterials for Application in Sensors, Catalysis and Emerging Technologies*, NASCENT, Project within the Marie Curie Research Training Network (Contract No. MRTN-CT-2006-033873). P.J.C. gratefully acknowledges the support by the Marie Curie Fellowship within the EC Project *Sensor Nanoparticles for Ions and Biomolecules* SNIB (MTKD-CT-2005-029554).

References

1. Hillberg AL, Brain KR, Allender CJ (2005) Adv Drug Deliv Rev 57:1875
2. Hulanicki A, Glab S, Ingman F (1991) Pure Appl Chem 63:1247
3. Theavenot DR, Toth TK, Durst RA, Wilson GS (1999) Pure Appl Chem 71:2333
4. Updike SJ, Hicks GP (1967) Nature 214:986
5. Nakamura H, Karube I (2003) Anal Bioanal Chem 377:446
6. Kröger S, Turner APF, Mosbach K, Haupt K (1999) Anal Chem 71:3698
7. Haupt K, Mosbach K (2000) Chem Rev 100:2495
8. Dickert FL, Hayden O (1999) Fresen J Anal Chem 364:506
9. Piletsky SA, Turner NW, Laitenberger P (2006) Med Eng Phys 28:971
10. Wulff G (1995) Angew Chem Int Ed Engl 34:1812
11. Andersson LI, Muller R, Vlatakis G, Mosbach K (1995) Proc Natl Acad Sci USA 92:4788
12. Mayes AG, Whitcombe MJ (2005) Adv Drug Deliv Rev 57:1742
13. Piletsky SA, Piletsky EV, Chen B, Karim K, Weston D, Barrett G, Lowe P, Turner APF (2000) Anal Chem 7:4381
14. Piletsky SA, Karim K, Piletsky EV, Turner APF, Day CJ, Freebairn KW, Legge C (2001) Analyst 126:1826

15. McCluskey A, Holdsworth CI, Bowyer MC (2007) Org Biomol Chem 5:3233
16. Piletsky SA, Turner APF (2002) Electroanalysis 14:317
17. Fernandez-Gonzalez A, Guardia L, Badia-Laino R, Díaz-García ME (2006) Trends Analyt Chem 25:949
18. Yano K, Karube I (1999) Trends Analyt Chem 18:199
19. Sellergren B (2001) Tech Instrum Anal Chem 23:362
20. Spivak DA, Simon R, Campbell J (2004) Anal Chim Acta 504:23
21. Hwang C-C, Lee W-C (2002) J Chromatogr A 962:69
22. Malitesta C, Losito I, Zambonin PG (1999) Anal Chem 71:1366
23. Merkoci A, Alegret S (2002) Trends Anal Chem 21:717
24. Blanco-Lopez MC, Lobo-Castanon MJ, Miranda-Ordieres AJ, Tunon-Blanco P (2004) Trends Anal Chem 23:36
25. Kriz D, Mosbach K (1995) Anal Chim Acta 300:71
26. Jenkins AL, Uy OM, Murray GM (1997) Anal Commun 34:221
27. McNiven S, Kato M, Levi R, Yano K, Karube I (1998) Anal Chim Acta 365:69
28. Haupt K, Noworyta K, Kutner W (1999) Anal Commun 36:391
29. Lanza F, Sellergren B (1999) Anal Chem 71:2092
30. Mullett WM, Martin P, Pawliszyn J (2001) Anal Chem 73:2383
31. Mullett WM, Lai EPC, Sellergren B (1999) Anal Commun 36:217
32. Lu Y, Li C, Zhang H, Liu X (2003) Anal Chim Acta 489:33
33. Manesiotis P, Hall AJ, Emgenbroich M, Quaglia M, De Lorenzi E, Sellergren B (2004) Chem Commun:2278
34. Burleigh MC, Dai S, Hagaman EW, Lin JS (2001) Chem Mater 13:2537
35. Matsui J (1998) Anal Commun 35:225
36. Lee JD, Greene NT, Rushton GT, Shimizu KD, Hong JI (2005) Org Lett 7:963
37. Striegler S, Tewes E (2002) Eur J Inorg Chem:487
38. Gong C, Wong KL, Lam MHW (2008) Chem Mater 20:1353
39. Liang C, Peng H, Bao X, Nie L, Yao S (1999) Analyst 124:1781
40. Kubo H, Nariai H, Takeuchi T (2005) Org Lett 7:359
41. Al-Kindy S, Badia R, Diaz-Garcia ME (2002) Anal Lett 35:1763
42. Proudfoot AT (2009) Clin Toxicol 47:89
43. Riskin M, Tel-Vered R, Willner I (2010) Adv Mater 22:1387
44. Muk S, Narayanaswamy R (2006) Anal Bioanal Chem 386:1235
45. Turkewitsch P, Wandelt B, Darling GD, Powell WS (1998) Anal Chem 70:2025
46. Cywinski P, Wandelt B, Danel A (2004) Adsorpt Sci Technol 22:719
47. Cywinski P (2006) Molecular and polymeric fluorescent pyrazoloquinoline sensors for nucleotides. PhD Dissertation, Technical University of Lodz
48. Cywinski P, Sadowska M, Danel A, Buma WJ, Brouwer AM, Wandelt B (2007) J Appl Polym Sci 105:229
49. Manesiotis P, Hall AJ, Sellergren B (2005) J Org Chem 70:2729
50. Kubo H, Nariai H, Takeuchi T (2003) Chem Commun:2792
51. Wolf C, Schmid RW (1990) J Liq Chromatogr 13:2207
52. Thanh NTK, Rathbone DL, Billington DC, Hartell NA (2002) Anal Lett 35:2499
53. Tong A, Dong H, Li L (2002) Anal Chim Acta 466:31
54. Wang W, Gao S, Wang B (1999) Org Lett 1:1209
55. Pestov D, Anderson J, Tepper G (2006) Proc SPIE Int Soc Opt Eng 6378:63780Y/1
56. Matsui J, Higashi M, Takeuchi T (2000) J Am Chem Soc 122:5218
57. Kugimiya A, Takeuchi T (2001) Biosens Bioelectron 16:1059
58. Southard GE, Van Houten KA, Murray GM (2007) Macromolecules 40:1395
59. Jenkins A, Uy OM, Murray GM (1999) Anal Chem 71:373
60. Jenkins AL, Yin R, Jensen JL (2001) Analyst 126:798
61. Jiahui L, Kending CE, Nesterov EE (2007) J Am Chem Soc 129:15911
62. Carlson CA, Lloyd JA, Dean SL, Walker NR, Edmiston PL (2006) Anal Chem 78:3537

63. Güney O, Cebeci FC (2010) J Appl Polym Sci 117:2373
64. Wandelt B, Mielniczak A, Turkewitsch P, Wysocki S (2003) J Lumin 102–103:774
65. Wandelt B, Turkewitsch P, Wysocki S, Darling GD (2002) Polymer 43:2777
66. Cywinski PJ, Sadowska M, Wandelt B, Buma WJ, Brouwer AM (2010) e-J Surf Sci Nanotech 8:293
67. Wandelt B, Sadowska M, Cywinski P, Hachulka K (2008) Mol Cryst Liq Cryst 486:257
68. Okumura T, Ariyoshi K, Hitomi T, Hirahara K, Itoh T, Iwamura T, Nakashima A, Motomura Y, Taki K, Suzuki K (2009) Toxin Rev 28:255
69. Pohanka M, Karasova JZ, Kuca K, Pikula J, Holas O, Korabecny J, Cabal J (2010) Talanta 81:621
70. Boyd JW, Cobb GP, Southard GE, Murray GM (2004) Johns Hopkins Apl Tech Digest 25:44
71. Stringer RC, Gangopadhyay S, Grant SA (2010) Anal Chem 82:4015
72. Li H, Li Y, Cheng J (2010) Chem Mater 22:2451
73. Wang YQ, Xu L, Zhang J, Lue RH (2009) Acta Chimi Sin 67:1475
74. Alvarez-Diaz A, Costa JM, Pereiro R, Sanz-Mendel A (2009) Anal Bioanal Chem 394:1569
75. Fang YJ, Yan SL, NB A, Liu N, Gao ZX, Chao FH (2009) Biosens Bioelectron 24:2323
76. Edmiston PL, Campbell DP, Gottfried DS, Baughman J, Timmers MM (2010) Sens Actuators, B 143:574
77. Taniwaki K, Hyakutake A, Aoki T, Yoshikawa M, Guiver MD, Robertson GP (2003) Anal Chim Acta 489:191
78. Raitman OA, Chegel VI, Kharitonov AB, Zayats M, Katz E, Willner I (2004) Anal Chim Acta 504:101
79. Lotierzo M, Henry OYF, Piletsky S, Tothill I, Cullen D, Kania M, Hock B, Turner APF (2004) Biosens Bioelectron 20:145
80. Li X, Husson M (2006) Biosens Bioelectron 22:336
81. Choi SW, Chang HJ, Lee N, Kim JH, Chun HS (2009) J Agric Food Chem 57:1113
82. Matsui J, Akamatsu K, Hara N, Miyoshi D, Nawafune H, Tamaki K, Sugimoto N (2005) Anal Chem 77:4282
83. Lin C-Y, Tai D-F, Wu T-Z (2003) Chem Eur J 9:5107
84. Tokareva I, Tokarev I, Minko S, Hutter E, Fendler JH (2006) Chem Commun:3343
85. Frasconi M, Tel-Vered R, Riskin M, Willner I (2010) Anal Chem 82:2512
86. Lavine BK, Westover DJ, Kaval N, Mirjankar N, Oxenford L, Mwangi GK (2007) Talanta 72:1042
87. Bompart M, De Wilde Y, Haupt K (2010) Adv Mater 22:2343
88. Kantarovich K, Tsarfati I, Gheber LA, Haupt K, Bar I (2009) Anal Chem 81:5686
89. Kantarovich K, Belmont AS, Haupt K, Bar I, Gheber LA (2009) Appl Phys Lett 94:194103.1
90. Lindner E, Buck RP, Kutner W, Inzelt G (2004) Pure Appl Chem 76:1139
91. Schumacher R (1990) Angew Chem Int Ed Engl 2:329
92. Buttry DA (1991) Application of Quartz Crystal Microbalance to Electrochemistry. In: Bard AJ (ed) Electroanalytical chemistry, vol 17. Marcel Dekker, New York, p 1
93. Buttry DA, Ward MD (1992) Chem Rev 92:1355
94. Sauerbrey G (1959) Z Phys 155:206
95. Morgan AD (1991) Surface-wave devices for signal processing. Elsevier, Amsterdam
96. Wohltjen H (1997) Acoustic wave sensors. Theory, design, and physico-chemical applications. Academic, San Diego
97. Thompson M, Stone DC (1997) Surface-launched acoustic wave sensors: chemical sensing and thin-film characterization. Wiley, New York
98. Haupt K, Mosbach K (1998) Trends Biotechnol 16:468
99. Komiyama M, Takeuchi T, Mukawa T, Asanuma H (2003) Molecular imprinting. Wiley-VCH, Weinheim
100. Fischerauer FDG, Forth P, Knauer U (1996) Proc IEEE Ultrason Symp 1:439
101. Ávila M, Zougagh M, Escarpa A, Rios A (2008) Trends Anal Chem 27:54
102. Uludag Y, Piletsky SA, Turner APF, Cooper MA (2007) FEBS J 274:5471

103. Dickert FL, Lieberzeit PA, Achatz P, Palfinger C, Fassnauer M, Schmid E, Werther W, Horner G (2004) Analyst 129:432
104. Ersöz A, Denizli A, Özcan A, Say R (2005) Biosens Bioelectron 20:2197
105. Wu A-H, Syu M-J (2006) Biosens Bioelectron 21:2345
106. Liu F, Liu X, Ng S-C, Chan HS-O (2006) Sens Actuators, B 113:234
107. Piacham T, Josell A, Arwin H, Prachayasittikul V, Ye L (2005) Anal Chim Acta 536:191
108. Cao L, Li SFY, Zhou XC (2001) Analyst 126:184
109. Percival CJ, Stanley S, Galle M, Braithwaite A, Newton MI, McHale G, Hayes W (2001) Anal Chem 73:4225
110. Sallacan N, Zayats M, Bourenko T, Kharitonov AB, Willner I (2002) Anal Chem 74:702
111. Liao H, Zhang Z, Li H, Nie L, Yao S (2004) Electrochim Acta 49:4101
112. Feng L, Liu Y, Tan Y, Hu J (2004) Biosens Bioelectron 19:1513
113. Matsuguchi M, Uno T (2006) Sens Actuators, B 113:94
114. Piletsky SA, Piletsky EV, Sergeyeva TA, Panasyuk TL, El'skaya AV (1999) Sens Actuators, B 60:216
115. Nomura T, Okuhara M (1982) Anal Chim Acta 142:281
116. Cooper MA, Singleton VT (2007) J Mol Recognit 20:154
117. Furtado LM, Su H, Thompson M, Mack DP, Hayward GL (1999) Anal Chem 71:1167
118. Percival CJ, Stanley S, Braithwaite A, Newton MI, McHale G (2002) Analyst 127:1024
119. Höök F, Ray A, Nordén B, Kasemo B (2001) Langmuir 17:8305
120. Jiang Y, Li P, Li SP, Wang YT, Tu PF (2007) J Pharm Biomed Anal 43:341
121. Fung YS, Wong YY (2001) Anal Chem 73:5302
122. Richert L, Lavalle P, Vautier D, Senger B, Stoltz J-F, Schaaf P, Voegel J-C, Picart C (2002) Biomacromolecules 3:1170
123. Karousos NG, Aouabdi S, Way AS, Reddy SM (2002) Anal Chim Acta 469:189
124. Dickert FL, Forth P, Lieberzeit P, Tortschanoff M (1998) Fresen J Anal Chem 360:759
125. Ji H-S, McNiven S, Ikebukuro K, Karube I (1999) Anal Chim Acta 390:93
126. Ji H-S, McNiven S, Lee K-H, Saito T, Ikebukuro K, Karube I (2000) Biosens Bioelectron 15:403
127. Bunte G, Hürttlen J, Pontius H, Hartlieb K, Krause H (2007) Anal Chim Acta 591:49
128. Suedee R, Intakong W, Dickert FL (2006) Talanta 70:194
129. Das K, Penelle J, Rotello VM (2003) Langmuir 19:3921
130. Yao W, Gao ZX, Cheng YY (2009) J Sep Sci 32:3334
131. Pietrzyk A, Suriyanarayanan S, Kutner W, Chitta R, D'Souza F (2009) Anal Chem 81:2633
132. Pietrzyk A, Suriyanarayanan S, Kutner W, Chitta R, Zandler ME, D'Souza F (2010) Biosens Bioelectron 25:2522
133. Pietrzyk A, Suriyanarayanan S, Kutner W, Maligaspe E, Zandler ME, D'Souza F (2010) Bioelectrochemistry 80:62
134. Pietrzyk A, Kutner W, Chitta R, Zandler ME, D'Souza F, Sannicolo F, Mussini PR (2009) Anal Chem 81:10061
135. Dryhurst G (1972) Talanta 19:769
136. Bongaers E, Alenus J, Horemans F, Weustenraed A, Lutsen L, Vanderzande D, Cleij TJ, Troost FJ, Brummer RJ, Wagner P (2010) Phys Status Solidi A 207:837
137. Trikka F, Yoshimatsu K, Ye L, Kyriakides DA (2008) FEBS J 275:373
138. Lakshmi D, Bossi A, Whitcombe MJ, Chianella I, Fowler SA, Subrahmanyam S, Piletska EV, Piletsky SA (2009) Anal Chem 81:3576
139. Spivak DA, Shea KJ (1998) Macromolecules 31:2160
140. Liang RN, Zhang RM, Qin W (2009) Sens Actuators, B 141:544
141. Diltemiz SE, Hur D, Ersoz A, Denizli A, Say R (2009) Biosens Bioelectron 25:599
142. Sun H, Mo ZH, Choy JTS, Zhu DR, Fung YS (2008) Sens Actuators, B 131:148
143. Zougagh M, Ríos A, Valcárcel M (2005) Anal Chim Acta 539:117
144. Ebarvia BS, Sevilla F III (2005) Sens Actuators, B 107:782
145. Ebarvia BS, Binag CA, Sevilla F III (2004) Anal Bioanal Chem 378:1331

146. Villamena AF, de la Cruz AA (2001) J Appl Polym Sci 82:195
147. Albano DR, Sevilla F III (2007) Sens Actuators, B 121:129
148. Syritski V, Reut J, Menaker A, Gyurcsányi RE, Öpik A (2008) Electrochim Acta 53:2729
149. Liang C, Peng H, Nie L, Yao S (2000) Fresen J Anal Chem 367:551
150. Prathish KP, Prasad K, Prasada Rao T, Suryanarayana MVS (2007) Talanta 71:1976
151. Sun H, Fung Y (2006) Anal Chim Acta 576:67
152. Kugimiya A, Takeuchi T (1999) Electroanalysis 11:1158
153. Kugimiya HYA, Takeuchi T (2000) Electroanalysis 12:1322
154. Otero TF, Cantero I (1995) J Electroanal Chem 395:75
155. Dickert FL, Lieberzeit PA (2007) Springer Ser Chem Sens Biosens 5:173
156. Shea KJ, Spivak DA, Sellergren B (1993) J Am Chem Soc 115:3368
157. Ávila M, Zougagh M, Escarpa A, Ríos Á (2007) Talanta 72:1362
158. Fu Y, Finklea HO (2003) Anal Chem 75:5387
159. D'Agostino G, Alberti G, Biesuz R, Pesavento M (2006) Biosens Bioelectron 22:145
160. Luo C, Liu M, Mo Y, Qu J, Feng Y (2001) Anal Chim Acta 428:143
161. Kutner W, Wang J, L'Her M, Buck RP (1998) Pure Appl Chem 70:1301
162. Céspedes F, Martinez-Fabregas E, Alegret S (1996) Trends Anal Chem 15:296
163. Alegret S (1996) Analyst 121:1751
164. Schöllhorn B, Maurice C, Flohic G, Limoges B (2000) Analyst 125:665
165. Piletsky SA, Piletsky EV, Elgersma AV, Yano K, Karube I, Parhometz YP, El'skaya AV (1995) Biosens Bioelectron 10:959
166. Dickert FL, Tortschanoff M, Bulst WE, Fischerauer G (1999) Anal Chem 71:4559
167. Kempe M (1996) Anal Chem 68:1948
168. Panasyuk TL, Mirsky VM, Piletsky SA, Wolfbeis OS (1999) Anal Chem 71:4609
169. Collinson MM (2002) Trends Anal Chem 21:31
170. Kobayashi T, Murawaki Y, Reddy PS, Abe A, Fujii N (2001) Anal Chim Acta 435:141
171. Ramanavicius A, Ramanaviciene A, Malinauskas A (2006) Electrochim Acta 51:6025
172. Okuno H, Kitano T, Yakabe H, Kishimoto M, Deore BA, Siigi H, Nagaoka T (2002) Anal Chem 74:4184
173. Pizzariello A, Stred'ansky M, Stred'anska S, Miertus S (2001) Sens Actuators, B 76:286
174. Johnson-White B, Zeinali M, Shaffer KM, Patterson CH Jr, Charles PT, Markowitz MA (2007) Biosens Bioelectron 22:1154
175. Cheng ZL, Wang EK, Yang XR (2001) Biosens Bioelectron 16:179
176. Piletsky SA, Piletskaya EV, Panasyuk TL, El'skaya AV, Levi R, Karube I, Wulff G (1998) Macromolecules 31:2137
177. Suedee R, Srichana T, Sangpagai C, Tunthana C, Vanichapichat P (2004) Anal Chim Acta 504:89
178. Thoelen R, Vansweevelt R, Duchateau J, Horemans F, D'Haen J, Lutsen L, Vanderzande D, Ameloot M, VandeVen M, Cleij TJ, Wagner P (2008) Biosens Bioelectron 23:913
179. Zhang Z, Liao H, Li H, Nie L, Yao S (2005) Anal Biochem 336:108
180. Kitade T, Kitamura K, Konishi T, Takegami S, Okuno T, Ishikawa M, Wakabayashi M, Nishikawa K, Muramatsu Y (2004) Anal Chem 76:6802
181. Zhou Y, Yu B, Shiu E, Levon K (2004) Anal Chem 76:2689
182. Weng C-H, Yeh W-M, Hob K-C, Lee G-B (2007) Sens Actuators, B 121:576
183. Kan X, Zhao Q, Zhang Z, Wang Z, Zhu J-J (2008) Talanta 75:22
184. Marx S, Zaltsman A, Turyan I, Mandler D (2004) Anal Chem 76:120
185. Huanga H-C, Huanga S-Y, Lin C-I, Lee Y-D (2007) Anal Chim Acta 582:137
186. Ulyanova YV, Blackwell E, Minteer SD (2006) Analyst 131:257
187. Belmont A-S, Jaeger S, Knopp D, Niessner R, Gauglitz G, Haupt K (2007) Biosens Bioelectron 22:3267
188. Bataillard P, Gardies F, Jaffrezic-Renault N, Martelet C, Colin B, Mandrand B (1988) Anal Chem 60:2374
189. Baranski AS, Krogulec T, Neison LJ, Norouzi P (1998) Anal Chem 70:2895

190. McNeil CJ, Athey D, Ball M, Ho WO, Krause S, Armstrong RD, Des Wright J, Rawson K (1995) Anal Chem 67:3928
191. Lasia A (1999) In: Conway BE, Bockris J, White RE (eds) Modern aspects of electrochemistry, vol 32. Kluwer Academic/Plenum Publishers, New York, p 143
192. Hedborg E, Winquist F, Lundström I, Anderson L, Mosbach K (1993) Sens Actuators, A 37:796
193. Panasyuk-Delaney T, Mirsky VM, Ulbritch M, Wolfbeis OS (2001) Anal Chim Acta 435:157
194. Panasyuk-Delaney T, Mirsky VM, Wolfbeis OS (2002) Electroanalysis 14:221
195. Yang L, Wei W, Xia J, Tao H, Yang P (2005) Electroanalysis 17:969
196. Zhang Z, Nie L, Yao S (2006) Talanta 69:435
197. Hutchins RS, Bachas LG (1995) Anal Chem 67:1654
198. Zhou Y, Yu B, Levon K (2003) Chem Mater 15:2774
199. Leung MKP, Chiu BKW, Lam MHW (2003) Anal Chim Acta 491:15
200. Alizadeh T, Akhoundian M (2010) Electrochim Acta 55:3477
201. Agostino GD, Alberti G, Biesuz R, Pesavento M (2006) Biosens Bioelectron 22:145
202. Sadeghi S, Fathi F, Abbasifar J (2007) Sens Actuators, B 122:158
203. Chen PY, Nien PC, Wu CT, Wu TH, Lin CW, Ho KC (2009) Anal Chim Acta 643:38
204. Ho K-C, Yeh W-M, Tung T-S, Liao J-Y (2005) Anal Chim Acta 542:90
205. Alizadeh T (2009) Electroanalysis 12:1490
206. Alizadeh T, Ganjali MR, Norouzi P, Zare M, Zeraatkar A (2009) Talanta 79:1197
207. Kindschy LM, Alocilja EC (2005) Biosens Bioelectron 20:2163
208. Gomez-Caballero A, Ugarte A, Sanchez-Ortega A, Unceta N, Goicolea MA, Barrio RJ (2010) J Electroanal Chem 638:246
209. Li JP, Zhao J, Wei XP (2009) Sens Actuators, B 140:663
210. Alizadeh T, Ganjali MR, Zare M, Norouzi P (2010) Electrochim Acta 55:1568
211. Patel AK, Sharma PS, Prasad BB (2009) Int J Pharm 371:47
212. Alizadeh T, Zare M, Ganjali MR, Norouzi P, Tavana B (2010) Biosens Bioelectron 25:1166
213. Gomez-Caballero A, Unceta N, Goicolea MA, Barrio RJ (2008) Sens Actuators, B 130:713
214. Mazzotta E, Picca RA, Malitesta C, Piletsky SA, Piletska EV (2008) Biosens Bioelectron 23:1152
215. Yoshimia Y, Ohdairaa R, Iiyamaa C, Sakaib K (2001) Sens Actuators, B 73:49

Top Curr Chem (2012) 325: 267–306
DOI: 10.1007/128_2010_100
© Springer-Verlag Berlin Heidelberg 2010
Published online: 18 November 2010

Chromatography, Solid-Phase Extraction, and Capillary Electrochromatography with MIPs

Blanka Tóth and George Horvai

Abstract Most analytical applications of molecularly imprinted polymers are based on their selective adsorption properties towards the template or its analogs. In chromatography, solid phase extraction and electrochromatography this adsorption is a dynamic process. The dynamic process combined with the nonlinear adsorption isotherm of the polymers and other factors results in complications which have limited the success of imprinted polymers. This chapter explains these problems and shows many examples of successful applications overcoming or avoiding the problems.

Keywords Fluoroquinolone · Mycotoxin · Nitrophenol · Propranolol · Thiabendazole · Triazine · Zearalenol

Contents

B. Tóth and G. Horvai (✉)
Department of Inorganic and Analytical Chemistry, Budapest University of Technology and Economics, Szt Gellert ter 4, H-1111 Budapest, Hungary
e-mail: george.horvai@mail.bme.hu

The majority of papers written on MIPs deal with their testing or application in a chromatographic, solid-phase extraction (SPE), or capillary electrochromatography (CEC) system. Many excellent reviews have been written in the last decade on such MIP systems [1–29]. These reviews discuss the large variety of templates used, the polymerization strategies, the formats of the separation system, and the optimization of the separation process. In earlier years there were only very few applications on real samples to be reported. This has changed recently and reviews are also becoming available on this topic [30].

This chapter does not intend to be another exhaustive review of all work done to date or in recent years. We rather attempt to direct attention to still existing problems and hope to contribute to a better appreciation of these problems. We believe that the slow progress of MIP methods in the direction of practical applications has been partly due to the lack of a clear and systematic discussion of some of the problems encountered. Two important topics of this kind are the peak shapes in MIP HPLC, SPE, and CEC, and the problems with defining MIP selectivity. Discussion of these topics requires appreciation of some elements of nonlinear chromatography, which will be presented in this chapter.

We also try in this chapter to look at MIPs from the point of view of a practicing analytical chemist who has a large inventory of tools other than MIPs available to solve his or her problems, and therefore looks critically at the MIP offer. On the other hand, we limit our discussion to analytical separations and do not consider technological applications of MIPs.

The chapter concludes with a thorough analysis of selected examples of MIP applications published recently.

1   The Possible Roles for MIPs in Modern Analytical Chemistry

Separation techniques have remained indispensable tools of the analytical chemist despite the great progress in selective measurement techniques like binding assays, enzymatic methods, or the use of tandem mass spectrometers. This is probably so because the typical task of today's analytical chemist is to determine multiple analytes at trace level in complex sample matrices. Often these tasks have to be solved in a very short time (days), including the method development and validation. Further on the method has to be rather robust, easily transferred to other laboratories, and the reagents should come from reliable commercial sources to ensure their availability in uniform quality over many years. The cost of method development should often be low enough to remain economical even if the number of samples to be analyzed will be limited to dozens or hundreds and not millions like in clinical chemistry.

These requirements are difficult to satisfy with, e.g., immunoassays. Instrumental techniques like MS/MS require sample clean-up and sometimes even an online coupled separation method. Thus separation techniques, first of all SPE, GC, and HPLC, remain indispensable tools of analysts. An expert in analytical separation can nearly always today solve the complex tasks described above. The large variety of commercial equipment and stationary phases provide him with the necessary tools in most cases. The question may arise, then, if he really needs MIP stationary phases. At this time there are many possible answers to this question and only the future will give the final answer.

MIPs have several features which may make them interesting for analytical chemists. A large body of literature shows that they are reasonably easy to synthesize for many interesting analytes, the synthesis is sufficiently reproducible, the MIPs obtained are stable for years, and the costs of production are low. The selectivity of binding is often predictable, at least in the sense that the template is better bound than its analogs or its enantiomer.

Where in analytical chemistry can these features be advantageous? Analytical chemists cannot always solve their problems with typical chromatographic or electrophoretic separations. In some of these cases they use affinity columns or affinity SPE. Affinity separations rely on reversible and very selective binding of the analyte to a biomolecule, e.g., antibody. Making the analyst's own preparation of affinity phases is not economical in most cases, so one has to rely on commercially available material. If this is not easily available the analyst may consider making an MIP, probably in the SPE format, because MIP preparations are fairly easy for any chemist.

A problem related to the previous one is the separation of a group of similar compounds from a complex sample matrix. A group of chemically similar compounds, like a family of drugs or pesticides, may include widely dissimilar members with respect to the properties on which typical separations are based. For example, the hydrophobicity of a drug and its metabolites often differ substantially despite relatively small changes in the overall structure of the molecule. In such situations the inherent shape selectivity of MIPs may be quite useful to achieve a group separation.

In practical life it may be important that not all chemists, and not even all analytical chemists, are experts in separation science. For these chemists it may be more economical to make an MIP than to try to solve their analytical problem by attempting chromatographic method development. But even the separations expert may encounter problems which are too complex to solve with a reasonable amount of effort, and then he may combine an MIP pre-separation with conventional techniques.

Much has been written about the usefulness of MIPs for chiral HPLC or chiral CEC separations. As long as one has an expert in chiral separation and can afford the expensive chiral HPLC columns which are commercially available, there seems to be little incentive to use MIPs for this purpose [25]. Chiral separations are today often required in the pharmaceutical industry, where the expertise and the financial resources are available to solve most problems. In the future, however, there may be more interest in chiral separation by other scientists, e.g., biochemists, and this could contribute to a wider use of MIPs in analytical chiral separations.

2 HPLC with MIP Columns

MIPs have been frequently used as stationary phases in HPLC. It is useful to distinguish two types of HPLC experiments with MIPs:

- Using HPLC to test new MIPs
- Using MIP HPLC columns to separate substances of interest

New MIPs are often tested by filling (or polymerizing) them in HPLC columns. Typically it is shown that the retention of the template on the MIP HPLC column is greater than on the control (NIP) column and it is also greater than the retention of some substances chemically related to the template. Such experiments – despite their widespread use – are of limited quantitative value as we shall show later in this chapter.

MIP HPLC columns have also been used to carry out analytically relevant separations. These fall into two categories: the separation of enantiomers from each other (usually in a very simple matrix) and the separation of the template from other substances in a complex sample. Due to the typically very wide peaks in MIP HPLC, such applications can become useful only in special cases, which will be presented in this chapter.

2.1 *Peak Tailing and Some Basics of Nonlinear Chromatography*

Any use of MIP HPLC columns carries the burden of unusually wide and tailing peaks. These features are most obvious with the template itself but other analytes also show it quite often. Many explanations have been proposed to explain this phenomenon, including the following:

- Too large MIP particles
- Irregular shape of MIP particles
- Inhomogeneity of binding sites

- Slow mass transfer
- Nonlinearity of the adsorption isotherm of the analyte on the MIP

All of these factors may indeed contribute to the problem, but their effects have rarely been separated and compared. There have been many successful ways found to make small, uniformly sized, spherical MIP beads (e.g., by precipitation polymerization or by making MIPs in the pores of porous, spherical particles or by multi-step swelling), but the width and asymmetry of the peaks could be hardly reduced in this way. Therefore the remaining three items of the above list (site inhomogeneity, slow mass transfer, nonlinear isotherm) appear to be more important. Isotherm nonlinearity has practically always been observed with MIPs, so its effects always need to be considered. Binding site inhomogeneity and isotherm nonlinearity are not independent from each other. Binding site inhomogeneity practically always results in a nonlinear binding isotherm. But even in the case of fully homogeneous binding sites the isotherm (in this case the Langmuir isotherm) becomes nonlinear when more than about one quarter of the binding sites is occupied. Thus, if the binding site concentration is low (e.g., less than 0.1 mol kg^{-1}) one has to consider problems with nonlinearity even in analytical (i.e., not preparative) separations.

The influence of isotherm nonlinearity is schematically shown in Fig. 1 using numerical simulation. The first part of the figure (Fig. 1a) shows a nonlinear isotherm. Figure 1b shows the corresponding chromatographic peak shapes at different injected concentrations of the analyte, assuming that the theoretical plate height (HETP) on the column is quite low (the plate number N is high). (The plate height or plate number needs to be determined independently by injecting a compound not retained on the column or retained only by nonspecific effects, like hydrophobicity.) Figure 1b clearly shows that the peaks are wide and asymmetrical. It also shows that the peak maximum's position depends very much on the concentration (or the amount) injected. Figure 1c shows the same peaks as Fig. 1b but this time with a much lower plate number on the column. The dominant role of isotherm nonlinearity is still obvious in this case, too.

The main conclusion from Fig. 1 is that the well-known ideas of linear chromatography cannot be used when the isotherm is not linear. For example, the position of the peak maximum should not be used for the determination of a retention factor k. As Fig. 1 shows, such a k value would depend on the sample concentration and thus it might not be used for characterizing the retention of the analyte independently from its concentration. The width of the peak also has little to do with the number of plates. It will also be shown later in this chapter that the separation of two analytes cannot be simply characterized by a selectivity factor α, which is calculated as the ratio of the corresponding k values.

Several papers have shown that isotherm nonlinearity is indeed an important factor in MIP HPLC [31–35]. The consequences will be dealt with in the next sections. Mass transfer kinetics, on the other hand, has not yet been sufficiently studied. Its role has been proven in some cases [36] but it is not known how general these observations are and how important the effects are generally.

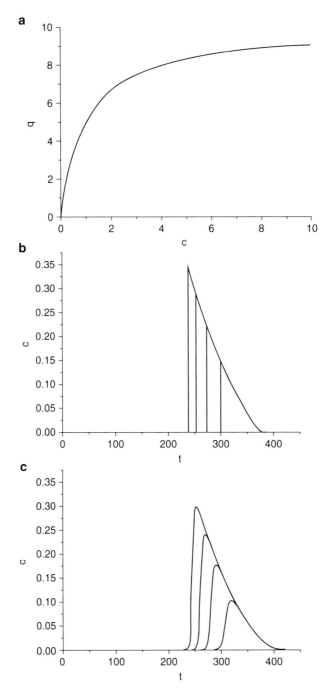

Fig. 1 Nonlinear effects in elution chromatography: (**a**) nonlinear adsorption isotherm; (**b**) the corresponding chromatographic peak shapes at $N = 10,000$ for different injected concentrations; (**c**) the corresponding chromatographic peak shapes at $N = 500$ for different injected concentrations. Isotherm: Langmuir type. Concentrations: 2.5, 5.0, 7.5, and 10.0 units. Length of column: 10 cm

2.2   Characterization of MIPs by HPLC

It has been shown above that HPLC is not ideally suitable for MIP characterization. Equilibrium batch adsorption experiments are probably much better suited for this purpose, but they are certainly tedious and conclusions are not so easily made from them, since one has to compare two isotherms (i.e., two curves) and not just two numbers.

Although HPLC (in the elution mode) is not giving us accurate characterization data (unless the whole peak shape is carefully analyzed and the kinetics is either fast or well understood), it is not without useful qualitative information. High retention values (compared to t_0) and long peak tails are signs of strong adsorption. Substantial differences in the peak position of two compounds (when identical amounts have been injected) show substantial differences in adsorption. Great differences of template retention on the MIP and the NIP columns, respectively, are a sign of great differences in adsorption on the two materials. These remarks justify to some extent the widespread use of HPLC for MIP characterization. However, quantitative conclusions should not be made in this simple way.

2.3   MIP HPLC for Analytical Separations

The problems with the wide, tailing peaks have prevented the widespread use of MIPs in HPLC. Recently, however, a number of studies have shown that MIPs can still have a place in the inventory of the chromatographic lab. While the more important MIP format for handling real samples appears to be SPE, use of MIP HPLC columns can also be useful.

The separation of a group of chemically related substances from complex samples is often required. An MIP column can be used when the MIP shows little selectivity between the members of the group but large selectivity against other substances. The unseparated peak of the group can then be collected and reanalyzed for its components. A nice example for such application will be shown at the end of this chapter in Sect. 7.1.

Other applications are possible when the usual inventory of analytical tools does not help. If a chiral separation is not easily solved by other means, it may be worth making an MIP for one of the enantiomers. It is worth noting that in cases where only the ratio of the two enantiomers is needed there should be less problems with peak tailing. This is because usually only the isotherm of the imprinted enantiomer is badly nonlinear. Thus the peak of the other, earlier eluting enantiomer is less tailing. The analytical information is already obtained from this first peak, so one may speed up such analyses substantially.

Other separations than chiral may also sometimes be too difficult on a single HPLC column and within reasonable time. In such cases a two-dimensional separation employing a conventional and an MIP column may be feasible. Several authors have used online MIP SPE (MISPE) precolumns in different column switching systems. Examples of this will be given later in this chapter as Sects. 7.2, 7.8, and 7.11. These applications may be considered as special cases of 2D-HPLC.

3 The Selectivity of MIPs

In analytical chemistry selectivity means that a method or tool is suitable to determine the concentration of a particular analyte independently from the composition of the sample matrix, at least in a wide range of possible sample compositions. The usual practical way to ascertain selectivity is to consider the chemistry behind the method and consider possible interferences in the range of samples to be expected.

Typically, interferences can be expected due to the presence of certain compounds in the sample. It is useful then to make an estimate of the maximum allowable concentrations of such compounds (individually and together) which do not bias the determination of the analyte more than to a certain limit. The maximum allowable concentrations of interferents usually depend on the analyte concentration to be determined, higher analyte concentrations allowing the presence of higher interferent concentrations. When the ratio of the analyte concentration to the maximum allowable concentration of an interferent is approximately independent of the analyte concentration, then this ratio (a "selectivity coefficient") may be considered as a measure of the selectivity of the method (or the instrument).

3.1 Selectivity Determined by HPLC

In the case of HPLC methods, selectivity is usually characterized by the ratio of the respective retention factors ("k" values) of two compounds. This "α" value has useful properties in linear chromatography, i.e., when the adsorption isotherms of both substances are linear, with slopes K_1 and K_2, these being the corresponding distribution ratios between the stationary phase and the eluent. In linear chromatography

$$\alpha = \frac{k_2}{k_1} = \frac{K_2}{K_1}.$$

Thus the selectivity "α" has a thermodynamic interpretation as the ratio of two distribution constants. Consequently "α" is itself a constant, independently from the injected concentrations of the analyte and the interferent, respectively.

The selectivity "α" in linear HPLC does not express selectivity in the same sense as we have defined the selectivity of an analytical method above. It is still a reasonable measure of analytical selectivity, because the resolution "R" of two neighboring peaks (in this case the analyte and the interferent peak, respectively) is directly proportional to $\alpha-1$. If $R > 1.5$ the two peaks are resolved to their baselines, so that the separate quantitation of the analyte is not disturbed by the presence of the interferent unless the latter is very much higher in concentration.

In the MIP literature the most widespread method for establishing the selectivity of an MIP for its template against an interferent has been to determine by HPLC the corresponding "α" value. In the section on MIP HPLC we have shown, however, that the "k" values determined in MIP HPLC depend on the injected sample concentration. Therefore they do not provide a distribution ratio. Indeed, the distribution ratio is obviously concentration dependent when the isotherm is not linear. It is not self-evident but it has been shown [35] that the ratio of the apparent "k" values (as determined from the peak maximum positions) for two compounds with nonlinear isotherms is not a constant and such an "α" value is not suitable to quantitate the separation selectivity.

Further on, in nonlinear chromatography the relationship between "α" and "R" does not exist in the form known from linear chromatography. The definition of "R" in linear chromatography occurs in terms of the Gaussian peak widths of the peaks but in nonlinear chromatography the peaks are by far not Gaussian (see Fig. 1). Thus even if "α" values determined from peak maxima of two peaks of similar injected concentrations may be used as rough measures of the (concentration dependent) selectivity of MIPs (an "α" being greater than 1 representing higher retention of the template compared to the interferent), one cannot easily estimate from this "α" how well the two substances are separated. The efficiency of separation depends to a great extent also on the tailing of the peak of the first eluting substance.

A very much neglected further topic about the chromatographic selectivity of MIPs is the role of competition. It is well known that compounds chemically similar to the template may compete for the MIP binding sites (see the chapter dealing with MIP binding assays). Therefore one may expect that, when the template is coinjected with a similar compound onto the HPLC column, these two compounds will compete for the binding sites of the MIP. Thus one can expect that the retention of each compound depends on the presence of the other. In other words, the separation cannot be predicted from the peak positions and shapes obtained with the two compounds injected separately. Yet it has been quite common with MIPs to determine their selectivities (usually as "α" values) from separate injections of template and interferent. Surprisingly, we have found [37] that the distorting effect of coinjection on the individual peaks is often rather small, particularly when the concentrations of both compounds are similar. In more complex situations, however, this is not true.

3.2 Selectivity Determined by Equilibrium Measurements

As we have seen, the HPLC determination of MIP selectivity is of limited value. The question naturally arises as to whether a better method is available. Since this is not the subject of this chapter, we mention only briefly that batch distribution measurements of the template (or of any other substance, separately or in mixture) appear to be useful due to the equilibrium nature of the batch method. The correct interpretation of the batch measurement results is, however, a further difficult question. For the characterization of selectivity researchers have often measured the individual isotherms of two substances. These isotherms have been fitted to one or another generally used isotherm equation (e.g., Langmuir, bi-Langmuir, Freundlich, etc.). The obtained parameters of these equations (e.g., the respective binding constants of two Langmuir isotherms) have been compared (divided) to express selectivity. This method neglects the possibility of competition, and its foundations are also not quite solid. It has often been observed (with MIPs as also with other sorbents) that the same measured isotherm fits different isotherm equations equally well. Thus the physical meaning of the fitted isotherm parameters is often questionable. Therefore selectivity values based on the constants' physical interpretation as equilibrium binding constants should be used with caution.

It may seem from the above discussion that it is impossible to use even batch isotherm measurements to design HPLC (or SPE) separations. This is not so, however, at least when the HPLC separation occurs under near-equilibrium conditions. Nonlinear chromatographic peaks can be simulated [38] once the corresponding isotherms have been measured. In this case one does not need a physical interpretation of the isotherm equation's constants; they can be regarded merely as interpolation factors. Separately measured isotherms of the two compounds are satisfactory in many cases because – as mentioned above – competition often has only minor influence on the separated peaks' position and shape.

MIP selectivity has often been related to observed differences of binding the template by the MIP and the NIP, respectively. This is in itself not a measure of the selective binding of the template by the MIP. The reason for using the MIP/NIP comparison is that a common model of MIPs assumes that the MIPs have two types of binding sites: selective and strong binding sites for the template and nonselective, weaker binding sites for interferents plus the template. Thus the extra binding on the MIP compared to the NIP would be due to the selective binding sites. This simple model may hold in some cases but it does not seem to be generally true.

A similar over-interpretation of measured data is to attribute high selectivity to MIPs which bind their templates very strongly. This is again based on the model with strong, selective sites for the template and weak, nonselective sites which bind the interferents. However, in a practical application of the MIP there may be other substances present in the sample, sometimes unrelated chemically to the template, which are also bound strongly. (Such substances have been used occasionally as "non-related" probes in competitive binding assays based on MIPs.) Therefore selectivity studies should be as extensive as possible.

4 Solid-Phase Extraction with MIPs (MISPE)

The ideal application for MIPs appears to be in SPE. SPE is generally used to improve the selectivity of analytical measurements by pre-separating the analytes from many other matrix components. In some cases the analytes are also preconcentrated on the SPE column.

Traditional (i.e., non-MIP) SPE sorbents are similar to HPLC stationary phases. The advantages of many of these materials are that they are widely available, well characterized, have high binding capacity, and show linear adsorption behavior. One may observe that just a few types are used in the majority of sample preparations, i.e., these materials are quite generic, and it is the wash and elution step which is varied according to the application. The generic nature of these materials is also a drawback because it reflects their limited selectivity.

Recently, in many demanding sample preparation situations, more selective sorbents have been used. Affinity type sorbents, particularly immunosorbents, have gained popularity in trace analysis, not only for biological macromolecules but also for small molecules, like aflatoxins. MIPs have properties resembling those of affinity phases and therefore they may find unique applications where other sample pretreatment methods are tedious. This relates both to the separation of a single analyte and to the separation of a group of related analytes from the sample matrix.

Considering the expected wider use of MIPs as SPE sorbents, we discuss below some general rules of thumb for developing and using MISPE methods.

4.1 MISPE Procedures

4.1.1 Preparation of the MIP

Methods for making an MIP with appropriate selectivity are the subject of other chapters. Here only the format of the MIP is mentioned. Crushed and sieved bulk MIPs have been widely used for MISPE but uniformly sized and spherical formats (obtained e.g., by precipitation polymerization) are more convenient. The cost of making the MIP is not negligible if the development work is included. However, this is still less than the cost of making a new affinity phase and the MIPs are stable, easily reproduced, and may even be reused, if so wished.

The most usual form for SPE is the cartridge format but membranes or high pressure precolumns may also be used. MISPE devices have also become commercially available.

4.1.2 Sample Loading

After adequate (and sometimes lengthy) cleaning of the MIP and conditioning with a sample-like solvent, one may load the sample on the SPE column. The samples may be aqueous or can be based on an organic solvent. With aqueous samples the retention of the analyte occurs mainly due to hydrophobicity and/or ionic charge, and is not particularly selective. With samples in organic solvents the loading step may be the first stage of the selective separation due to the selectivity of interactions with the MIP. On the other hand, the breakthrough of the analyte may also occur unless its interaction with the MIP is very strong in the organic solvent used. This often limits the maximum volume of sample that can be applied.

4.1.3 Washing

The washing step is the most crucial one in SPE procedures. The goal is to let the selective binding forces retain the analyte while most other compounds should be washed off.

An SPE cartridge may be considered as a short chromatographic column with a very low number of theoretical plates. In MISPE this is still true but the peak shapes are also influenced by the nonlinearity of the isotherms. As shown in Fig. 1, plate number and nonlinearity both have their peak broadening effects. In the case of MISPE both effects are comparable due to the very low plate numbers. These remarks allow the conclusion that the analyte needs to be bound very strongly on the MISPE column (i.e., with a high distribution ratio at the given concentration) so that a few bed volumes of washing solvent should not elute it at all. In view of the expected wide peaks the retention of interferents should be substantially lower than that of the analyte. Luckily the peaks of the interferents are usually much less tailing than that of the template. Moreover the front of the template peak is relatively sharp. These features are quite useful because in MISPE one needs only the complete elution of the interferents before the template begins to be eluted. Preliminary experiments on an MIP HPLC column (and comparison with the NIP column) may help to find the suitable conditions. Wash solutions consisting of methanol, with acetic acid or dichloroacetic acid added, are quite common.

It has become a standard procedure to test how much the washing step removes the interferents and how little of the analyte is removed during washing. This test is useful to support the method development, but often it is based merely on a trial and error approach, so that any changes in the method or the sample matrix would require a reoptimization.

A difficult problem during the wash step is the presence of water. Some MIPs are working satisfactorily with partly aqueous wash solvents but the template retention of many MIPs is greatly reduced by small concentrations of water present. Samples can still be applied as aqueous solutions because at high water concentrations the template (and many other substances) are retained due to their hydrophobicity. After application of an aqueous sample the wash step may occur with pure non-aqueous solvent. However, the small amount of water remaining on the cartridge from the sample may suffice to destroy retention or at least to make its efficiency badly reproducible. Air drying of the cartridge has been found useful but in some cases this is not sufficient and the reproducibility of the method suffers. Applying the aqueous sample to a less selective sorbent at first, such as C18 or the NIP, and transferring the adsorbed substances to the MIP with a nonaqueous solvent has been found a useful solution to the problem [39]. Even better is, of course, to use MIPs working in partly aqueous solutions. Recent research has produced a variety of such

MIPs. These are made by adding a hydrophilic monomer to the polymerization mixture, e.g., HEMA, PETRA, or methacrylamide. Some urea polymers have also been found useful. In some cases, however, the hydrophilized MIPs were less satisfactory, perhaps due to interactions during the polymerization between the hydrophilic and the functional monomer.

4.1.4 Elution

The last step is elution of the analyte with a solution which can release the analyte from the binding sites. The stronger the eluent the less eluent volume is needed and the more concentrated the resulting eluate will be with respect to the analyte. If the eluting solvent is not strong enough, a subsequent evaporation step may be used to concentrate the analyte. Thus in MISPE elution there is no serious problem with the tailing of the template peak.

A difficult problem with MISPE in trace analysis has been the bleeding of the template from the cartridge. This is probably due to template molecules being buried during polymerization under the MIP surface and later diffusing slowly but steadily to the surface. Bleeding is a disadvantage of MISPE against affinity SPE. Recently microwave assisted extraction or accelerated solvent extraction have been used to clean the MIP sufficiently before use and thereby to reduce the bleeding of the MIP. Analyte bleeding may be completely overcome by carrying out the imprinting with an analog compound of the analyte. Due to cross-reactivity (i.e., limited selectivity) the MIP so obtained may work well for the analyte while it bleeds only the analog used for templating. The selectivity of the MIP may suffer, however.

4.2 MISPE with Real Samples

A recent review [30] has demonstrated that in the last 5 years there has been a trend in the MISPE literature to go away from mere proof-of-principle demonstrations towards applications in real samples. The authors found 110 original papers on MISPE applications in real samples in this period of time. These have dealt with four main types of samples: environmental, biological, food, and drug. The MIPs used were mainly prepared by noncovalent methods and were used in off-line mode. The review gives a nice tabulated summary of all these papers with information about the templates, sample matrices, and detection methods.

MISPE is often followed by an HPLC determination but there have also been interesting instances of going on directly to detection, e.g., by electrochemical or spectroscopic methods.

It is interesting to mention that occasionally one may use batch MIP extraction instead of the MISPE, i.e., instead of the column or cartridge format, even with real samples [40].

5 MIP Membranes for Separation

MIPs can be prepared in membrane or film formats, unsupported or supported by an inert matrix. There are many possible varieties and also various potential applications for MIP membranes. Probably the most important application is in sensors. In this chapter we deal only with applications for separation and particularly those based on adsorption.

One way to apply MIP membranes for separations is to construct them as selectively permeable membranes. Since a membrane with large pores (compared to the molecular size of the template and the interferents) is necessarily easily permeable for all these substances, selective permeability may only be expected from nonporous membranes or membranes with very small pores. Work with selectively permeable nonporous MIP membranes has been limited and the underlying mechanisms may hardly be categorized as adsorption. Therefore a more specific recent review (M. Ulbricht in [3], p 455) should be consulted in this respect.

Porous MIP membranes are much closer to our topic. These are similar to SPE disks which are commercially available and commonly used in analytical laboratories. Such disks allow the collection of low concentration analytes from large volumes of sample. This type of problem is often encountered in environmental analysis. The disks have high flow permeability and show fast adsorption kinetics so that large sample volumes can be extracted with them in a short time. One may consider these disks as extremely short and wide chromatographic or SPE columns. The purpose may be mainly the concentration of the analyte(s) but separation may also be achieved between substances differing rather much in their distribution coefficients.

Porous MIP disks may be made, e.g., by incorporating MIP particles in a mesh-like inert network or by polymerizing MIPs in the pores of a porous inert membrane. The MIP so generated may either fill the pores (without clogging the membrane since the MIP is itself porous) or the MIP may form a thin film on the pore walls of the inert substrate.

The possibility of making such MIP membrane disks has been well demonstrated yet we do not know of real-life analytical applications. This may be related to the low selective capacity of the MIPs, the slow adsorption kinetics, and the known bleeding of MIPs which may be disturbing in environmental ultratrace analysis. A further problem is that the matrix of the large sample volume (typically aqueous) may not be compatible with the MIP separation.

An alternative and relatively simple approach to the same analytical problem is to carry out the large volume sample extraction with a commercial SPE disk, e.g., C18 disk, and then elute the analyte from the C18 disk with a smaller volume of suitably selected solvent onto an MISPE column [39].

Recently, porous MIP membranes have been made to accelerate MIP development [41]. MIPs prepared in this format are easily cleaned after preparation. Parallelization of preparation and testing is also easily done in multiwell filter plates.

6 Capillary Electrochromatography with MIPs

CEC has recently become an alternative to HPLC. A capillary is filled or its internal wall covered with a porous sorbent. The free volume remaining in the capillary is filled with an electrolyte. High voltage (on the order of ten kV) is applied across the length of the capillary. Sample plugs are introduced at one end. Sample components are carried to the other end due to electro-osmosis and – in the case of ions – also electrophoresis. In CEC the more important effect is electro-osmosis, which is essentially a flow mechanism of the electrolyte solution without the need for applied pressure. The separation of the sample components occurs mainly due to phase distribution between the stationary phase and the flowing electrolyte. Thus CEC is very similar to HPLC in a packed capillary except that the flow is not pressure driven and that ionic analytes undergo electrophoresis additionally to phase separation.

Electro-osmotic flow depends largely on the zeta potential arising at the interface of the stationary phase with the electrolyte. For this potential to develop, the stationary phase has to carry some amount of ionic or ionizable groups at its surface. When MIPs are used as stationary phase in CEC their composition has to be designed with this point in mind besides the usual criteria for an MIP. Some of the more usual MIP compositions, like MIPs made with the functional monomer methacrylic acid, satisfy this criterion quite well. It is also important that the stationary phase is well retained in the capillary. Frits at the ends of the packing are difficult to make in capillaries, and therefore monoliths covalently bound to the (prederivatized) surface of the capillary are advantageous. The preparation of MIP monoliths has been mostly promoted in the CEC field.

The main argument for making MIP CEC is to combine the selectivity of the MIPs with the high separation efficiency of CEC. This argument appears to fail, however, if the adsorption isotherm of the MIP is nonlinear, which seems to be the rule. In the case of nonlinear isotherms, the peak shapes depend mainly on the isotherm, particularly so if the separation system is otherwise very efficient (has low theoretical plate height, see Fig. 1). In the case of ionized analytes the situation is more complex. If an ionized analyte is not adsorbed at all on the MIP, then it is separated only due to electrophoresis, and its peak will not be widened due to the nonlinear effect. In this case, however, the MIP is merely behaving like an inert porous material. In intermediate cases an ionized analyte may participate in both separation mechanisms and for this case we do not have exact predictions of the peak shape.

The problem that reduction of the plate heights will not considerably improve the peak shapes on MIP phases has perhaps not been recognized immediately when MIP CEC was developed. A recent (2006) review by Ch. Nilsson and S. Nilsson [13] concludes, however: "So far all verified imprint-based CEC separations showed peak tailing, due to polydispersity of the imprinted receptor sites, resulting in different affinity and poor mass transfer." Although we do not agree with the exact wording (because the isotherm may be nonlinear even if all sites are the same and because the cause of tailing on MIPs has perhaps more to do with the nonlinear isotherm than with slow mass transfer), this statement made by one of the prime developers of MIP CEC seems to justify our conclusion made above. A further important statement on the same problem has been made in a 2005 review by E. Turiel and A. Martin-Esteban [14]. These authors say that nearly all reviewed MIP CEC methods have been directed toward the separation of racemic mixtures. Separation of "completely different" analytes could only be demonstrated using the partial-filling MIP CEC technique (see below) with mixtures of MIPs or with MIPs made towards two templates. The apparent reason behind their observation is the wide tailing of the template peak (which does not prevent chiral separations where the template is the later eluting enantiomer).

One should mention, however, that MIP CEC has the potential to be carried out in long capillaries because the mobile phase is not driven by pressure. Longer columns lead to better separations in nonlinear chromatography too. In some cases capillaries about 1 m long have been used. The problem with long capillaries may be the change of the electrolyte, which is usually done by applying pressure. Another observed trend in MIP CEC has been just the opposite: short capillaries (around 10 cm effective length) and high field strengths (up to 850 kV cm^{-1}) have been used to achieve rather fast (less than 1 min) chiral separations. Such short separation systems may indeed find their way into micro total analysis systems.

It is worth mentioning here that comparisons between the efficiency of different MIP separation systems like two HPLC systems or two CEC systems, or an HPLC system with a CEC system, are quite difficult when the adsorption isotherms are nonlinear. One of the typical difficulties is that the phase ratios in the two systems may be different. The effect of phase ratio on the separation and particularly on the achievable optimum separation is a complex question even in linear chromatography. In nonlinear chromatography this is really difficult and also burdened by the differences between the isotherms of the two compounds to be separated. The complexity of this matter has been mostly overlooked in the MIP literature and the visual comparison of two separations in rather different systems, operated under very different conditions, has frequently lead to statements declaring one technique better than the other.

Much ingenuity has gone into the technical development of MIP CEC systems. These have been nicely reviewed in 2007 by Liu et al. [20]. Here only one technique is mentioned: the use of MIP (nano)particles in the background electrolyte. Similarly to the better known micellar electrokinetic chromatography (MEKC), where a micellar solution is employed in the electrophoresis capillary, one may disperse a solid adsorbent in the electrolyte as well. To avoid settling of the particles these should be nanometric in size. The particles need to be charged so that they migrate with a speed different from the electro-osmotic flow. Analytes distribute between the electrolyte and the solid particles and therefore their effective velocity will be a weighted average of the electro-osmotic velocity and the particle velocity. The weighting factor depends on the phase ratio and the analyte's distribution coefficient.

MIP nanoparticles (prepared for example by precipitation polymerization) may also be used in this technique. To avoid problems with optical detection in the presence of particles a partial filling method has been employed [42]. Here the charge of the MIP particles has to be such that their overall velocity is slower than the electro-osmotic flow velocity. A plug of MIP particles is injected followed by the sample plug. The analytes are carried by the electro-osmotic flow through the particle plug, where they are differentially slowed down due to partition. Finally, the analytes leave the particle plug at its front end, so that they arrive – the components being already separated – at the detector before the particle plug itself reaches the detector. Thus there is no interference from the particles in the detector.

There are many advantages (but also some disadvantages) of using particles dispersed in the electrolyte as pseudostationary phase. There is no particle packing procedure needed. Frits are not used. The solid phase is always refreshed so that carryover from complex samples is no problem.

An interesting variant of the dispersed particles technique has been to use the partial filling method with a particle plug consisting of two differently imprinted MIPs [42]. The goal was to separate members of two chemically rather different groups of compounds. The adsorption of the members of the first group was considerably less than the adsorption of the second group on their respective MIPs. Cross-adsorption of one group on the other MIP was negligible. Thus the members of the first group arrived at the detector – separated from each other by their MIP – before the second group which was itself separated into its members. In a second variant of the same technique a single MIP was used. This MIP has been imprinted with a mixture of the two chemically rather different templates used to make the two MIPs in the previous case. The final result on the separation was similar to the previous one. These techniques are really masterful but the difficulties of optimizing such systems (including optimization of the MIP compositions) should not be overlooked.

7   Selected Examples

After the general discussion of our topic it is useful to look at a few examples. The literature on MIPs as employed in HPLC, SPE, CEC, and membrane separations is vast and steadily increasing. Therefore examples need to be selected and there is little chance to show all interesting examples in a short space. Thus our selection is necessarily very subjective and far from comprehensive. As mentioned earlier, there exist excellent reviews on the topic from the last decade which cover quite thoroughly some narrower time spans or narrower subfields.

7.1   Example 1 [43]

This recent paper shows an excellent example of how MIPs can be used for practical analysis both in the HPLC column and the MISPE format. The high selectivity of the MIP for a group of antibacterial compounds (similar to the template) has been used in the HPLC format for a fast screening of soil samples. A more thorough analysis of the positive samples can then be made by a reversed phase HPLC method which employs sample clean-up on a MISPE.

In recent years antibiotics residues have accumulated in different environmental compartments. Fluoroquinolone antibiotics are strongly adsorbed by soil, and therefore their extraction requires exhaustive procedures, leading to complex extracts due to coextraction. Despite this the authors succeeded in selectively detecting the fluoroquinolones from crude soil extracts directly injected onto an MIP HPLC column and using UV detection [Fig. 2 (Fig. 6 of the original)]. Recoveries close to 100% were obtained for five fluoroquinolones. The analysis time was only a few minutes, so this method was found suitable for screening soil samples for the presence of fluoroquinolones. The individual fluoroquinolones were not resolved on the column.

For the determination of the individual fluoroquinolones the authors developed a MISPE clean-up method followed by HPLC–UV on a C18 column. Figure 3 (Fig. 7 of the original) shows the substantial clean-up effect due to MISPE. Recoveries for five fluoroquinolones tested were around 80%. The selectivity of the MISPE clean-up

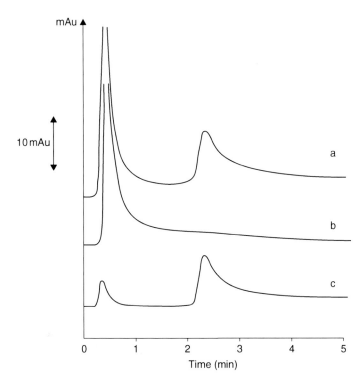

Fig. 2 Chromatograms obtained in the analysis of (**a**) a soil crude extract spiked with 0.5 μg g^{-1} of the fluoroquinolone antibiotic ciprofloxacin (CIP), (**b**) non-spiked soil extract, and (**c**) a standard solution of 0.5 μg g^{-1} of CIP prepared in methanol [43]

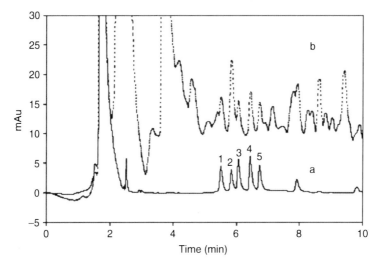

Fig. 3 Chromatograms obtained in the analysis of a soil sample extract spiked with 0.1 μg g^{-1} of the selected fluoroquinolones (**a**) with and (**b**) without performing MISPE. Peak assignment: 1 = enoxacin (ENO), 2 = norfloxacin (NOR), 3 = ciprofloxacin (CIP), 4 = danofloxacin (DAN), 5 = enrofloxacin (ENR) [43]

proved to be sufficient to differentiate between fluoroquinolones and the closely related quinolones. The authors compared this method with alternative existing methods based on conventional SPE clean-up and HPLC–MS. They found that the sensitivity and the selectivity of the new MISPE–HPLC–UV method were similar or even better than of the existing methods.

One should note that, according to the authors of this paper (submitted in April 2007), only three other papers had dealt before with using selective MIP HPLC columns for determinations from real samples.

7.2 Example 2 [44]

This example shows a comparison of a noncovalently and a semicovalently imprinted MIP in MISPE. It also shows the development of an online coupled MISPE–HPLC system for environmental analysis of river water.

Caro et al. made a noncovalent and a semicovalent MIP, respectively, for 4-nitrophenol (4-NP). Their purpose was to extract selectively 4-NP from river water in the presence of ten other phenolic pollutants. The MIPs were used in an online MISPE procedure employing dichloromethane as wash solvent. Figure 4 (Fig. 1 of the original) shows the experimental setup. The pump on the left is used to deliver the conditioning solutions, the sample, and the wash solvent. The two pumps on the right deliver the organic and the aqueous components of the HPLC

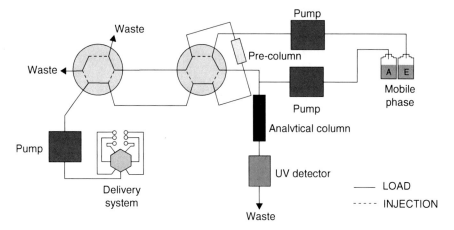

Fig. 4 Set-up for online MISPE (see text) [44]

eluent, respectively. The organic phase of the HPLC eluent also serves as the eluent of the MISPE column. The aqueous part of the HPLC eluent is mixed with the organic one only after the MISPE column. This interesting setup allows the direct elution of the cleaned sample from the MISPE onto the HPLC column. Thus the usual incompatibility of the MISPE elution solvent with the HPLC conditions is circumvented.

Figure 5 (Fig. 2 of the original) shows that with both MIPs the increase of wash solvent (dichloromethane) volume has led to cleaner eluates for 4-NP but also the peak area of 4-NP decreased substantially. The chromatograms of the eluates were cleaner when the non-covalent MISPE was used while the retention of 4-NP was better with the semicovalent MISPE.

The analysis of river water samples with this system was further complicated by the interference from humic acids. This interference could be eliminated by adding some Na_2SO_3 to the samples. Figure 6 (Fig. 4 of the original) shows the respective chromatograms. Note that the removal of humic acids has been solved in a different way in another online MISPE system by Koeber et al. [45]. These authors had used a restricted access material (RAM) precolumn. The RAM material retains the low molecular weight analytes while the high molecular weight interferents are easily washed off.

One should note that the MISPE system served two goals here. First, the 10-mL samples were substantially preconcentrated on the MIP. Second, the levels of the other phenolic pollutants were selectively diminished against 4-NP by using the optimized washing procedure. While this example is educative it is not quite clear why the selective wash was needed. Figure 5 (Fig. 2 of the original) shows that the selectivity of HPLC alone was sufficient to determine 4-NP.

A further interesting feature in this paper has been how the increase of wash solvent volume influenced the retention of 4-NP on the MISPE column when mixtures of 11 phenols were used as the sample. For example, in the case of the

Fig. 5 Chromatograms obtained by online MISPE with the non-covalent 4-NP imprinted polymer of 10 mL standard solution (pH 2.5) spiked at 10 µg L^{-1} with each phenolic compound. (**a**) Without washing step, and (**b–d**) with washing step using 0.1, 0.2, and 0.3 mL of dichloromethane, respectively: (1) phenol (Ph), (2) 4-nitrophenol (4-NP), (3) 2,4-dinitrophenol (2,4-DNP), (4) 2-chlorophenol (2-CP), (5) 2-nitrophenol (2-NP), (6) 2,4-dimethylphenol (2,4-DMP), (7) 4-chloro-3-methylphenol (4-C-3-MP), (8) 2-methyl-4,6-dinitrophenol (2-M-4,6-DNP), (9) 2,4-dichlorophenol (2,4-DCP), (10) 2,4,6-trichlorophenol (2,4,6-TCP) [44]

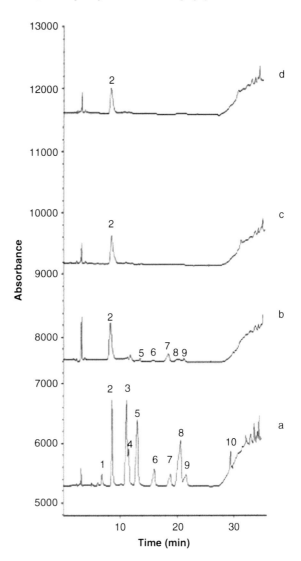

semicovalent MIP the 4-NP retention was 78%, 52%, 51%, and 50% after washing with 0, 0.2, 0.3, and 0.5 mL dichloromethane, respectively. Note that after a first sharp decrease the retention remains nearly constant. The authors attributed this to the elimination of the weaker, nonselective bonds by 0.2 mL dichloromethane. This explanation is unlikely unless 4-NP would produce peak splitting in an MIP column when eluted with dichloromethane. Such peak splitting being unusual, a more likely explanation is that the first 0.2 mL of dichloromethane removed the other phenols (as evidenced by the data of Table 2 in the paper). These compounds may have competed for the binding sites with 4-NP and thus reduced its retention. When they

Fig. 6 Chromatograms
obtained by online MISPE
with the non-covalent 4-NP
imprinted polymer (P1) of
10 mL Ebro river water
(pH 2.5) spiked at 10 μg L^{-1}
with each phenolic
compound. (**a**) With washing
step using 0.2 mL of
dichloromethane and (**b**) with
addition of Na$_2$SO$_3$ to the
washing step. Peak
designation as in the
preceding figure [44]

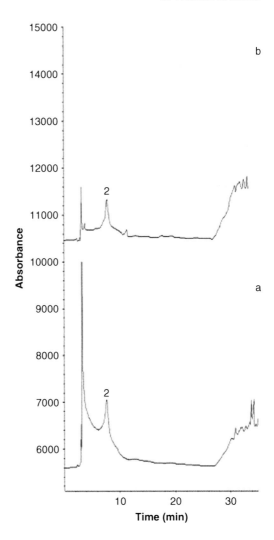

were removed the competition disappeared and the remaining 4-NP was not further
eluted by dichloromethane.

7.3   Example 3 [46]

This work shows that simple cations in real samples may interfere with MISPE
when the MIP is made with carboxylic functionality on it. The coauthors of this
paper had earlier participated in a work [47] where a triazine MIP was used to
analyze water samples. Water samples had been directly applied to the MIP. The

triazines were retained by hydrophobic forces. A subsequent wash with a small amount of porogen had removed the interferences while the analytes were selectively retained due to the prevalence of specific binding in porogen. The recovery of the analytes had been low, however. In the paper discussed here, the method has been improved by applying an acid wash after the sample application. This was found necessary because it has been found that Ca^{2+} ions present in tap and mineral water samples occupied the carboxylic sites by ion exchange and thereby reduced the retention of the triazines.

The same protocol could also be used to clean up an effluent form the textile industry. Figure 7 (Fig. 8 of the original) shows that the investigated triazines were retained on the MIP and the baseline was quite clean. An alternative method using conventional hydrophobic SPE produced a much worse baseline. Phenylurea herbicides were also spiked to the samples to check the selectivity of the procedure for triazines. The MIP proved selective (i.e., did not retain the phenylureas) while the traditional sorbent was not selective and also retained the phenylureas.

There are also some other features worth mentioning about this protocol. The authors dried the MIP between the aqueous and organic (dichloromethane) washes. This step was thought necessary because the two solvents are immiscible. However,

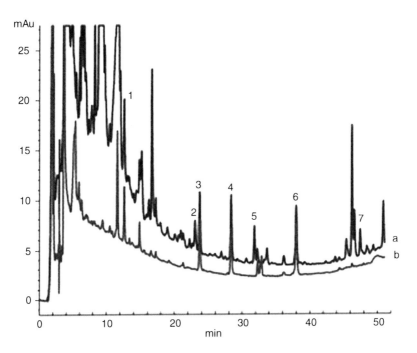

Fig. 7 Chromatograms obtained after the preconcentration of 50 mL of a diluted industrial effluent spiked at 1 μg L^{-1} with a mixture of triazines and phenylureas through a styrene-divinylbenzene (SDB) SPE cartridge (**a**) and through the terbutylazine MIP cartridge (**b**). Compounds: (1) deethylatrazine (DEA), (2) monuron, (3) deethylterbutylazine (DET), (4) atrazine, (5) diuron, (6) terbutylazine, (7) neburon. UV detection at 220 nm [46]

in other papers authors had applied immiscible phases without air drying. We believe that drying also helps to avoid eventual retention drops caused by the presence of water traces (remaining from the aqueous wash) in the organic wash solvent.

It is also interesting that the authors preferred 5 mL of dichloromethane wash to 2 mL because they found that the MIP/NIP retention ratio was much higher with 5 mL. This ratio was indicative of the ratio of retention (for a particular substance) by selective vs nonselective binding forces. The authors did not show, however, that this increased selectivity was really necessary in real samples. On the other hand, by going to 5 mL from 2 mL of wash, the recovery of three triazines (those with CH_3S groups) drastically dropped (below 20%) and recovery of all other triazines decreased by about 10–12%. This means that, with the 5 mL of dichloromethane wash, all analytes began to break through and the protocol may have lost from its robustness.

7.4   Example 4 [48]

This very interesting paper compares MISPE both with SPE on a "classical" polymeric sorbent (SDB) and with the clean-up on an immunosorbent. The comparison is made with real samples: grape juice and soil extract, respectively. Figure 8 (Fig. 5 of the original) shows that the MISPE cartridge and the immunosorbent cartridge are both very useful in cleaning up a soil sample for the HPLC–UV analysis of triazines. Actually the baseline is even cleaner with the MISPE sample pretreatment. Figure 9 (Fig. 4 of the original) shows that the SDB sorbent is practically useless in cleaning up the grape juice sample for low level triazines detection. The MISPE treatment produces a much cleaner baseline but even this result is barely sufficient to quantitate the triazines. One reason for this may be the low recovery of the analytes (estimated by us from the figure as ca. 15%).

It has been considered a disadvantage of MISPE that aqueous samples often need to be reconstituted in a suitable organic solvent before being applied to the cartridge. The examples of this paper show that in several practical problems the samples need to be concentrated to dryness anyway, even for the classical methods. In such cases the need for solvent change is not an extra burden.

This work shows very nicely the problems which may be encountered with different types of substances in MISPE. A series of triazines and triazine metabolites were investigated for their retention when using different sample application solvents. All investigated compounds were retained on the MISPE column when using dichloromethane. However, the retention of some substances was not due to a good fit into the imprinted sites. These substances were strongly retained even on the non-imprinted control polymer (NIP) in contrast to the others. This effect could be attributed to strong H-bonded interactions of the OH or primary amino groups of these substances with the acidic functional groups of the polymer.

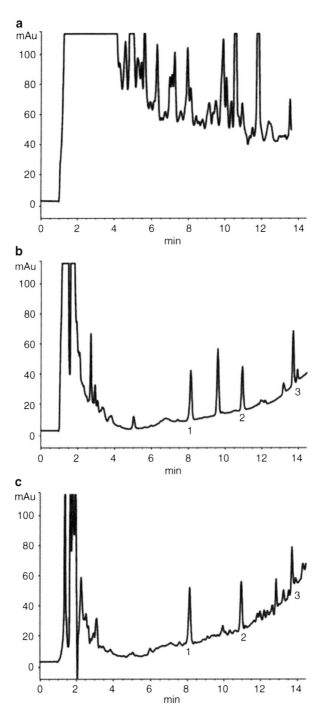

Fig. 8 Chromatograms obtained after the injection of a soil extract containing 20 ng g^{-1} of triazines (**a**) without and (**b**) with a clean-up on the terbutylazine MIP and (**c**) on the anti-triazines immunosorbent. (1) atrazine; (2) simazine; (3) terbutylazine. UV detection at 220 nm [48]

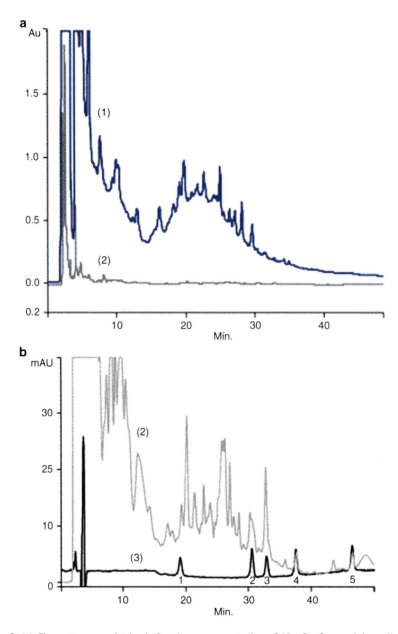

Fig. 9 (**a**) Chromatograms obtained after the preconcentration of 10 mL of grape juice spiked at 10 μg L^{-1} of a mixture of triazines through the styrene-divinylbenzene (SDB) sorbent: (1) without and (2) with a clean-up on the ametryn MIP. (**b**) Chromatograms resulting from the clean-up on MIP (2) and of the direct injection of standard solution containing 15 ng of each compound (3). Note the ca. 500-fold scale difference between (**a**) and (**b**)! Peak assignment: (1) deethylatrazine (DEA); (2) simazine; (3) deethylterbutylazine (DET); (4) atrazine; (5) terbutylazine. UV detection at 220 nm [48]

To reduce the retention of interfering compounds by non-selective mechanisms, the authors added 1% methanol to the dichloromethane. As a result only an OH containing compound was bound on the NIP column, all others were completely washed off. With the same solvent mixture the MIP retained 10 out of 13 investigated compounds with higher than 97% efficiency.

These results show that in a MISPE procedure one may encounter substances with different patterns of behavior. In the ideal case the template (or the template and a group of its analogs which need to be carried on for further analysis) are retained 100% during the sample application and the wash step. All other substances (or at least those which interfere with further analysis) should be fully removed. In the non-ideal case some interferents will not be removed by a particular protocol. In this case it is useful to observe the behavior of these interferents on the NIP. If they are retained by the NIP, too, then their retention is likely due to non-specific binding. In this case a better choice of the solvents in the protocol may help to eliminate the interferent. The danger here is that the new protocol may also reduce the retention of the template and reduce its recovery. It is therefore interesting to note that even in this good paper it has not been shown that the addition of 1% methanol to the dichloromethane sample was really necessary with the real sample matrices.

In this paper we also find an interesting experiment. The authors wished to determine the capacity of their terbutylazine MIP. They had filled an HPLC column (50 × 4.6 mm) with the MIP and passed through this column various volumes of atrazine solutions of different concentrations. The eluent was dichloromethane: methanol (99:1). Figure 10 (Fig. 3 of the original) shows how much atrazine was retained on the column (i.e., did not break through). In most cases the retention is less than 100% (i.e., the points do not fall on the dotted curve). In the authors' opinion the increase of sample volume reduces the number of accessible binding sites and thereby decreases the capacity. This is not a clear explanation and we believe the situation is much simpler. If one dissolves a certain amount of substance (shown on the abscissa of the figure) in various volumes one obtains lower concentrations at higher volumes. On the other hand, in all points where breakthrough occurred, the whole column became equilibrated with the original (pumped) sample concentration. So at higher volume the sample concentration was lower and consequently the equilibrium bound concentration on the column was less. Therefore the retained amount was also less. From this explanation it follows that the retained amount in these experiments should be the same every time the solution concentration was the same and breakthrough occurred. Indeed, one can read from the figure that this conclusion is quite well satisfied.

This discussion on the MISPE column capacity shows that it is not easy to define in a "scientific" manner the capacity of MISPE cartridges. Most published protocols apply the samples as large volumes and then employ a different solvent for the wash. Sample matrix components may also affect retention. Therefore we believe it is more practical to study the capacity for the full protocol and with real samples. If under these conditions the analyte(s) are fully retained, the capacity is sufficient and one may even try to reduce the MIP quantity used.

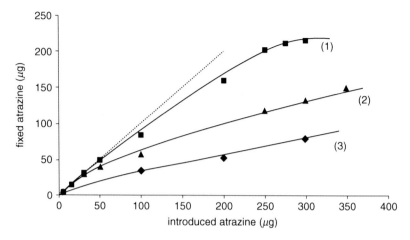

Fig. 10 Curves of capacity obtained after the percolation of: (1) 5 mL, (2) 10 mL, and (3) 25 mL of a dichloromethane–methanol (99:1) mixture spiked with increasing amount of atrazine on the terbutylazine MIP (130 mg). The *dotted line* corresponds to a slope of 1 meaning an extraction recovery of 100% [48]

7.5 Example 5 [49]

This example shows the development of the washing step of a MISPE procedure. Le Moullec et al. had developed a MISPE method for the separation of ethyl methylphosphonic acid (EMPA), a compound collected from the environment (soils) to obtain evidence of the prohibited production and use of nerve agents. Soil samples were extracted with pressurized hot water. The extracts were evaporated to dryness. This allowed the residue to be dissolved in acetonitrile, a suitable solvent for applying the sample to the MISPE cartridge. The wash solution (5 mL) consisted of acetonitrile with varying amounts of methanol. Figure 11 (Fig. 4 of the original) shows that addition of methanol made the wash solvent "stronger." This effect could be observed both on the NIP and the MIP. The difference between the two polymers was greatest at about 7% methanol concentration. This result has been interpreted in such a way that the nonselective binding sites of the MIP (which are also present on the NIP) release the template at lower methanol concentrations, while the stronger, imprinted sites release it only above 7% methanol.

This method development strategy (i.e., comparison of the MIP with the NIP) has often proved useful but two potential pitfalls need to be considered. First, the underlying model assumption may be wrong. The difference between the MIP and the NIP binding is not always due to the appearance of the selective binding sites on the MIP. For example, one would expect results similar to those in Figure 11 (Fig. 4 of the original) if the MIP had the same kind of binding sites as the NIP but at higher concentration. Therefore methods developed using the MIP/NIP comparison strategy need to be thoroughly tested by crosschecking the selectivity of the method in real samples.

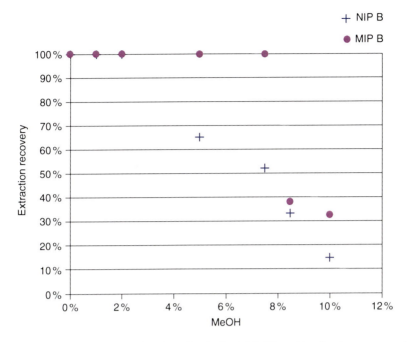

Fig. 11 Effect of the proportion of methanol in the 5 mL MeCN washing fractions on extraction recovery of ethyl methylphosphonic acid (EMPA) on the MIP and the NIP, respectively (The polymers have been named B to differentiate them from another preparation A in the same paper) [49]

The second problem to note is related to the sharp drop of extraction recovery noticed with the MISPE cartridge at about 8% methanol in the wash solvent. This drop is very likely due to the chromatographic breakthrough of the template from the MISPE column. It is then dangerous to work close to the 7% methanol concentration because the breakthrough volume is likely to depend greatly on the concentration of the analyte, the sample matrix, the quality of SPE packing, etc. (In the paper the "optimized" wash occurred with 10 mL of 5% methanol. In this case 87% of EMPA was retained. This wash was probably also close to the dangerous range).

7.6 Example 6 [50]

This paper shows a nice example for solving an important analytical problem using MISPE. Mycotoxins and particularly zearalenone (ZON) and *trans*-α-zearalenol (α-ZOL) present an everyday problem in food analysis. Existing sample clean-up techniques have different drawbacks. Liquid–liquid extraction is characterized by poor accuracy and high detection limits. SPE sorbents are not sufficiently selective. Immunosorbents have high costs per sample.

Although MIPs can overcome these problems, no appropriate MIPs had been developed before, due to the high cost and the toxicity of these mycotoxins. The authors succeeded in synthesizing a good ZON mimicking template and made with it an MIP of suitable binding characteristics. This MIP has been applied for the analysis of cereal and swine feed samples.

The MISPE method of the authors uses a pressurized liquid extraction (also called accelerated solvent extraction) extract of the samples. This acetonitrile–methanol extract had to be dried and redissolved in acetonitrile because methanol was not compatible with the MISPE. The redissolution volume had to be relatively large (5 mL) due to solubility problems. The samples were loaded on the MIP, washed with 1 mL of water, and eluted with 3×1 mL of methanol/phosphoric acid (95:5 v/v) solution. The eluate was evaporated to dryness and redissolved in 1 mL of methanol. Final analysis was carried out by HPLC using fluorescence detection.

The results obtained with a variety of real samples (swine feed, wheat, corn, barley, rye, and rice) are rather convincing. Recoveries around 90% have been reported in all cases with consistently less than 6% standard deviation. The accuracy for a corn reference material deviated from 100% only insignificantly. The detection limits were similar to LC-MS and immunoaffinity methods, and well below the maximum allowed limits in the EU.

Figure 12 (Fig. 2 of the original) shows the comparison of a wheat extract without clean-up and with clean-up on the MIP and the NIP, respectively. The general cleaning effect of the MIP is nicely shown, but it is not easy to judge only from this figure whether the method without clean-up would not have satisfied the requirements for this determination. It is also somewhat unclear why the elution step was done with phosphoric acid. Although the authors state that methanol and methanol/acetic acid (97:3 v/v) were slightly less satisfactory, these eluents would not have left much acid back after evaporation. Probably the phosphoric acid did not cause problems in HPLC because only 8 μL of the 1 mL redissolved sample was injected.

7.7   Example 7 [51]

This recent work is a preliminary study to compare MISPE with classical SPE for determination of benzodiazepines from hair samples for forensic purposes. Benzodiazepines are common prescription tranquilizers. They have also been widely abused and have contributed to many drug-related deaths.

The authors had prepared an MIP imprinted with diazepam, a frequently used benzodiazepine. The MIP did not differentiate between the template and a series of analogs, so those were also retained on the MIP cartridge. The MISPE sample clean-up was followed by HPLC–MS–MS analysis to quantify the drugs. Both the MISPE and a classical SPE procedure were successfully validated. Recoveries of

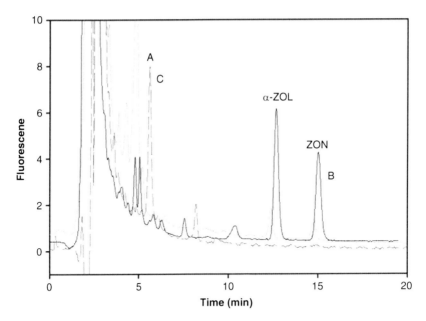

Fig. 12 Chromatograms of a pressurized liquid extraction (PLE) wheat extract (5 g sample spiked with 100 ng g^{-1} of ZON and α-ZOL): (**a**) without and (**b**) with clean-up on the MIP. (**c**) Chromatograms resulting from clean-up on NIP [50]

nine investigated substances were typically above 80%, with a few exceptions on either cartridge. The authors also compared the performance of both methods on ten postmortem hair samples from individuals whose blood tests were positive. Many of the measured values in these samples were below the respective lower detection limits. In a number of cases, however, the SPE method produced considerably higher values than the MISPE method. The authors could show by mass spectrometry that this was due to matrix interferences in the classical SPE method.

7.8   Example 8 [52]

This paper from 2007 shows that with recent developments in making aqueous compatible MIPs it has become possible to determine drugs in blood plasma samples by HPLC–UV essentially by direct injection and using a rather simple MIP precolumn system.

The authors have reviewed the major steps of development of online MISPE systems in drug analysis from biological samples. This development started with the first use of MISPE in 1994 by one of the coauthors [53]. An efficient online system was later introduced [45] where a RAM precolumn and an MIP precolumn were used in series to remove large molecules and small molecular

weight interferences, respectively. A further milestone of development has been
when a RAM style MIP precolumn was used which had merged the two functions
of the previous system [54, 55].

In the paper discussed here, partially hydrophobic MIPs containing 2-hydro-
xyethyl methacrylate (HEMA) have been used in an online MIP precolumn.

Plasma samples containing the analytes bupivacaine and ropivacaine were
diluted 1:1 with citrate buffer (containing the internal standard ethycaine). Then
20 μL of this solution was injected and transferred to the MIP precolumn with a
water:ethanol 90:10 v/v solution. The same solution washed the MIP for 4 min after
which the analytes were transferred to the analytical C18 column by backflushing
the MIP precolumn using 1% formic acid in water:acetonitrile 90:10. Separation of
the analytes occurred by raising the acetonitrile content of the last solution to 30%
in 5 min. The resulting chromatograms are shown in Fig. 13 (Fig. 6 of the original).
Further investigations have proved that this simple method has sufficient recovery.

When an MIP with exactly the same composition as that described above
has been made using a different porogen (1,1,1-trichloroethane instead of toluene),
the results were different in the sense that the corresponding NIP gave too
high recoveries. This has been interpreted as reduced selectivity for the analytes.
Figure 14 (Fig. 7a of the original) shows the chromatograms obtained of plasma
samples. In our view the quality of the chromatograms obtained with the two MIPs
is quite similar. It would be interesting to see a comparison of both MIPs using
many blank plasma samples. This would better show the differences in selectivity
than a comparison with the respective NIPs.

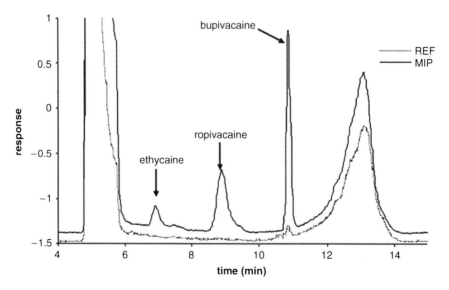

Fig. 13 Chromatograms of the online extraction of a human plasma sample spiked with bupivacaine and ropivacaine at 75 μmol L^{-1} and ethycaine at 40 μmol L^{-1} on the MIP and the reference
polymer REF, both prepared using toluene as the porogen [52]

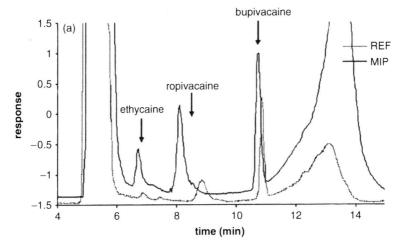

Fig. 14 Chromatograms of the online extraction of a human plasma sample spiked with bupivacaine and ropivacaine at 75 μmol L^{-1} and ethycaine at 40 μmol L^{-1} on the MIP and the reference polymer REF, both prepared using 1,1,1-trichloroethane as the porogen [52]

7.9 Example 9 [42]

This is a unique paper in many ways. It shows an example of partial filling CEC, which has been described earlier in this chapter.

It also presents the application of a mixture of two MIPs for the simultaneous chiral separation of two different drugs, propranolol and ropivacaine. Last but not least, the authors have prepared an MIP with incorporation of two different templates, *S*-propranolol and *S*-ropivacaine. This "multiply templated" MIP was also used in the partial filling CEC technique and sufficiently separated the enantiomers of both compounds in a single run.

Figure 15 (Fig. 3 of the original) shows the results obtained with the MIP mixture while Fig. 16 (Fig. 4B of the original) shows those obtained with the multiply templated MIP.

7.10 Example 10 [56]

This very recent paper is apparently the first instance when MIP CEC has been used as a routine methodology for the determination of environmental pollutants in real (food) samples.

The authors prepared a monolithic MIP CEC column in 100 μm i.d. fused-silica capillary, imprinted with thiabendazole (TBZ). This compound is a commonly used post-harvest fungicide to control diseases during storage and distribution of fruits

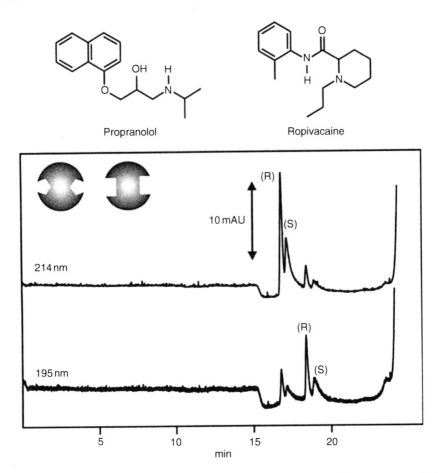

Propranolol   Ropivacaine

Fig. 15 Separation of ropivacaine and propranolol enantiomers using an MIP plug composed of (S)-ropivacaine MIP and (S)-propranolol MIP. The capillary was 100 cm in total length and 91.5 cm in effective length. The electrolyte contained acetonitrile/2 mol L^{-1} acetic acid adjusted to pH 3 by the addition of triethanolamine (90/10, v/v). The separation voltage was 15 kV, and the capillary column was thermostated to 60 °C. The MIPs were injected hydrodynamically at 50 mbar for 6 s each, and the sample was composed of 50 µmol L^{-1} *rac*-propranolol (first eluting) and *rac*-ropivacaine injected electrokinetically at 16 kV for 3 s. Detection was performed at 214 (*top*) and 195 nm (*bottom*) [42]

like lemons and oranges. Existing methods for determination of TBZ at the maximum residue limits are tedious. The MIP CEC method presented in this paper allowed identification (due to MIP selectivity) and quantification (by UV detection) of TBZ from crude citrus peel or pulp extracts without any clean-up and within 6 min. Figure 17 (Fig. 7 of the original) shows the electrochromatograms for orange. There were no peaks detected at the retention time of TBZ in the non-spiked samples. Recoveries higher than 85% and RSDs less than 6% were achieved.

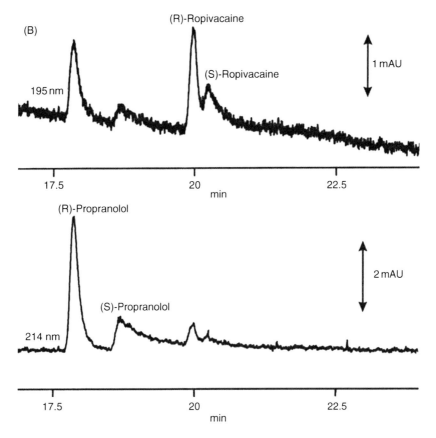

Fig. 16 Enantiomer separation of propranolol and ropivacaine on multiply templated MIP nanoparticles. Separation conditions: the electrolyte contained acetonitrile/2 mol L^{-1} acetic acid adjusted to pH 3 using triethanolamine (90/10, v/v). The capillary was 100 cm in total length, and 91.5 cm in effective length. The separation voltage was 10 kV, and the capillary column was thermostated to 60 °C. MIP nanoparticles suspended in electrolyte were injected hydrodynamically at 50 mbar for 12 s and the sample, 50 µmol L^{-1} *rac*-propranolol and *rac*-ropivacaine, was injected electrokinetically at 8 kV for 3 s. Detection was performed at 214 and 195 nm for propranolol and ropivacaine, respectively [42]

The linear range was wider than two orders of magnitude. The limits of detection and quantification were suitable to satisfy current legislation.

7.11 Example 11 [57, 58]

These two very recent papers deal with a novel method of MISPE, the lab-on valve format. Here renewable portions of MIP are dosed into the loop of an injection valve. The full MISPE and HPLC process evolves automatically. Besides a detailed description of the technique used, these papers are also important from other aspects.

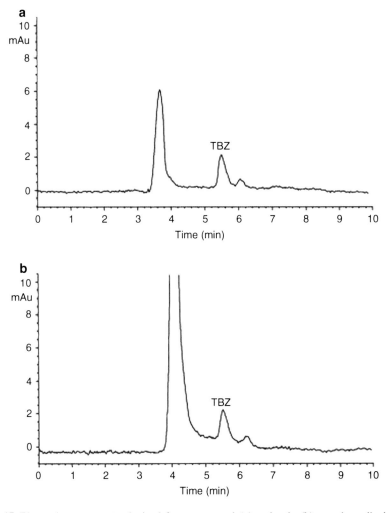

Fig. 17 Electrochromatograms obtained for orange peel (**a**) and pulp (**b**) samples spiked with thiabendazole (TBZ) (5 mg kg^{-1}) by the MIP-CEC method. Voltage: 2 kV, temperature: 60 °C, electrolyte: citric acid/acetonitrile (4%, w/v) [56]

One is the use of commercially available MIPs. According to the web page of the main commercial source for MIPs, the open literature on the applications of their MIPs seems to be mostly coming from the manufacturer's lab or from labs closely collaborating with them. The works cited in this example appear to be made by an independent laboratory. The other interesting aspect of these papers is the extensive testing of the proposed methods on real samples and also on standard reference materials. Earlier in this review we mentioned that unexpected matrix effects may change unpredictably the recovery of the analyte in a MISPE procedure. Testing on different sample types and preferably on standard reference samples can make MISPE methods more credible.

8 Conclusions and Outlook

In a book published in 2005 ([3], page 13), Andrew G. Mayes wrote an interesting interview with three pioneers of MIPs: Günter Wulff, Ken Shea, and Klaus Mosbach. Their views on the future of MIPs in separation were somewhat divergent. Shea, for instance, considered MISPE a possible niche application while he did not trust that commercial applications of MIPs for chromatography, including chiral separations, would become a reality. Wulff hoped that future research may improve MIPs, so that even production scale separations for high value products would become feasible with MIPs. Mosbach was confident that many commercial applications will appear in the coming years, e.g., in MISPE or in enantiopolishing of products.

Our impression has been that unclear interpretations of the tailing peaks on MIP columns have slowed down the development of the field. It is of course a question whether, with a better understanding of this phenomenon, one can find more real-life applications for MIPs.

Some of the more recent examples presented in this chapter show, however, that several researchers have learned how to live with this problem and succeeded in finding areas where MIP or CEC columns and MISPE cartridges or online pre-columns can be used with similar or better results than competing technologies.

Acknowledgment The authors are indebted to Maria Cruz Moreno-Bondi for a thorough discussion on experiences with MISPE.

References

1. Sellergren B (ed) (2001) Molecularly imprinted polymers: man-made mimics of antibodies and their applications in analytical chemistry, 1st edn. In: Techniques and instrumentation in analytical chemistry, vol 23. Elsevier, Amsterdam
2. Komiyama M (2003) Molecular imprinting: from fundamentals to applications. Wiley-VCH, Weinheim
3. Yan M, Ramström O (2005) Molecularly imprinted materials: science and technology. Marcel Dekker, New York
4. Sellergren B (1999) Polymer- and template-related factors influencing the efficiency in molecularly imprinted solid-phase extractions. TrAC, Trends Anal Chem 18(3):164–174
5. Andersson LI (2000) Molecular imprinting for drug bioanalysis – a review on the application of imprinted polymers to solid-phase extraction and binding assay. J Chromatogr B 739 (1):163–173
6. Lanza F, Sellergren B (2001) The application of molecular imprinting technology to solid phase extraction. Chromatographia 53(11–12):599–611

7. Andersson LI (2001) Selective solid-phase extraction of bio- and environmental samples using molecularly imprinted polymers. Bioseparation 10(6):353–364
8. Lanza F, Sellergren B (2001) Molecularly imprinted extraction materials for highly selective sample cleanup and analyte enrichment. Adv Chromatogr 41:137–173
9. Masque N, Marce RM, Borrull F (2001) Molecularly imprinted polymers: new tailor-made materials for selective solid-phase extraction. TrAC, Trends Anal Chem 20(9):477–486
10. Martin-Esteban A (2001) Molecularly imprinted polymers: new molecular recognition materials for selective solid-phase extraction of organic compounds. Fresenius J Anal Chem 370(7):795–802
11. Sellergren B (2001) Imprinted chiral stationary phases in high-performance liquid chromatography. J Chromatogr A 906(1–2):227–252
12. Alexander C et al (2006) Molecular imprinting science and technology: a survey of the literature for the years up to and including 2003. J Mol Recognit 19(2):106–180
13. Nilsson C, Nilsson S (2006) Nanoparticle-based pseudostationary phases in capillary electrochromatography. Electrophoresis 27(1):76–83
14. Turiel E, Martin-Esteban A (2005) Molecular imprinting technology in capillary electrochromatography. J Sep Sci 28(8):719–728
15. Tamayo FG, Turiel E, Martin-Esteban A (2007) Molecularly imprinted polymers for solid-phase extraction and solid-phase microextraction: recent developments and future trends. J Chromatogr A 1152(1–2):32–40
16. Turiel E, Martin-Esteban A (2004) Molecularly imprinted polymers: towards highly selective stationary phases in liquid chromatography and capillary electrophoresis. Anal Bioanal Chem 378(8):1876–1886
17. Pichon V, Haupt K (2006) Affinity separations on molecularly imprinted polymers with special emphasis on solid-phase extraction. J Liq Chromatogr Related Technol 29(7–8):989–1023
18. Haginaka J (2008) Monodispersed, molecularly imprinted polymers as affinity-based chromatography media. J Chromatogr B Anal Technol Biomed Life Sci 866(1–2):3–13
19. Pichon V (2007) Selective sample treatment using molecularly imprinted polymers. J Chromatogr A 1152(1–2):41–53
20. Liu ZS et al (2007) Molecularly imprinted polymers as a tool for separation in CEC. Electrophoresis 28(1–2):127–136
21. Quaglia M, Sellergren B, De Lorenzi E (2004) Approaches to imprinted stationary phases for affinity capillary electrochromatography. J Chromatogr A 1044(1–2):53–66
22. Spegel P, Schweitz L, Nilsson S (2003) Molecularly imprinted polymers in capillary electrochromatography: recent developments and future trends. Electrophoresis 24(22–23): 3892–3899
23. Nilsson J, Spegel P, Nilsson S (2004) Molecularly imprinted polymer formats for capillary electrochromatography. J Chromatogr B Anal Technol Biomed Life Sci 804(1):3–12
24. Caro E et al (2006) Application of molecularly imprinted polymers to solid-phase extraction of compounds from environmental and biological samples. TrAC, Trends Anal Chem 25(2):143–154
25. Maier NM, Lindner W (2007) Chiral recognition applications of molecularly imprinted polymers: a critical review. Anal Bioanal Chem 389(2):377–397
26. Wei ST, Jakusch M, Mizaikoff B (2006) Capturing molecules with templated materials – analysis and rational design of molecularly imprinted polymers. Anal Chim Acta 578(1): 50–58
27. Nicholls IA et al (2009) Theoretical and computational strategies for rational molecularly imprinted polymer design. Biosens Bioelectron 25(3):543–552
28. Pichon V, Chapuis-Hugon F (2008) Role of molecularly imprinted polymers for selective determination of environmental pollutants – a review. Anal Chim Acta 622(1–2):48–61
29. Dias ACB et al (2008) Molecularly imprinted polymer as a solid phase extractor in flow analysis. Talanta 76(5):988–996

30. He CY et al (2007) Application of molecularly imprinted polymers to solid-phase extraction of analytes from real samples. J Biochem Biophys Meth 70(2):133–150
31. Sajonz P et al (1998) Study of the thermodynamics and mass transfer kinetics of two enantiomers on a polymeric imprinted stationary phase. J Chromatogr A 810(1–2):1–17
32. Chen YB et al (1999) Influence of thermal annealing on the thermodynamic and mass transfer kinetic properties of D- and L-phenylalanine anilide on imprinted polymeric stationary phases. Anal Chem 71(5):928–938
33. Miyabe K, Guiochon G (2000) A study of mass transfer kinetics in an enantiomeric separation system using a polymeric imprinted stationary phase. Biotechnol Prog 16(4):617–629
34. Toth B, Laszlo K, Horvai G (2005) Chromatographic behavior of silica-polymer composite molecularly imprinted materials. J Chromatogr A 1100(1):60–67
35. Toth B et al (2006) Nonlinear adsorption isotherm as a tool for understanding and characterizing molecularly imprinted polymers. J Chromatogr A 1119(1–2):29–33
36. Kim H, Kaczmarski K, Guiochon G (2006) Isotherm parameters and intraparticle mass transfer kinetics on molecularly imprinted polymers in acetonitrile/buffer mobile phases. Chem Eng Sci 61(16):5249–5267
37. Tóth B, Felinger A, Horvai G. In preparation (unpublished)
38. Guiochon G et al (2006) Fundamentals of preparative and nonlinear chromatography, 2nd edn. Academic, Boston, xiv, 975 p
39. Pap T et al (2002) Effect of solvents on the selectivity of terbutylazine imprinted polymer sorbents used in solid-phase extraction. J Chromatogr A 973(1–2):1–12
40. Bui BTS, Merlier F, Haupt K (2010) Toward the use of a molecularly imprinted polymer in doping analysis: selective preconcentration and analysis of testosterone and epitestosterone in human urine. Anal Chem 82(11):4420–4427
41. Ceolin G et al (2009) Accelerated development procedure for molecularly imprinted polymers using membrane filterplates. J Comb Chem 11(4):645–652
42. Spegel P, Schweitz L, Nilsson S (2003) Selectivity toward multiple predetermined targets in nanoparticle capillary electrochromatography. Anal Chem 75(23):6608–6613
43. Turiel E, Martin-Esteban A, Tadeo JL (2007) Molecular imprinting-based separation methods for selective analysis of fluoroquinolones in soils. J Chromatogr A 1172(2):97–104
44. Caro E et al (2002) Non-covalent and semi-covalent molecularly imprinted polymers for selective on-line solid-phase extraction of 4-nitrophenol from water samples. J Chromatogr A 963(1–2):169–178
45. Koeber R et al (2001) Evaluation of a multidimensional solid-phase extraction platform for highly selective on-line cleanup and high-throughput LC-MS analysis of triazines in river water samples using molecularly imprinted polymers. Anal Chem 73(11):2437–2444
46. Chapuis F et al (2003) Optimization of the class-selective extraction of triazines from aqueous samples using a molecularly imprinted polymer by a comprehensive approach of the retention mechanism. J Chromatogr A 999(1–2):23–33
47. Ferrer I et al (2000) Selective trace enrichment of chlorotriazine pesticides from natural waters and sediment samples using terbuthylazine molecularly imprinted polymers. Anal Chem 72(16):3934–3941
48. Chapuis F et al (2004) Retention mechanism of analytes in the solid-phase extraction process using molecularly imprinted polymers – application to the extraction of triazines from complex matrices. J Chromatogr B Anal Technol Biomed Life Sci 804(1):93–101
49. Le Moullec S et al (2006) Selective extraction of organophosphorus nerve agent degradation products by molecularly imprinted solid-phase extraction. J Chromatogr A 1108(1):7–13
50. Urraca JL, Marazuela MD, Moreno-Bondi MC (2006) Molecularly imprinted polymers applied to the clean-up of zearalenone and alpha-zearalenol from cereal and swine feed sample extracts. Anal Bioanal Chem 385(7):1155–1161
51. Anderson RA et al (2008) Comparison of molecularly imprinted solid-phase extraction (MISPE) with classical solid-phase extraction (SPE) for the detection of benzodiazepines in post-mortem hair samples. Forensic Sci Int 174(1):40–46

52. Cobb Z, Sellergren B, Andersson LI (2007) Water-compatible molecularly imprinted polymers for efficient direct injection on-line solid-phase extraction of ropivacaine and bupivacaine from human plasma. Analyst 132(12):1262–1271
53. Sellergren B (1994) Direct drug determination by selective sample enrichment on an imprinted polymer. Anal Chem 66(9):1578–1582
54. Haginaka J, Sanbe H (2000) Uniform-sized molecularly imprinted polymers for 2-arylpropionic acid derivatives selectively modified with hydrophilic external layer and their applications to direct serum injection analysis. Anal Chem 72(21):5206–5210
55. Sanbe H, Haginaka J (2003) Restricted access media-molecularly imprinted polymer for propranolol and its application to direct injection analysis of beta-blockers in biological fluids. Analyst 128(6):593–597
56. Cacho C et al (2008) Molecularly imprinted capillary electrochromatography for selective determination of thiabendazole in citrus samples. J Chromatogr A 1179(2):216–223
57. Oliveira HM et al (2010) Exploiting automatic on-line renewable molecularly imprinted solid-phase extraction in lab-on-valve format as front end to liquid chromatography: application to the determination of riboflavin in foodstuffs. Anal Bioanal Chem 397(1):77–86
58. Boonjob W et al (2010) Online hyphenation of multimodal microsolid phase extraction involving renewable molecularly imprinted and reversed-phase sorbents to liquid chromatography for automatic multiresidue assays. Anal Chem 82(7):3052–3060

Top Curr Chem (2012) 325: 307–342
DOI: 10.1007/128_2010_93
© Springer-Verlag Berlin Heidelberg 2010
Published online: 14 October 2010

Microgels and Nanogels with Catalytic Activity

M. Resmini, K. Flavin, and D. Carboni

Abstract Molecular imprinting has grown considerably over the last decade with more and more applications being developed. The use of this approach for the generation of enzyme-mimics is here reviewed with a particular focus on the most recent achievements using different polymer formats such as microgels and nanogels, beads, membranes and also silica nanoparticles.

Keywords Molecular imprinting · Nanogels · Microgels · Enzyme-mimic · Catalysis

Contents

M. Resmini (✉), K. Flavin, and D. Carboni
School of Biological and Chemical Sciences, Queen Mary University of London, Mile End Road, London, E1 4NS, UK
e-mail: m.resmini@qmul.ac.uk

Nature has promoted enzymes' evolution with the aim of optimising the binding ability of the active site towards the transition of a reaction, and therefore favouring its stabilisation [1]. For this reason enzymes' catalytic pockets have evolved in order to perform this task with a range of functional groups capable of interacting selectively with the substrate. These interactions are responsible for the molecular recognition of the substrate and their nature is strictly dependent upon the amino acid residues in the active site and the chemical structure of the substrate. However, it is important to note that, in order to reach a degree of molecular recognition adequate enough for catalysis using artificial macromolecular systems, it is essential that key enzymatic features are mimicked.

Two different approaches, catalytic antibodies (Ab) and molecularly imprinted polymers (MIPs), have been investigated with the aim of achieving this target and the next section will provide some general outlines.

1 Catalytic Antibodies

Antibodies are Y-shaped proteins, belonging to the class of the immunoglobulins, that are produced by the immune system in response to the presence of an antigen, a molecular structure that is not recognised as part of the body. This characteristic was exploited to develop antibodies with catalytic properties by immunising animals with haptens mimicking an intermediate or the transition state of the reaction (TSA). In this case, the resulting antibody will be able to catalyse the corresponding reaction by inducing a conformational change in the reaction substrate, thus allowing it to fit its catalytic pocket. This process is similar to that occurring in an enzyme active site, where the enzyme–substrate complex is formed. As a consequence of this conformational change, an antibody–substrate complex is generated that lowers the activation energy required to form the transition state, thereby facilitating the evolution of the TS into product [2].

Although very successful in term of catalytic activity and selectivity, the use of catalytic antibodies has raised some issues with the long and complex methodology needed to prepare them. Most importantly, they present the same limitations of narrow working ranges of pH and temperature and the impossibility of their use in organic solvents shown by the enzymes due to their protein nature. A significant and valuable alternative technique was developed by applying a similar approach to polymeric matrices, leading to the preparation of molecular imprinted polymers with catalytic activity.

The aim of this chapter is to give a brief overview of the molecularly imprinted catalysts reported up to approximately the turn of the millennium, followed by a more concise review of the literature thereafter. In addition to catalysis, that is cases in which a reaction TSA or intermediate are used as the template, imprinted polymers capable of aiding chemical transformations will also be discussed. In these cases the reaction substrate or product are often used as the template in order to control the regio- or stereochemistry of the reaction.

2 Synthesis and Catalysis by Imprinted Polymers: Background

The idea of generating recognition sites in polymeric materials, as a result of a molecular memory obtained by imprinting highly organised structures with a template molecule, was initially proposed by Linus Pauling and developed further. By 1931, Polyakov, from Kiev, observed some unusual properties in the material obtained while preparing silica particles. He added some solvent additives to the dry silica and let them dry again for about a month, after which the silica was thoroughly washed with hot water. By studying the adsorption properties of the new material, he discovered that this presented higher absorption properties towards those additives previously used, as if the material had a "memory" of its structure "imprinted" in its three-dimensional polymeric matrix [3].

The first experimental application of this concept was reported in the seminal work of Dickey, who, in 1949, stated that silica, adsorbed with methyl orange, showed preferential absorption properties towards the same structure. Dickey, hypothesising the mechanism with which the specific adsorption was generated, invoked Pauling's antibody formation theory: "This mechanism is the same as that proposed by Pauling for the formation of antibodies with use of antigen molecules as a template." [4].

These results were not fully exploited until two milestone papers, from Gunter Wulff on the use of imprinted polymers to resolve racemates [5], and Klaus Mosbach describing a polymer preparation with clear rebinding characteristics towards phenyl-alanine ethyl ester [6], were published, giving a clear boost to the field. The authors showed that, in analogy with Pauling's model for antibodies generation, chemical structures could be used as templates to generate specific recognition sites by "*imprinting*" a molecular memory in polymer matrices.

Molecular imprinting allows the generation of specific three-dimensional cavities in polymer matrices by using a template molecule around which functional monomers and cross-linker are self-assembled in a pre-polymerisation state. Following polymerisation and template removal, the polymer matrix is left with the free three-dimensional cavities capable of rebinding the molecule, or others structurally very similar, used for the imprinting.

Figure 1 illustrates schematically the molecular imprinting process. The imprinting is performed using a template molecule (Fig. 1a), around which the functional

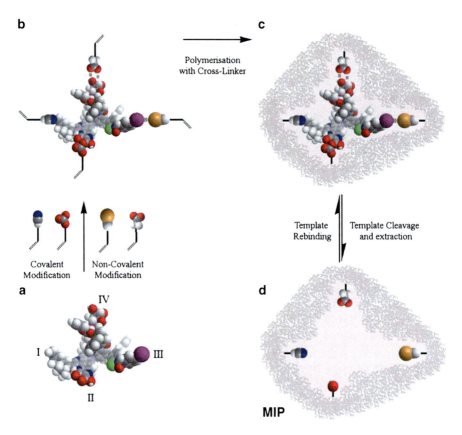

Fig. 1 Schematic representation of the molecular imprinting procedure

monomers and cross-linker self-assemble to generate the pre-polymerisation complex (Fig. 1b). The polymerisation leads to an imprinted polymer still containing the template inside the cavities (Fig. 1c). The removal of the template makes available an imprinted polymer with active cavities (Fig. 1d).

The choice of the co-monomers and cross-linker used together with the template will have a very important role in determining the physical properties of the final materials and, ultimately, the rebinding and catalytic properties of the polymers.

3   Catalytic Molecular Imprinting: The Early Works

The molecular imprinting approach, chosen to obtain enzyme mimics, has a lot in common with the generation of catalytic antibodies. The immunisation, in fact, allows generation of binding pocket in the catalytic antibodies as an immune response elicited from a hapten [7]. By following the same theoretical approach,

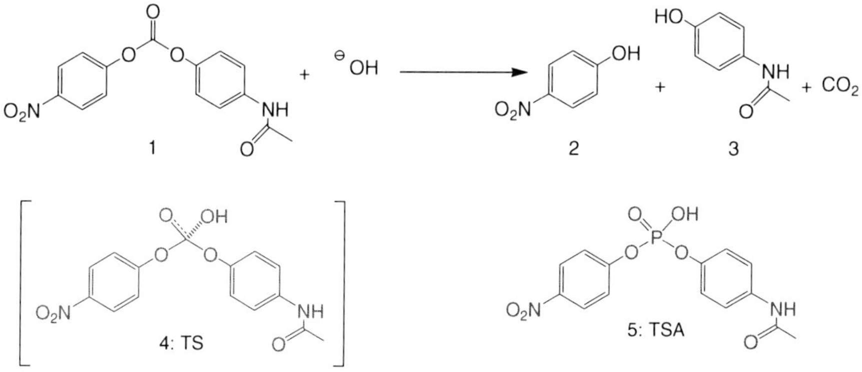

Scheme 1 Hydrolysis of diphenyl-carbonate (**1**) to give 4-nitro-phenol (**2**), 4-acetamido-phenol and carbon dioxide. The diphenyl-phosphate (**4**) was used as TSA to mimic the tetrahedral transition state formed during the hydrolysis

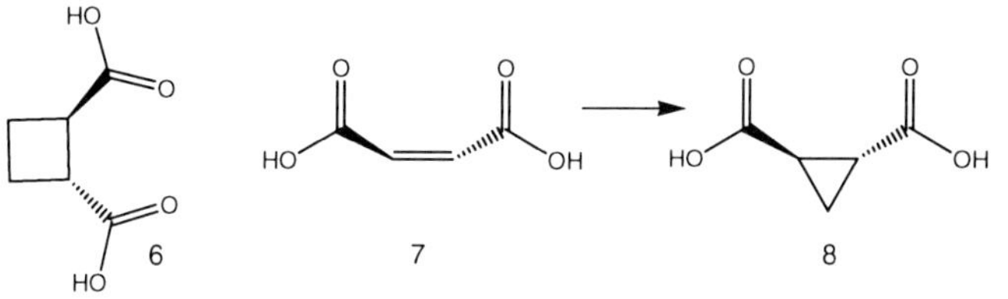

Scheme 2 Polymer imprinted with the structure (**6**) were able to convert the achiral substrate (**7**) in the chiral substrate (**8**)

if a synthetic polymer is cast around a template molecule, either a TSA or an intermediate analogue of a chemical reaction, then the "imprinted" cavities should be able to catalyse the corresponding reaction.

An example of this imprinting approach is illustrated in Scheme 1, where a polymer with hydrolytic properties towards the diphenyl-carbonate (**1**) was imprinted with the corresponding phosphate template (**4**), which represents a TSA for the hydrolytic reaction. This structure, in fact, mimics the tetrahedral intermediate formed during the hydrolysis of the carbonate, and therefore it allows imprinting a cavity with functional groups placed in the right spatial position.

The first reported attempt of using MIPs to control the stereochemical course of a reaction dates back to 1980, when the two research groups of Neckers and Shea published, simultaneously, examples of bulk polymers able to control the formation of the product by using a chiral template. Shea et al. reported that bulk polymers imprinted with stereochemically pure (−)-*trans*-1,2,cyclobutane-dicarboxyilic acid (**6**) were able to keep a molecular memory of the asymmetry of the template [8]. In fact, this was transferred to an achiral substrate, such as fumaric acid (**7**), inducing a diastereoselective methylation, which led to *trans*-1,2,cyclopropane-dicarboxyilic

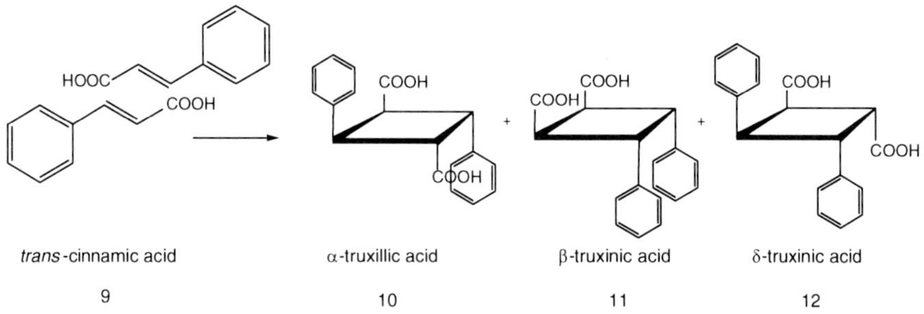

Scheme 3 Photodimerisation of *trans*-cinnamic acid (**9**) and three of the possible diastereomers, the α-truxillic acid (**10**), the β-truxinic acid (**11**) and the δ-truxinic acid (**12**)

acid (**8**) only as a result of the shape of the cavity where the methylation took place [8] (Scheme 2).

Neckers et al. reported the synthesis of bulk polymers imprinted with one of the three possible isomeric products of the photodimerisation of *trans*-cinnamic acid (**9**). This polymer was able to convert more than 50% of the *trans*-cinnamic acid in δ-truxinic acid (**12**), an isomer that is never formed with the catalyst in solution. This demonstrated how the cavity of the imprinted polymer could control the stereochemistry of the reaction in a direction which did not occur at all using the free catalyst in solution (Scheme 3) [9].

Leonhardt and Mosbach in 1987 [10] described the first attempt to synthesise a polymer by combining the molecular imprinting approach with the insertion, in the imprinted cavity, of polymerisable imidazole units (**13**), already known to be responsible for the esterolytic activity of some enzymes, with the aim of generating a polymer with an enzyme functional mimicry. The pyridine-derivative (**14**) was utilised, as a substrate analogue template, instead of the true substrate (**15**), a *p*-nitro-phenol-ester of a Boc-amino acid, to avoid hydrolysis in the pre-polymerisation stage. The imprinting complex was generated by formation of a Co^{2+} complex involving complexation of two polymerisable imidazole units and the substrate analogue (Fig. 2). After removal of both the template and the metal, the polymers were incubated with a solution of the *p*-nitro-phenol ester and the rates of hydrolysis were recorded, showing activity two to three times higher than the corresponding non-imprinted polymers. Further experiments also demonstrated the presence of substrate selectivity and catalytic turnover.

Wulff and his collaborators reported, in 1989, the preparation of imprinted polymers able to perform enantioselective synthesis [11, 12]. The imprinting complex was prepared by reacting 3,4-di-hydroxy-phenyl-alanine methyl-ester (L-DOPA methyl ester) (**16**) with the 4-vinyl-salicylaldehyde (**17**) to form the corresponding Schiff's base (**18**), which was further reacted with the 4-vinyl-phenyl-boronic acid (**19**) to afford to the corresponding ester (**20**) (Scheme 4). The imprinting complex obtained was then polymerised and the template removed. The resulting polymers were incubated with sodium glycinate to allow formation

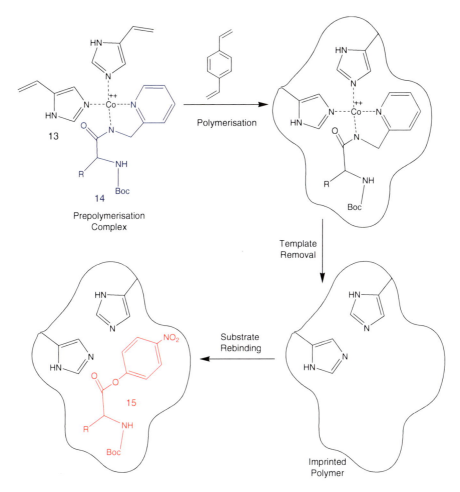

Fig. 2 Use of imidazole units (**13**) binding the pyridine derivative (**14**) to generate imprinted polymers with esterolytic activity towards 4-nitro-phenyl-esters (**15**)

of the corresponding imine, which was further reacted with a solution of Ni acetate to form the catalytic metal complex.

This complex was reacted with acetaldehyde to give the desired product with 36% *ee* as a unique result of the asymmetric arrangement of the functional groups in the imprinted cavity. Although in this case the imprinting complex was not designed to mimic a transition state analogue, the authors nevertheless suggested that this result showed how important it was to mimic the TSA instead of the final product.

A few years later, in 1993, Akermark et al. reported the preparation of imprinted polymers capable of reducing selectively a 3-17-steroidal diketone (**21**) to the corresponding alcohol with high control of the stereochemistry [13]. The polymers were prepared attaching a polymerisable unit via an ester linkage to the desired steroidal product in position 17 (Fig. 3). The resulting compound was then used as a

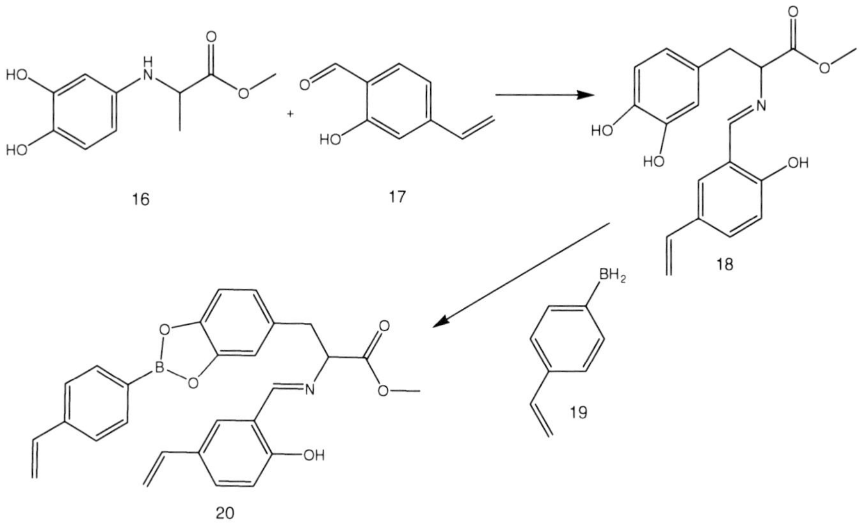

Scheme 4 Imprinting approach used by Wulff et al. for the preparation of a polymer capable of performing enantioselective synthesis. The L-DOPA (**16**) is reacted with salicylaldehyde (**17**) to give the corresponding Schiff's base (**18**) that reacts with borane (**19**) to generate the imprinting complex (**20**)

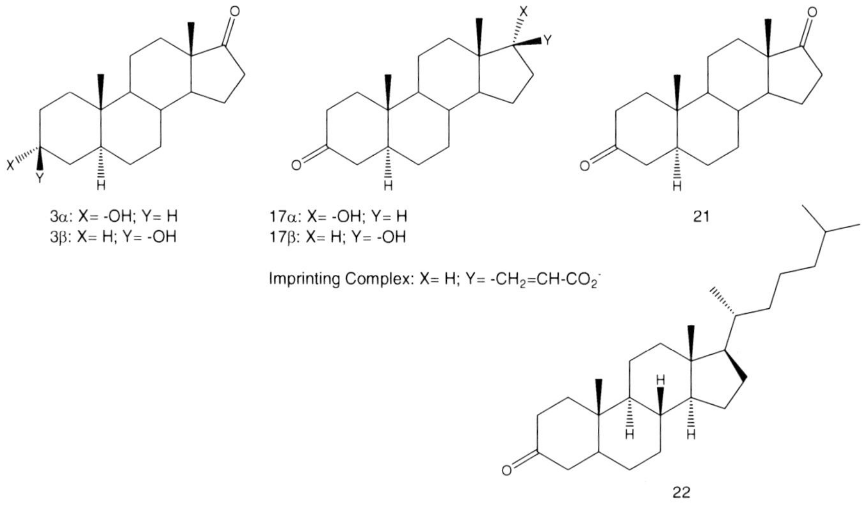

3α: X= -OH; Y= H
3β: X= H; Y= -OH

17α: X= -OH; Y= H
17β: X= H; Y= -OH

21

Imprinting Complex: X= H; Y= -CH₂=CH-CO₂⁻

22

Fig. 3 Structures of the templates, the 3-17-steroidal diketones (**21**) and the natural cholestan-3-one (**22**)

templating agent during the polymer synthesis, after which it was removed using LiAlH₄ to leave the free imprinted cavities. The polymers, after a second treatment with LiAlH₄ to generate the active hydride species inside the cavities, were suspended in THF at room temperature and allowed to react with an excess of substrate, the 3-17-steroidal diketone. The analysis of the product of the reaction showed a complete preference for the reduction at position 17 whereas in solution, or using a non-imprinted polymer, the completely opposite outcome was preferred, with a 99% reduction of the ketone in position 3.

Further experiments, made using a template selective for position 3, permitted a high stereocontrol of the reaction involving the natural substrate cholestan-3-one (**22**). This was reduced to the corresponding, less readily available, cholestanol 3α-OH, with a ratio 3α-OH/3β-OH $= 72/28$. The corresponding result for a hydride reduction in solution usually gave 3α-OH/3β-OH $= 10/90$.

3.1 MIPs Based on TSA

The affinity of an enzyme's active site for the transition state of the reaction is an essential feature for the achievement of the catalytic activity. The use of the transition state analogue approach has been extensively used for the generation of catalytic antibodies, starting from the early results obtained by Lerner and his group in 1986 [14]. The same approach has also been used successfully to generate synthetic receptors in polymer matrices using the casting procedure of the molecular imprinting.

The first example of polymer imprinting using a TSA has to be ascribed to Mosbach and collaborators in 1989 [15]. Since the esterolytic activity toward p-nitro-phenol-acetate (**23**) was the target, the imprinting complex was designed by taking into account that the transition state of the esterolytic reaction had to be tetrahedral instead of planar (**24**), as for the acetate substrate (Scheme 5). For this purpose the p-nitro-phenol-methyl-phosphonate (**25**) was chosen as transition state analogue (TSA). This was held in place during the polymerisation stage using a Co^{2+} complex made using two imidazole units, with a technique previously introduced by the same group [10] (see also Fig. 2). After removal of the template,

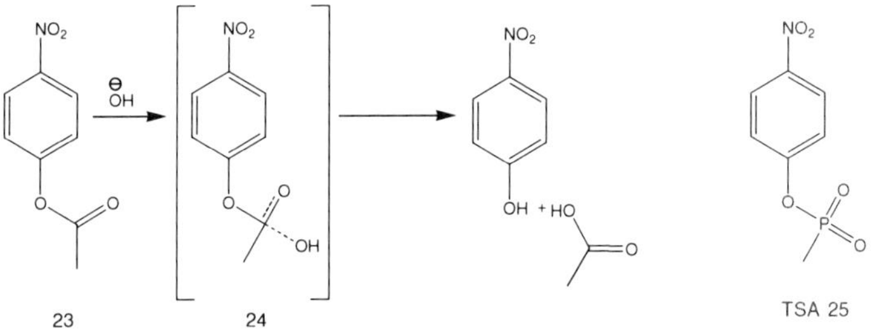

Scheme 5 The polymers were imprinted with the TSA (**25**) to mimic the tetrahedral transition state (**24**) generated during the hydrolysis of the 4-nitro-phenol-acetate (**23**)

the imprinted polymer (MIP) was shown to be 60% catalytically more active than the corresponding non-imprinted polymer (NIP). Moreover, it was proved that the template was rebound from the active cavity and acted as specific site inhibitor, demonstrating that, despite the small rate enhancement observed, the cavity was able to stabilise the transition state of the reaction, therefore acting as a true enzyme mimic.

Following this first attempt of Mosbach and co-workers to imprint using a TSA, the number of examples reported in the literature in the following years flourished with reports of various esterolytic polymers based on the same type of transition state analogue. Ohkubo et al. carried out a complete kinetic analysis of this system, proving that the imprinted polymer behaved like an enzyme mimic increasing the rate of the reaction sevenfold compared to the background at pH = 7 [16]. This paper was immediately followed by another from the same group, describing a water soluble polymer imprinted with phosphonate TSA able to increase the rate of hydrolysis of (L)-leucine ester (**26**) up to ninefold at pH = 7 [17]. In 1995 the same group developed both insoluble and water-soluble polymers based on the catalytic hystidyl group (**27**) for the hydrolysis of Z-(L)-leucine ester, using a non-covalent imprinting strategy [18]. The water-soluble polymer showed a rate increase of 12-fold compared with the background whilst the insoluble polymer increased the rate 3-fold. Both polymers demonstrated substrate specificity and template inhibition. This result showed the importance of the transport properties in the catalytic activity of the imprinted polymers.

In 1994 Shea et al. reported the preparation of gel-like imprinted polymers with enantioselective esterolytic activity toward the Boc-D-phenyl-alanine p-nitrophenol ester (**28**) [19]. The polymers were prepared using a covalent approach, rather than metal complexes or non-covalent interactions, by attaching the catalytic phenol-imidazole unit to the TSA phosphonate via ester linkage (**29**). The imprinted polymer, containing the catalytic unit (**30**), showed little selectivity toward the D-enantiomer used for the imprinting.

Inspired by the great rate acceleration, by a factor of 10^3–10^4, obtained using phosphonate derivatives as TSA for the preparation of esterolytic antibodies [20],

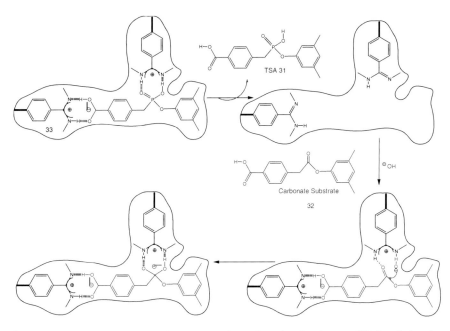

Scheme 6 System used by Wulff to catalyse the hydrolysis of carbonate (**32**) imprinting the cavities with the TSA phosphonate (**31**)

Wulff et al., in 1997, developed a more promising strategy involving this type of TSA and using a stoichiometric non-covalent imprinting [21]. The system was based on polymerisable *N,N'*-di-ethyl-amidine units, used as functional monomers, which showed a strong association constant towards both the phosphonate ester (**31**), a TSA used as template, and the substrate, the carbonate 4-(2-(3,5-dimethyl-phenoxy)-2-oxoethyl)benzoic acid (**32**). The amidine functional monomers (**33**) were used in a ratio of 2:1 to the template. In fact, the first unit interacts with the carboxylic acid of the template functioning as an anchor, whilst the second unit acts as a catalyst, stabilising the incipient transition state of the esterolysis (Scheme 6).

The result of this approach was a 100-fold increase in the hydrolytic activity of the imprinted polymer compared with the background at pH = 7.6. As a control, another polymer was made using a complex between amidine and benzoate, showing a surprisingly 20-fold increase in the hydrolysis of the substrate. The authors also reported a kinetic investigation of the TSA-imprinted and the benzoate-imprinted polymers, in addition to the free catalyst in solution. Although the ratio substrate/catalyst is not specified, and therefore the steady-state conditions could not be verified, the authors claimed for the two polymers a Michaelis–Menten kinetic behaviour, with a higher profile for the TSA-imprinted polymer. On the other hand, the free catalyst in solution showed, as expected, a linear dependence of the rate from the substrate concentration. The TSA also showed a moderate selectivity towards its "own" substrate.

Since the first report on the use of the phosphonate as a TSA for esterolytic reactions, a similar approach has been used by other researchers for the synthesis of MIPs with catalytic activity. Following this approach, Mosbach and co-workers, in 1993, made the first attempt to generate an imprinted polymer able to perform a dehydrofluorination reaction [22]. The system was based on the TSA (**34**) already used for the generation of catalytic antibodies for the same reaction [23], where a secondary amine was used to deprotonate an amino acid residue in order to form a carboxylate able to catalyse a β-elimination reaction. The methacrylic acid (MAA) was reacted with *N*-benzyl-isopropylamine and ethylene-di-methacrylate (EDMA) was used as cross-linking agent. After template removal the polymer cavities containing the carboxylate residues in the right spatial disposition were incubated with the substrate, 4-fluoro-4-(*p*-nitrophenyl)-2-butanone (**35**). The dehydrofluorination was carried out in various solvent and the best rate enhancement was observed in methylene dichloride where the imprinted polymer resulted in a 2.4 times increase compared to the non-imprinted and 3.4 times compared to the background in generating the unsaturated compound (**36**).

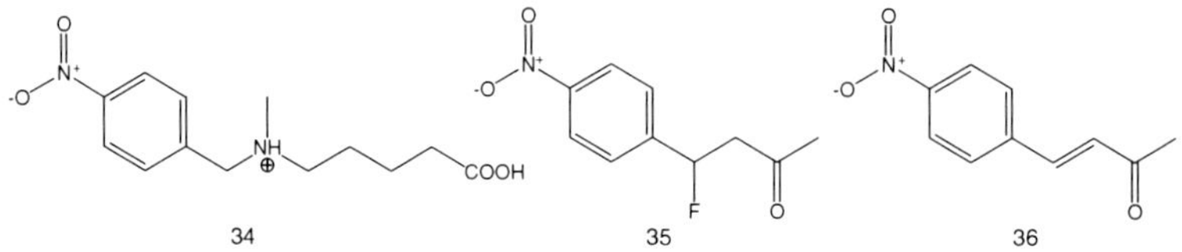

At the same time another example of catalytic polymers for the β-elimination was reported by Shea and co-workers [24]. This group adopted a different approach, using different dicarboxylic acids, such as benzylmalonic acid (**37**), as template, in order to hold the amino functionalities (**38**) in the right places (Fig. 4).

The target substrate was the same as for the previous report (**35**). The control polymer was prepared using acetic acid as the template. The imprinting efficiency was reported to be 3.5. Interestingly, an imprinted polymer made using a

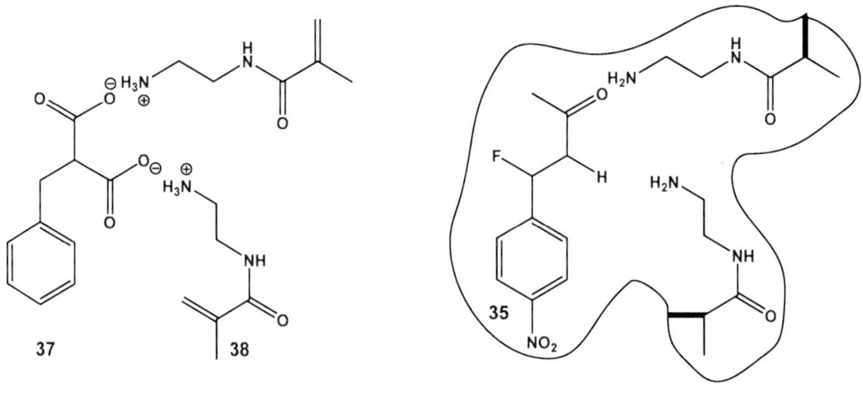

Fig. 4 Formation of the imprinting complex between benzyl–malonic acid (**37**) and the amino functionalities (**38**) and the product rebinding operated by the imprinted polymer (*right*)

dicarboxylic acid with the two groups far away showed a catalytic activity no better than the control polymer, suggesting that the position of the two amino functions present in the imprinted cavities played a big role in directing the reaction.

Catalysis of the C–C bond formation is a significant challenge, certainly greater than hydrolysis, in terms of energy requirements. This might in part explain the reasons why, in the field of catalytic molecular imprinting, there are very few examples of polymers able to catalyse this type of reaction. Matsui et al. in 1996 reported one of the first examples of C–C bond formation catalysed by an imprinted polymer [25]. The system was designed to mimic an Aldolase type II, which is a metallo-enzyme that owes its activity to a Zn ion in the catalytic centre able to promote the reaction. The catalytic centre of the imprinted polymer was based on a metal complex described by two molecules of pyridine coordinating a Co(II) ion. The reaction between acetophenone (**39**) and benzaldehyde (**40**) to give the corresponding condensation product, the chalcone (**41**), was chosen as a target.

The polymer was imprinted with dibenzoyl-methane (**42**), a 1,3-diketone able to be held in place by formation of a complex with the Co(II) ion, and structurally similar to the product of the cross-aldol condensation. A schematic representation of this approach is given in Scheme 7.

The imprinted polymer was able to increase the reaction rate by eightfold over the reaction in solution but only twofold with respect to the non-imprinted polymer.

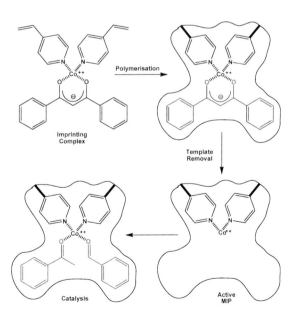

Scheme 7 Schematic representation of the imprinting and catalytic approach used to prepare the Aldolase II mimic

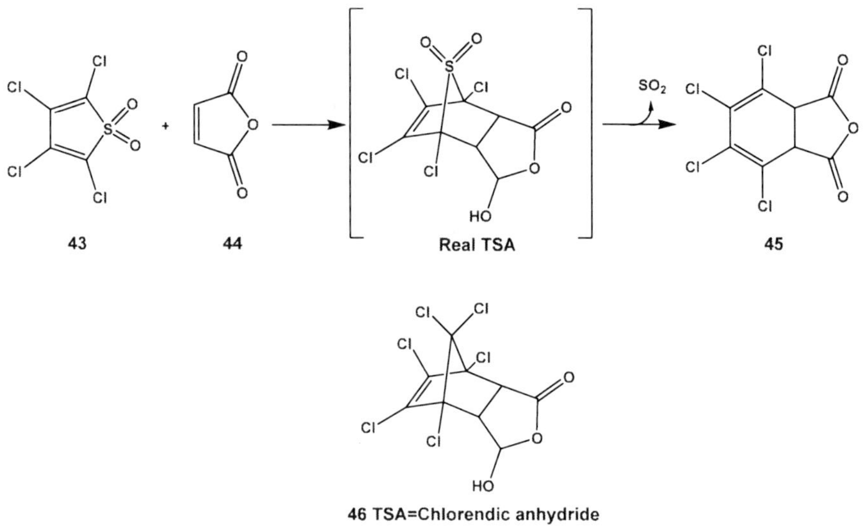

Scheme 8 Imprinting approach used to catalyse the reaction between tetrachlorothiophene dioxide (**43**) and maleic anhydride (**44**) to give (**45**). The chlorendic anhydride (**46**), representing the TSA, is used as the template for imprinting.

Moreover, it necessitated 100 °C for several weeks to achieve complete conversion of the substrate.

In 1997 the same group developed the first imprinted polymer able to catalyse a Diels–Alder reaction between tetrachlorothiophene dioxide (**43**) and maleic anhydride (**44**) to give the product (**45**). The imprinting strategy was inspired by previous work carried out by Hilvert et al. in 1989 for the development of catalytic antibodies with Diels–Alder capabilities [26]. The chlorendic anhydride (**46**) was used as a template because of its structural analogy with the transition state of the reaction (TSA). The resulting imprinted polymer showed a Michaelis–Menten behaviour and a ratio k_{cat}/k_{uncat} equal to 270 (Scheme 8).

This first part of the chapter was intended to give the reader an overview of the first examples of catalytic imprinted polymers based on the use of the TSA as template. However, a more detailed discussion about imprinted polymers with synthetic and catalytic properties can be found in the reviews by G. Wulff et al. [27, 28] and by C. Alexander et al. [29] The second part of this chapter will focus on the most successful examples of molecular imprinted polymers with catalytic activity that have been reported in the last decade.

3.2 Imprinted Polymers for Applications in Synthesis

Whitcombe et al., in 1999, described the synthesis of a bulk polymer capable of selectively protecting two of the three hydroxyl groups in a steroid based compound, leaving the third one available for functionalisation [30]. The system was based on the covalent imprinting approach involving a steroid template possessing two hydroxyl groups, which were allowed to react with two polymerisable boronaphthalide monomers, making boronic ester linkages. After polymerisation and template removal, the imprinted polymer was incubated with a steroidal substrate containing three hydroxy groups and a selective acylation was performed on the "free" alcohol. The reaction was successfully carried out with a regioselectivity ratio of 23:1 for the desired monoester product. The same reaction performed on the non-imprinted polymer led to a reversed ratio of 1:100, proving the relevance of having functional groups in the right spatial position inside an imprinted cavity.

Mosbach and collaborators, in 2001, described an innovative approach to drug discovery using polymers imprinted with a biologically active template [31]. The approach, called anti-idiotypic for the similarity with anti-idiotypic antibodies in the immune-response, can be used to create synthetic receptors able to generate inhibitors or receptor antagonists by exploiting the complementarity with the cavity. The imprinted cavity promotes preferentially the formation of compounds with high affinity, which can later be evaluated for the inhibitory activity and the more active selected for further analysis.

The authors reported the preparation of polymers imprinted with kallikrein (**47**) as a template, a known tissue inhibitor, which, after template removal, were used to allow direct coupling between a di-chloro-triazine (**48**) and a series of aromatic amines. The first substrate was used to "resynthesise" the compound that was used as template and the yield of this was four times higher than with the corresponding control polymer. Moreover, the same reaction performed in free solution in the same conditions gave no product. The polymer was then tested against the other substrates to verify whether it was possible to synthesise compounds with slightly different properties. This experiment resulted in compounds with lower yields, 21% for one and 34% for another, whilst using a bulkier substrate it was not possible to obtain any product.

47 kallikrein   48 R_1 = -H, -OH
                                                                             R_2 = -H, -C(O)-OCH$_3$

The year after, the same group developed a system based on the same anti-idiotypic approach and a similar template possessing a chiral centre, in order to investigate the possibility of controlling the enantioselectivity of the reaction uniquely by the direct moulding of the cavity. In this case the polymer was able to give 67% ee as opposed to the control polymer, which produced a virtually racemic mixture [32].

Nicholls and co-workers, in 2004, reported the preparation of imprinted polymers for the stereoselective reduction of (−)-menthone (**49**) to diastereomeric products (−)-menthol (**50**) and (+)-neomenthol (**51**) [33]. The imprinting strategy made use of a covalent approach to prepare a template based on two molecules of (−)-menthol coupled with one of fumaric acid by means of ester linkages. After polymerisation the (−)-menthol was removed by prolonged exposure to an excess of LiAlH₄, which cleaved the menthol and subsequently transformed the alcohol residues in reducing agents. The polymer was able to reduce selectively the (−)-mentone, shifting the diastereomeric ratio to 1:1 in favour of the unflavoured (+)-neomenthol and altering the natural outcome of the reaction that would give a ratio of 2:1 in favour of the (−)-menthol. Although interesting, these results were not as significant as those obtained by Akermark and co-workers, who were able to reverse the regioselectivity of a steroid substrate reduction [13].

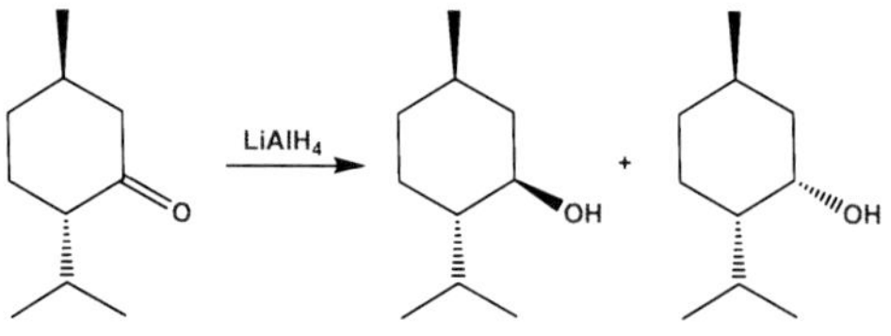

In 2006, Ye et al. developed a non-covalently imprinted polymer capable of catalysing a regioselective 1,3-dipolar cycloaddition between azides and alkynes [34]. The polymer was imprinted using the *anti*-isomer (**54**), obtained by reacting the ethyl propiolate (**52**) and 2-aminobenzyl-azide (**53**), and methacrylic acid as functional monomer. After template removal, the same reaction catalysed by the imprinted polymer produced an amount of product 6.7 times higher than the control polymer and surprisingly 1.6 times higher than that of the polymer-free solution. Moreover, 94% of the product obtained with the imprinted polymer was the *anti*-isomer against the 75% of the control polymer and 69% of the polymer-free solution. This suggested that the cavities imprinted with the product could increase the regioselectivity of the cycloaddition only by directing the spatial disposition of the reactants (Scheme 9).

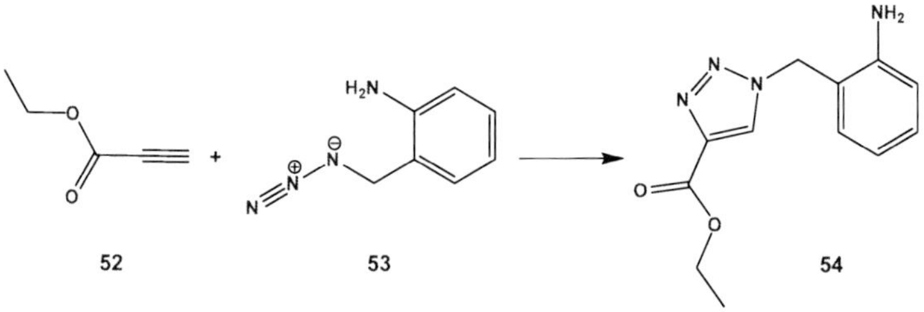

Scheme 9 Preparation of the imprinted polymer for the catalysis of 1,3-dipolar cycloaddition between azides and alkynes obtained using the ethyl propiolate (**52**) and the azide (**53**) to generate the imprinted imprinting complex (**54**)

Meng and Sode [35, 36] also reported the preparation of imprinted polymers to be used as "reaction vessels" for the transesterification of p-nitro-phenol acetate (**23**) and hexanol. The MIP showed eightfold increased activity when compared to the NIP.

After giving a brief overview of some examples of molecular imprinting for synthetic purposes, the attention will now focus on the description of the most significant examples of catalytic polymers appeared in the literature in the last decade. During this time the molecular imprinting field has seen significant advances in terms of materials technology with new polymeric formats used, in addition to the traditional bulk polymers. The following sections, covering the use of catalytic imprinted polymers, will be divided into subsections characterised primarily by the polymer format used and by the type of reaction (hydrolytic or C–C formation).

4 Catalysis by 'Bulk' Acrylate Imprinted Polymers

The bulk polymeric format, characterised by highly cross-linked monolithic materials, is still widely used for the preparation of enzyme mimic despite some of its evident drawbacks. This polymerisation method is well known and described in detail in the literature and has often be considered the first choice when developing molecular imprinted catalysts for new reactions. The "bulk" polymer section is presented in three subsections related to the main topics covered: hydrolytic reactions, carbon–carbon bond forming reactions and functional groups interconversion.

4.1 Hydrolytic Polymers

One of the most investigated type of reaction in the field of catalytic imprinted polymers, as indicated by the large number of publications available, is certainly ester hydrolysis. In particular, a great deal of work has been carried out on systems inspired by hydrolytic enzymes since 1987. In 2000, Shea et al. [37] reported the preparation of enantioselective imprinted polymers for the hydrolysis of N-*tert*-butoxycarbonyl phenylalanine-p-nitrophenyl ester (**55**), using a system already developed by the same group in 1994 [19]. The system was inspired by the natural hydrolytic enzyme chymotrypsin and polymerisable imidazole units (**27**) were used as functional monomers coupled via ester linkages to a chiral phosphonate (**56**), analogue of (D)- or (L)-phenyl-alanine. After template removal, the imprinted polymers showed selectivity towards the hydrolysis of the enantiomer with which they were imprinted. The ratio of the rate constants, k_D/k_L, was 1.9 for the polymer imprinted with the D-enantiomer and k_L/k_D was 1.2 for that imprinted with the L-enantiomer. Moreover, the imprinted polymer showed a 2.5-fold increase in the rate of the reaction when compared with the control polymer, imprinted with a benzoyl-substituted imidazole, and a 10-fold increase when compared with the imidazole monomer in solution.

55   56   57

The use of polymerisable imidazole as functional monomer has been largely used for the preparation of hydrolytic imprinted polymers from other groups as well. Ohkubo et al. [38], for instance, described the preparation of imprinted polymers with esterolytic activity towards N-dodecanoyl leucine-p-nitrophenyl ester (57) using a system similar to the Shea one but based on a stoichiometric non-covalent approach. The imprinting was carried out with a phosphonate TSA and the activity of the resulting imprinted polymer was twice the non-imprinted.

Another example of imidazole-based hydrolytic imprinted polymer was also reported by Goto and collaborators [39], who developed a new methodology called "surface molecular imprinting technique". The imprinted polymer was prepared in a water/oil emulsion by using an oleyl-imidazole (58), as host molecule, and a substrate analogue, the N-Boc-L-hystidine (59), as imprinting guest molecule. These were held together inside water droplets as Co(II) complexes. The radical polymerisation of the olefinic tails of the host molecules produced a porous "bulk" polymer due to the presence of the water droplets. Following removal of the imprinting guest, the free imprinted cavities will be left with the host units in a suitable position to rebind and hydrolyse the substrate. The authors claimed for the polymers an enzymatic behaviour and comparison of the two $V_{maxMIP}/V_{maxNIP} = 1.8$, demonstrate the imprinting efficiency of the system.

The group of Sode also reported the use of an imidazole derivative for the preparation of an imprinted polymer with phosphotriesterase (PTE) activity towards the pesticide metabolite Paraoxon, diethyl 4-nitro-phenolphospate (60) [40]. The authors synthesised a non-imprinted polymer catalyst based on a Zn(II)-imidazole complex, which was claimed to increase 105-fold the rate of the reaction compared to the spontaneous hydrolysis. The corresponding imprinted polymer, obtained by imprinting with the substrate analogue D4NP (diethyl (4-nitrobenzyl)phosphonate) (61), showed little increase in the acceleration rate compared to the non-imprinted polymer. The authors demonstrated that using a Zn(II) complex, present in the natural PTE enzyme, instead of a Co(II) complex, the activity increased fivefold, although presenting an imprinting efficiency of only 1.3.The Zn(II)-based system was also used to construct an amperometric sensor for the Paraoxon, whose detection limit was established in 0.1 mM.

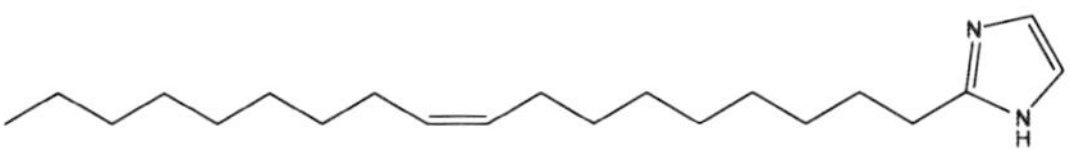

58

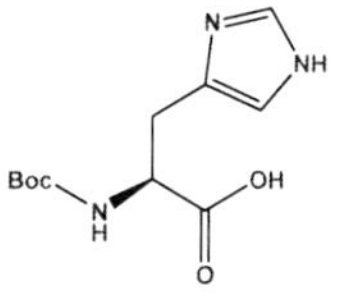

59 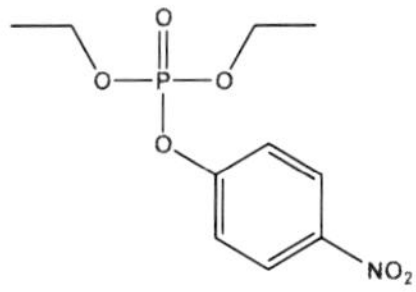   60 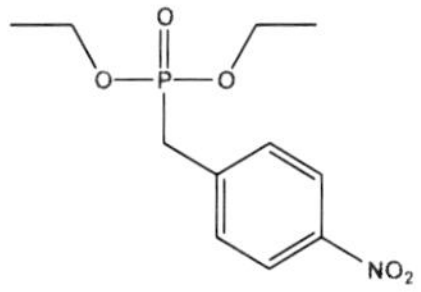   61

The preparation of imprinted polymers based on imidazole was also recently reported by Li et al. [41], who used essentially the same system as Mosbach to investigate the effect of the monomer-template ratio on catalytic specificity.

A different approach was used by Emgenbroich and Wulff [42] to develop imprinted polymers with enantioselective esterase activity. The system was based on the use of an amidinium functional monomer (**33**), already developed earlier by the same group, but in this case a chiral phosphonate (**62**) was used as imprinting TSA in order to catalyse the hydrolysis of the corresponding chiral ester (**63**). The polymer imprinted with the L-enantiomer was able to enhance the esterolytic activity 325-fold when compared with the background and 80 times compared to the non-imprinted polymer. The ratio between the two rates constant, $k_{cat-L}/k_{cat-D} = 1.4$, can be taken as a measure of the enantioselective efficiency of the reaction.

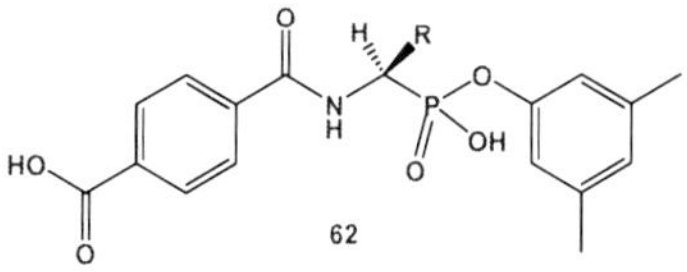

62   63

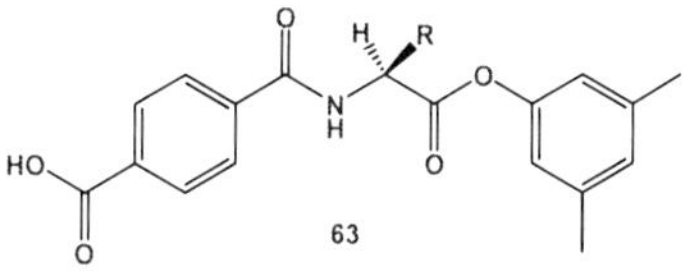

R= isobutyl, isopropyl   R= isobutyl, isopropyl

More recently the same group reported the preparation of polymers with even higher activity based on functional monomers containing amidinium units and Zn(II) (**64**) [43], inspired by the natural metallo-enzyme carboxypeptidase A. The functional monomer is constituted by the amidinium group, which mimics the guanidium moiety present in the arginine and by a trialkyl-amine unit able to hold the metal in proximity of the amidinium function and the phosphate TSA. This strategic design allowed generation of a hydrolytic imprinted polymer with an impressive pseudo-first order rate constant 3,200-fold higher than the background reaction when Zn(II) was used. In addition, the imprinting effect, calculated as ratio of k_{impr}/k_{contr}, was equal to 61.

With the aim of increasing the stability of the metal complex involved in the catalytic cycle, the authors substituted the Zn(II) (**64**) with the Cu(II) complex (**65**) [44]. This in fact not only forms a better complex but is also able to increase the nucleophilicity of the OH^- on which depends the rate of the hydrolytic reaction. As a result of this modification, the new imprinted polymer showed an extraordinary 8,000-fold increase in the acceleration rate if compared with the uncatalysed reaction. The reaction performed on a carbonate substrate containing pyridyl nitrogen (66), allowing a tighter binding with the catalytic unit, presented an enhanced ratio on both the uncatalysed and control polymer catalysed reactions.

64   65   66

A further evolution of this system has recently been published by Liu and Wulff [45], in which one amidine unit was linked with on both sides of the trialkyl-amine, in order to have a complex coordinating a single atom of Cu(II) in close proximity to two molecules of substrate interacting with the amidines (67). As a consequence of this structural modification the imprinted polymer was able to accelerate the rate of the reaction by an extraordinary factor of 410,000 when compared with the background. This is the highest rate enhancement ever achieved for an imprinted polymer.

67

It seems clear from this brief overview that, since the seminal work of Mosbach in 1987 [10], imprinted polymers with hydrolytic activity have been going through important developments, which have led to some interesting results. The discussion will therefore move to examine the use of the same format in the application of the imprinting technology to the carbon–carbon forming reactions.

4.2   Carbon–Carbon Bond Forming Polymers

The formation of a new carbon–carbon bond needs activation energy generally higher than that involved in a carbon–heteroatom bond breakage, and therefore the requirements for a catalyst, aiming to promote this kind of reaction, are stronger if compared with those for hydrolysis. The generation of molecularly imprinted catalyst is not exempted from this issue and, as a consequence, the number of imprinted polymers with carbon–carbon bond capabilities is so far limited. Apart from a few examples of Diels–Alder reactions [46], the majority of the catalytic imprinted polymers, so far, make use of metal complexes to lower the energetic barrier needed to catalyse the new C–C bond formation. The purpose of this section is to provide the reader with a general idea of the type of work done in the past decade on this topic.

In 2001, Cammidge et al. [47] reported the preparation of an MIP for Suzuki cross-coupling between *p*-bromoanisole (**68**) and phenyl-boronic acid (**69**) to give the corresponding 4-methoxy-biphenyl (**70**).

The polymer was imprinted with a metal complex (**71**) where two polymerisable triphenyl-phospine ligands and a catecholate, as a template, coordinated a Pd(II) ion. Although the polymer was made with only 25% of cross-linker, the polymer matrix was rigid enough to preserve the imprinted shape and, as a result, allow the polymer to achieve up to 81% of yield instead of the 56% maximum yield obtained with the catalyst prepared from the commercial ligand. In addition, the imprinted polymer showed consistently higher yields (76–81%) even after being re-used five times, whilst the homogeneous catalyst showed a drop in the yield from 56% to 45% when recycled, proving the enhanced activity and re-usability of the imprinted catalyst.

On the basis of this report, in 2003, Gagnè and collaborators described an imprinting system involving the use of the crown ether to improve Cammidge's system [48]. The polymer was prepared by imprinting the polymerisable bis-triphenyl-phosphine-Pd(catecolate) together with a polymerisable 1:1 complex formed by the 4'-vinylbenzo-18-crown-6 ether and *n*-butylamine (**72**).

The introduction of the crown ether served the purpose of coordinating the cation of the base, such as K_2CO_3, used to promote the reaction. The activity of the new imprinted polymer was compared with Cammidge's polymer and was greatly enhanced by the use of the crown ether and was also dependent on the nature of the cation used. In all cases the activity was higher than Cammidge's polymer but the best result was obtained using K_2CO_3 as a base, since the new polymer showed a conversion 2.5 times higher.

Another interesting example of C–C bond formation can be found in reports from Li and co-workers who, in 2004, described the preparation of an MIP with peroxidase-like activity capable of dimerising the homovanillic acid (HVA) (73) [49]. In this case a polymer was prepared by using the HVA substrate, instead of a TSA, as a template and a haemin unit as catalytic centre (74). The polymerisation was carried out in the presence of acrylamide and vinyl-pyridine in order to add extra functionalities aiding substrate recognition. The imprinted polymer showed an enzyme-like activity, as confirmed by adherence to the Michaelis–Menten saturation model, and it was inhibited by ferulic acid (75), a structural analogue of the substrate, which is also capable of inhibiting the natural peroxidase.

A successive report from the same group [50] discussed the influence of molecular recognition on the substrate binding, and therefore catalytic activity, in the peroxidase system previously developed. The authors reached the conclusion that the haemin played a role not only in the catalytic cycle but it was also essential in the molecular recognition of the substrate, by cooperating with the other co-monomers, 4-vinyl-pyridine and acrylamide. Moreover, imprinting efficiency was demonstrated by showing that the catalytic activity of the MIP was enhanced 7.6 times with respect to the NIP.

On the basis of a catalytic system previously developed by the same group, Nicholls and collaborators [51] reported the preparation of an imprinted polymer for enantioselective formation of a C–C bond with properties of a metallo-enzyme aldolase type II. Polymers were imprinted using the two enantiomers of a 1,3-diketone, the (1S,3S,4S)-(75), and the corresponding (1R,3R,4R)-(75), together with two 4-vinyl-pyridine held in place by a Co(II). The cross-aldol condensation between camphor (**76**) and benzaldehyde (**40**) to give (**77**) was chosen as a target reaction with the aim of controlling the enantioselectivity of the reaction using imprinting technology.

75   76   77

Both imprinted polymers showed an enhancement in the catalytic activity that was about 50-fold higher than the control polymer (P0) and turnover of the catalytic cavities was also demonstrated. However, when comparison was made with a polymer containing Co(II) but which was not imprinted with the template (P1), the rate acceleration dropped to about fourfold. In addition, the control of the enantioselectivity of the reaction was very low. In fact, the polymer, imprinted with the diketone derived from the *R*-camphor, was able to catalyse the reaction, between the *S*-camphor and benzaldehyde, with an acceleration rate almost identical to that obtained with the polymer imprinted with the opposite enantiomer. The rate enhancement between the two polymers was in fact equal to 1.04.

Earlier this year the same group described the preparation of novel imprinted polymers with Diels–Alder cycloaddition capabilities [52]. The imprinting system was inspired by the same approach used by Gouverneur et al. for the preparation of catalytic antibodies for Diels–Alder cycloaddition [53]. The reaction between 1,3-butadiene carbamic acid benzylester (**78**) and *N,N'*-dimethylacrylamide (**79**) to give the corresponding *endo*- and *exo*-products (**80**) was chosen as a target. The imprinted polymer made it possible to enhance 20-fold the rate of the reaction when compared with the background.

78   80

So far we have discussed only the use of molecular imprinting for the synthesis of enzyme-like catalytic polymers with hydrolytic and carbon–carbon bond forming properties. However, in the past decade, a number of research groups have worked on another important application, the functional groups interconversion. The next section will describe some of the most significant reports that have appeared in the literature during the past decade.

5   Functional Groups Interconverting Polymers

Severin and Polborn, in 2000, reported the development of imprinted polymers capable of performing regioselective hydrogenation of ketones to alcohols [54]. The imprinting strategy was based on the use of the first characterised organometallic TSA, the (η^6-arene)Ru(II) complex (**81**), with a diphenylphosphinato ligand acting as a pseudo-substrate, and capable of catalysing the reduction of benzophenone (**82**) via hydrogen transfer. After removal of the diphenylphosphinato ligand, operated by displacement with chloro ligand, the free shape-selective cavity (**83**) was able to catalyse the reaction. When the 2-propanol was used instead of formic acid as reducing agent, the imprinted polymer demonstrated rate acceleration up to seven times higher than the non-imprinted polymer, prepared using the complex but without the substrate. In particular, when a reaction was carried out over a pool of seven aromatic ketones, including the benzophenone, a clear preference of the imprinted polymer towards reduction of the benzophenone to diphenylmethanol (**84**) was demonstrated. This result proved the ability of the imprinted polymer in differentiating between substrates that are structurally very similar.

Moreover, when the reaction was performed with a multifunctional substrate, such as 4-acetyl-benzophenone (**85**), the yield in diaryl methanol was 1.3 times higher than that derived from the reduction of the acetyl group. On the other hand, when the non-imprinted polymer was used, the regioselectivity was reversed, with the yield of reduced acetyl being 1.7 times higher than the diarylmethanol.

In a subsequent report, in 2005 [55], the same group described the preparation of imprinted polymer capable of oxidising alcohols and alkanes with 2,6-dichloropyridine N-oxide (**86**) without mineral acid activation. The polymer was imprinted with a ruthenium porphyrin complex (**87**) using the diphenylmethanamine (**88**) as pseudo-substrate template in order to achieve a shape of the cavity complementary to the substrates, diphenylmethane (**89**) and diphenylmethanol (**84**). The reaction, carried out with the imprinted polymer on the diphenylmethanol as substrate, showed a rate enhancement 2.5 higher than with the non-imprinted polymer. In the same conditions, but with diphenylmethane and anthracene as substrates, the results are even higher with a rate enhancement of 6.4 for the first and 15.7 for the second. These results were possibly due to the fact that in the alkane oxidation the imprinting effect is manifested twice (alkane ⇒ alcohol and alcohol ⇒ ketone) with a greater influence on the overall reaction rate.

An alternative approach to the oxidation of alcohols to ketones was also reported by Shea et al., who incorporated a nitroxide catalyst into a polymeric matrix [56]. A polymerisable 2,2,6,6-tetramethylpiperidine (90) was derivatised as N-allyl-amine (91), which was removed after polymerisation, leaving a catalytically active nitroxide (92) able to form stable free radicals, thereby efficiently catalysing the reaction of oxidation with yields ranging from 55 to 88%.

Another interesting application of molecular imprinting to functional groups interconversion was described in 2004 by Nicholls and co-workers [57], who developed an enzyme mimic to convert α-keto-acid into α-amino-acid, taking inspiration from the natural transaminase. The reaction between pyridoxamine (**93**) and phenylpyruvic acid (**94**) to give phenylalanine (**95**) was selected as target. The polymer was imprinted using a chiral oxazine-based TSA (**96**) as a template and methacrylic acid as a functional monomer. After polymerisation and template removal, the polymer was found to follow a Michaelis–Menten saturation model and to undergo inhibition as a result of template rebinding. In addition, the imprinted polymer showed a rate enhancement of 15-fold when compared with the background solution and threefold with the non-imprinted polymer, which showed a fivefold increase in comparison with the background. Another important feature of this system was represented by the induction of chirality during the transammination, which resulted in a 32% ee with the imprinted polymer as opposed of a racemic mixture obtained with non-imprinted polymer.

An interesting example, although it cannot be described as functional inter-conversion but rather as isomerisation, was reported by Motherwell and collabora-tors in 2004 [58]. In this case a polymer was imprinted with a *trans*-carvyl amine (**97**), which was used as TSA, for the isomerisation of α-pinene oxide (**98**) to *trans*-carveol (**99**) that was obtained with 45% yield.

Takeuchi et al. reported the preparation of an imprinted polymer for the conver-sion of the herbicide atrazine (**100**) in atraton (**101**), a less toxic compound, by conversion of an atrazine chloride into methoxy [59, 60]. After polymerisation and removal of the template, analysis of the imprinted polymer showed saturation kinetics, suggesting an enzyme-like behaviour of the polymer.

Having described few examples of catalytic imprinted polymer based on the "bulk" polymer format, the attention will now move to describe catalytic systems based on different types of polymer formats, often developed with the intention of addressing some of the issues presented by bulk.

6   Catalytic Beads, Particles and Membranes

The previous section illustrated some of the most significant work regarding the topic of catalytic molecular imprinting in the past decade using the "bulk" polymer format. Although the target was often envisaged in imprinted polymers with capabilities inspired by natural enzymes, the "bulk" format might nevertheless have hampered the catalytic activity of the systems because of some obvious drawbacks linked with this format. In view of this fact, new polymeric materials have been developed, which has resulted in a more homogeneous size distribution and sites homogeneity, as opposed to the heterogeneous and irregularly shaped insoluble particles resulting from the grinding of the "bulk" polymers. These in fact, inter alia, affect the solubility, the load capacity and, overall, the reproducibility of the results.

With the aim of overcoming the major issues of the bulk format, new polymeric materials, such as beads, particles and membranes, were developed, prompting the application of these new formats to well known imprinting systems.

Wulff and collaborators, for instance, reported the preparation of TSA imprinted beads for the hydrolysis of carbonate and carbamate [61, 62], exploiting the amidine (**33**) functional monomer previously developed by the same group and successfully applied to the bulk format [63]. The polymers were prepared using a suspension polymerisation that produced beads with sizes in the range 8–375 μm, depending on the polymerisation conditions. The pseudo-first order reaction rate of the imprinted beads (k_{MIP}/k_{soln}) was enhanced by a factor of 293 for the carbonate hydrolysis and 160 for the carbamate, when compared with the background.

While comparing the pseudo-first order constant for the same polymers with the relative non-imprinted polymers (k_{MIP}/k_{NIP}),the rate increase dropped to 24 for carbonate and 11 for carbamate. The corresponding bulk polymers showed better results since the rate enhancement, due to the imprinted polymer, was 588 for the carbonate hydrolysis and 1,435 for the carbamate. However, when comparing the imprinted with the non-imprinted bulk, the ratio dropped to 10 for carbonate and 5.8 for carbamate, suggesting that the higher selectivity showed by the beads could be due to an enhanced accessibility of the active sites compared to the bulk.

Following the first reports in the literature of catalytic imprinted beads, a number of authors also reported applications of this polymer format to several imprinting systems. Busi et al. [64] reported the preparation of catalytic active beads for the Diels–Alder reaction using a TSA as a template. Jakubiak and co-workers developed imprinted beads for the oxidation of phenols based on a Cu(II) complex as catalytic centre [65]. Say and collaborators described the synthesis of microbeads also based on a Cu(II) complex with esterase activity towards paraoxon (**60**), a potent nerve agent [66]. The imprinted beads enhanced the rate of reaction over the non-imprinted polymer by a factor of 40, as resulted from the ratio of the corresponding k_{cat}.

Brüggemann and co-workers have applied the molecular imprinting technique to several polymeric materials with the aim of addressing some of the issues related to the "bulk" polymeric format. The examples range from the synthesis of imprinted membranes for the catalysis of a dehydrofluorination reaction [67], to the preparation of MIP-shells for a Diels–Alder reaction [68, 69]. In the first case an imprinting approach similar to that previously reported by Mosbach et al. with "bulk" polymers was used [70], whilst in the second the MIP-shells were generated by acrylate polymerisation around a TSA template immobilised on the surface of silica nanoparticles. In the latter, the MIP-shells were obtained by dissolution of the silica nanoparticles, which resulted in a porous imprinted material with active sites located very close to the shells-surface. Interestingly, the authors claim that a threefold increase in rate enhancement was achieved from the MIP-shells in comparison with the relative control shells, as compared to the corresponding bulk systems where both imprinted and non-imprinted polymers showed the same rate.

7 Catalytic Microgels and Nanogels

A large part of the literature related to catalytic molecular imprinting describes the synthesis of insoluble polymers with a high degree of cross-linking. The applications of these materials, characterised by a low degree of flexibility due to the rigidity of their structures, are hampered by the level of cross-linking, which can affect the transport properties and therefore their turnover. On the other hand, microgels seem to be attractive materials to be used in molecular imprinting, especially in catalysis, because their particular features (good solubility in both organic and aqueous solvents, stability in a broad range of pH and temperatures, good degree of flexibility, rigidity to preserve strong recognition properties and high surface/volume ratio) are expected to lead to an increased efficiency and to a higher rate enhancement, overcoming the major drawbacks of the bulk polymers.

The following two sections will describe how these materials have been exploited so far to improve the activity of systems, previously developed as a bulk or newly designed, creating molecularly imprinted microgels with hydrolytic activity or C–C bond formation capability.

7.1 Hydrolytic Microgels

The first example that describes imprinted catalytic microgels was reported in 2004 by Resmini et al. [71], which described the synthesis of soluble acrylamide-based microgels with hydrolytic activity towards 4-acetamidophenyl 4-nitrophenyl carbonate (102). The imprinting strategy was based on the corresponding phosphate TSA (103). Literature data demonstrated that arginine (104) and tyrosine (105) are

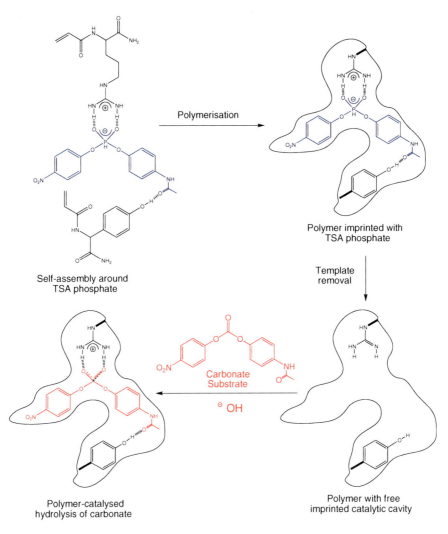

Fig. 5 Imprinting strategy based on the TSA (**103**). The phosphate template mimics the tetrahedral intermediate deriving from the hydrolysis of the analogue carbonate, imprinting therefore a cavity with a shape complementary to the real transition state of the reaction

responsible for improved catalytic activity in hydrolytic reactions, and therefore they were used as functional monomers after making them polymerisable by adding an acryloyl functionality. The polymers have been made in DMSO using a high dilution radical polymerisation initiated by AIBN (**106**), and acrylamide (**107**) as backbone monomer, 70 wt% N,N'-ethylenebisacrylamide (**108**) as cross-linker and $C_m = 1.5\%$ (Fig. 5).

A full kinetic characterisation of the imprinted (Pol397) and not imprinted (Pol396) polymers was carried out in a 9:1 solution of DMSO and Tris–HCl buffer. The kinetic profile showed a Michaelis–Menten behaviour with an 8.9 value of $V_{max}/K_m k_{uncat}$, with a significant enhancement of an order of magnitude over the uncatalysed reaction. The generation of a selective three-dimensional cavity providing substrate selectivity has been demonstrated using Pol397 to catalyse the reaction with the 2-nitro isomer of the substrate, which turned out to be almost nine times less active than the 4-nitro isomer. In 2005 the same group reported a more detailed study related to this catalytic system [72], where microgels with different percentages of cross-linker (from 70 to 90%) were fully characterised using a variety of techniques to determine their chemico-physical properties, including relative molecular mass, M_r, coil density and particle size.

A step further in the synthesis of enzyme-like microgels was made in 2006 by G. Wulff and co-workers [73] who reported a new procedure to obtain molecularly imprinted catalysts in the form of soluble, single-molecule nanogels of defined structure. The imprinting strategy was based on the system previously developed by the same group for bulk polymers [74]. After investigating different polymerisation methodologies using ethylene dimethacrylate (**109**) and methyl methacrylate (**110**) with high dilution radical polymerisation, a new procedure, called '*post dilution method*', was developed to increase the rigidity of the structure without increasing the monomer concentration. The macrogelation is avoided, stopping the polymerisation at high concentration just before that occurs, and largely diluting the solution with polymerisation solvent to keep the concentration of monomer below C_m. In this way the density of the particles increases but the polydispersity becomes lower and the catalytic activity improves. This method allowed synthesis of nanogels very similar to natural enzymes because of their solubility and average molecular weight (40 kDa), and decreased the number of active sites to one per particle. Kinetic characterisation of these nanogels were carried out with two equivalents of cavities per one equivalent of substrate, which resulted in a very high k_{impr}/k_{sol} value of 291 and k_{impr}/k_{contr} value of 18.5 for the imprinted nanogel **ING4**, where trimethylolpropane–trimethacrylate (TRIM) (**111**) was positively used as cross-linker.

7.2   Microgels in C–C Bond Formation

The hydrolysis reaction is generally less demanding in terms of energy required to convert reagents into products since it presupposes a bond breakage which is entropically favoured. Reactions involving creation of a new bond are very challenging, especially a C–C bond with a new chiral centre, given that two atoms have to approach and join together in the right spatial configuration. This partly explains why, looking at the plethora of publications related to catalytic molecular imprinting, only a small percentage is dedicated to C–C bond formation, the reason being the difficulty in designing a system capable of catalyse a reaction with high level of selectivity.

The first attempt in using a microgel system to catalyse a C–C bond formation dates from 2007, when Yuanzong Li et al. [75] reported the synthesis and characterisation of soluble nanogels mimicking a peroxidase-like catalyst for the oxidation (dimerisation) of HVA (73) under aqueous conditions, using a catalytic system previously developed by the same group for bulk polymers; see Sect. 4.2 [49]. The soluble nanogels were made using the Fe-porphyrin (*hemin*) (74), acrylamide (107) and 4-vinylpiridine as functional monomers and HVA, the substrate, as template with a solution polymerisation method initiated by AIBN. Chemo-physical characterisation of these polymers by GPC, TEM, DLS and ESEM established an average size of 200 nm in DMSO and 160 nm in a mixture of DMSO–buffer. Kinetic characterisations performed in DMSO–Tris–HCl buffer (5:95), where the polymers are completely homogeneous, resulted in a sevenfold increased activity compared with the bulk and threefold increase compared with the same polymer in 100% buffer. These results indicate that the improvement in catalytic activity can be attributed mainly to the size decrease of the enzyme-like polymer and therefore to the homogeneity of the solution. This example is indicative of how the microgels materials lead to a serious improvement even using a more sophisticated system such as those catalysing bond formation, not just bond breakage.

In 2008 Resmini et al. [76] presented their work on the synthesis of novel molecularly imprinted nanogels with Aldolase type I activity in the cross-aldol reaction between 4-nitrobenzaldehyde and acetone. A polymerisable proline derivative was used as the functional monomer to mimic the enamine-based mechanism of aldolase type I enzymes. A 1,3-diketone template, used to create the cavity, was designed to imitate the intermediate of the aldol reaction and was bound to the functional monomer using a reversible covalent interaction prior to polymerisation. Soluble imprinted nanogels were prepared by using a high-dilution radical polymerisation, which was followed by template removal and estimation of active site concentrations by monitoring the release of 4-nitrophenolate due to the acetylation of proline units with 4-nitrophenyl acetate. Analysis by DLS and TEM confirmed an average particle size of 20 nm for the nanogel preparations, comparable to the size of natural enzymes, and an average molecular mass ranging from 258 kDa to 288 kDa, as determined by GPLC using polymethylmethacrylate standards. The polymers were found to be soluble in DMF, DMSO and mixtures thereof, giving rise to homogeneous solutions. The kinetic characterisation of both imprinted and non-imprinted nanogels was carried out with catalyst concentrations between 0.7 mol% and 3.5 mol%. Imprinted nanogel MIP-AS147 was found to have a k_{cat} value of 0.26×10^{-2} min^{-1}, the highest value ever achieved with imprinted nanogels catalysing C–C bond formation. Comparison of the catalytic constants for both imprinted nanogel MIP-AS147 and non-imprinted nanogel NIP-AS133 gave a ratio of $k_{catMIP}/k_{catNIP} = 20$, which is indicative of good imprinting efficiency and highlights the significance of the template during the imprinting process. Analysis of the stereoselectivity of the reaction catalysed by nanogels gave 62% ee, which is comparable with that of the "free" catalyst. This work is the first and so far the only example of an imprinted nanogel capable of catalysing the cross-aldol reaction. The kinetic data obtained provide additional evidence that the use of microgels as materials for molecular imprinting reduces sensibly the gap between the natural and artificial enzymes in terms of transport and molecular recognition properties and gives hope for further improvement in specific applications as tailor-made catalysts capable of complementing natural enzymes.

8   Imprinting in Silica

The concept of imprinting in silica stretches back to the initial work performed by Dickey in the late 1940s [4] and, although it is often directly compared to the analogous technique of imprinting in acrylates, it is generally considered as a more complex approach for the preparation of antibody and enzyme mimics. The complexity of this process is a consequence of the large number of factors that influence the properties and structure of the final materials. In addition to this, preparation of these materials is performed in aqueous or alcoholic environments, which, in the case of non-covalent, and also moisture sensitive covalent, template-monomer interactions, can be extremely problematic. A number of authors have, however, despite the complexities of the method, reported the successful preparation of silica-imprinted catalysts. This section describes those reports and is subdivided based on the different polymeric formats employed, which includes imprinting in "bulk" silica, surface imprinting and imprinting in silica nanoparticles.

8.1 "Bulk" Silica

Although there have recently been a number of publications which report the imprinting of "bulk" silica or sol–gel materials, most of these have been used for purposes other than catalysis. Katz and Davis, however, reported the successful imprinting of "bulk" microporous silica, which was subsequently used for the catalysis of the Knoevenagel condensation reaction between malononitrile and isophthalaldehyde [77]. Here a covalent imprinting approach was used in which a substituted aromatic template was bound to one, two or three silane precursors via carbamate linkages and polymerised with tetraethoxysilane. This approach was used in order to create active sites where aminopropyl groups, following template removal, are situated in the appropriate position within the cavity. The carbamate linkages were utilised to protect the amine groups during polymerisation in order to prevent basic catalysis occurring during the sol-gel process. Although, characterisation of the imprinted materials was of the highest calibre, very little was reported on their catalytic behaviour. It was simply stated that the material imprinted with a template bound to two silane precursors was an active base catalyst for the condensation reaction.

In order to overcome some of the problems generally associated with the preparation of imprinted catalysts in "bulk" materials, such as poor mass transfer properties and non-uniform polymer matrices, Iwasawa and co-workers report the development of a technique for the production of catalytically active sites at the surface of commercially available Ox.50 silica microspheres [78–81]. Initially a catalytically active Rhodium dimer was coordinatively attached to the silica surface, followed by subsequent coordination of trimethyl phosphite (template) to the metal centre. The template was chosen because of its structural analogy to the half-hydrogenated species present during the hydrogenation of 3-ethyl-2-pentene. In order to produce a cavity, the silica surface was further built up around the template-catalyst complex by hydrolysis and subsequent condensation of tetramethoxysilane with the surrounding silica matrix. The material containing imprinted cavities, relative to the non-imprinted material (the catalyst supported at the surface of the silica), showed a 35-fold enhancement in catalytic activity for the hydrogenation of 3-ethyl-2-pentene.

8.2 Silica Nanoparticles

Markowitz et al. developed a different approach, again in an attempt to overcome some of the inherent difficulties that arise when imprinted "bulk" materials are used as catalysts [82]. Here, the authors used a template-directed method to imprint an α-chymotrypsin TSA at the surface of silica nanoparticles, prepared with a number of organically modified silanes as functional monomers. Silica particle formation was performed in a microemulsion, where a mixture of a non-ionic surfactant and the acylated chymotrypsin TSA (with the TSA acting as the headgroup at the surfactant water interface) were used as a method of creating a cavity capable of hydrolysis of benzoylarginine-*p*-nitroanalyde, a trypsin substrate. Unexpectedly, the particles were highly selective for the D-isomer of the substrate, even though the imprint molecule had the L-isomer configuration.

In a subsequent study, Markowitz and co-workers investigated the effect that the addition of functional silanes had on the catalytic activity of the surface imprinted nanoparticles [83]. It was suggested that a variation of the basicity of the functional monomers affected initial rates of hydrolysis and that imprinted particles prepared with mixtures of functional monomers demonstrated a cooperative effect promoting catalytic activity. Later the same group used a template-directed method for the imprinting of a hydrolysis product of Soman (nerve agent) at the surface of silica nanoparticles [84]. Again, a number of different functionalized silane precursors were used and the binding characteristics of the imprinted particles were investigated. The results demonstrated that the imprinted nanoparticles displayed a significantly higher degree of specificity for the imprint molecule than structurally related organophosphates.

9 Conclusions

The field of MIPs with catalytic activity has evolved considerably over the last decade, as more polymeric formats are being exploited and new combinations of strategically designed monomers and templates are used. The development of artificial enzyme-mimics, although successful in some cases, is still a considerable challenge. The achievement of high catalytic activities coupled with very good imprinting efficiencies requires in-depth knowledge of the polymeric matrices, close evaluation of the experimental parameters and detailed kinetic characterisations.

Acknowledgements The authors wish to thank the EU (MCA-RTN-033873), Queen Mary University of London and GlaxoSmithKline for financial support.

References

1. Kirby AJ (1996) Angew Chem Int Ed 35:707–724
2. Schultz PG, Lerner RA (1995) Science 269:1835–1842
3. Polyakov MV (1931) Zh Fiz Khim 2:799–805
4. Dickey FH (1949) Proc Natl Acad Sci USA 35:227–229
5. Wulff G, Sarhan A (1972) Angew Chem Int Ed 11:341
6. Andersson L, Sellergren B, Mosbach K (1984) Tetrahedron Lett 25:5211–5214
7. Schultz PT, Yin J, Lerner RA (2002) Angew Chem Int Ed 41:4427–4437
8. Shea K, Thompson E, Pandey S, Beauchamp P (1980) J Am Chem Soc 102:3149–3155

9. Damen J, Neckers D (1980) J Am Chem Soc 102:3265–3267
10. Leonhardt A, Mosbach K (1987) React Polym 6:285–290
11. Wulff G, Vietmeier J (1989) Macromol Chem Phys 190:1717–1726
12. Wulff G, Vietmeier J (1989) Macromol Chem Phys 190:1727–1735
13. Bystroem S, Boerje A, Akermark B (1993) J Am Chem Soc 115:2081–2083
14. Tramontano A, Janda KD, Lerner RA (1986) Science 234:1566–1570
15. Robinson DK, Mosbach K (1989) J Chem Soc Chem Commun:969–970
16. Ohkubo K, Urata Y, Hirota S, Honda Y, Fujishita Y, Sagawa T (1994) J Mol Catal A Chem 93:189–193
17. Ohkubo K, Urata Y, Honda Y, Nakashima Y, Yoshinaga K (1994) Polymer 35:5372–5374
18. Ohkubo K, Funakoshi Y, Urata Y, Hirota S, Usui S, Sagawa T (1995) J Chem Soc Chem Commun:2143–2144
19. Sellergren B, Shea K (1994) Tetrahedron Asymmetry 5:1403–1406
20. Schultz PG (1989) Angew Chem Int Ed 28:1283–1295
21. Wulff G, Gross T, Schonfeld R (1997) Angew Chem Int Ed 36:1961–1964
22. Muller R, Andersson LI, Mosbach K (1993) Macromol Rapid Commun 14:637–641
23. Shokat KM, Leumann CJ, Sugasawara R, Schultz PG (1989) Nature 338:269
24. Beach J, Shea K (1994) J Am Chem Soc 116:379–380
25. Matsui J, Nicholls IA, Karube I, Mosbach K (1996) J Org Chem 61:5414–5417
26. Hilvert D, Hill K, Nared K, Auditor M (1989) J Am Chem Soc 111:9261–9262
27. Wulff G (1995) Angew Chem Int Ed 34:1812–1832
28. Wulff G (2002) Chem Rev 102(1):1–27
29. Alexander C, Andersson HS, Andersson LI, Ansell RJ, Kirsch N, Nicholls IA, O'Mahony J, Whitcombe MJ (2006) J Mol Recognit 2:106–180
30. Alexander C, Smith C, Whitcombe M, Vulfson E (1999) J Am Chem Soc 121:6640–6651
31. Mosbach K, Yu Y, Andersch J, Ye L (2001) J Am Chem Soc 123:12420–12421
32. Yu Y, Ye L, Haupt K, Mosbach K (2002) Angew Chem Int Ed 41:4459–4463
33. Dahlstrom J, Shoravi S, Wikman S, Nicholls IA (2004) Tetrahedron Asymmetry 15: 2431–2436
34. Zhang H, Piacham T, Drew M, Patek M, Mosbach K, Ye L (2006) J Am Chem Soc 128: 4178–4179
35. Meng Z, Yamazaki T, Sode K (2004) Biosens Bioelectron 20:1068–1075
36. Meng Z, Sode K (2005) J Mol Recognit 18:262–266
37. Sellergren B, Karmalkar RN, Shea KJ (2000) J Org Chem 65:4009–4027
38. Sagawa T, Togo K, Miyahara C, Ihara H, Ohkubo K (2004) Anal Chim Acta 504:37–41
39. Toorisaka E, Uezu K, Goto M, Furusaki S (2003) Biochem Eng J 14:85–91
40. Meng Z, Yamazaki T, Sode K (2003) Biotechnol Lett 25:1075–1080
41. Zhang D, Li S, Li W, Chen Y (2007) Catal Lett 115:169–175
42. Emgenbroich M, Wulff G (2003) Chem Eur J 9:4106–4117
43. Liu JQ, Wulff G (2004) Angew Chem Int Ed 43:1287–1290
44. Liu JQ, Wulff G (2004) J Am Chem Soc 126:7452–7453
45. Liu JQ, Wulff G (2008) J Am Chem Soc 130:8044–8054
46. Liu X, Mosbach K (1997) Macromol Rapid Commun 18:609–615
47. Cammidge AN, Baines NJ, Bellingham RK (2001) Chem Commun:2588–2589
48. Viton F, White PS, Gagné MR (2003) Chem Commun:3040–3041
49. Cheng Z, Zhang L, Li Y (2004) Chem Eur J 10:3555–3561
50. Cheng Z, Li Y (2006) J Mol Catal A Chem 256:9–15
51. Hedin-Dahlström J, Rosengren-Holmberg JP, Legrand S, Wikman S, Nicholls IA (2006) J Org Chem 71:4845–4853
52. Kirsch N, Hedin-Dahlström J, Henschel H, Whitcombe M, Wikman S, Nicholls IA (2009) J Mol Catal B Enzym 58:110–117
53. Gouverneur VE, Houk KN, de Pascual-Teresa B, Beno B, Janda KD, Lerner RA (1993) Science 262:204–208

54. Polborn K, Severin K (2000) Chem Eur J 6:4604–4611
55. Burri E, Ohm M, Daguenet C, Severin K (2005) Chem Eur J 11:5055–5061
56. Anderson C, Shea K, Rychnovsky S (2005) Org Lett 7:4879–4882
57. Svenson J, Zheng N, Nicholls IA (2004) J Am Chem Soc 126:8554–8560
58. Motherwell WB, Bingham MJ, Pothier J, Six Y (2004) Tetrahedron 60:3231–3241
59. Takeuchi T, Ugata S, Masuda S, Matsui J, Takase M (2004) Org Biomol Chem 2:2563–2566
60. Yane T, Shinmori H, Takeuchi T (2006) Org Biomol Chem 4:4469–4473
61. Strikovsky AG, Kasper D, Grun M, Green BS, Hradil J, Wulff G (2000) J Am Chem Soc 122:6295–6296
62. Strikovsky A, Hradil J, Wulff G (2003) React Funct Polym 54:49–61
63. Wulff G, Gross T, Schönfeld R (1997) Angew Chem Int Ed 36:1962–1964
64. Busi E, Basosi R, Ponticelli F, Olivucci M (2004) J Mol Catal A Chem 217:31–36
65. Jakubiak A, Kolarz BN, Jezierska J (2006) Macromol Symp 235:127–135
66. Say R, Erdem M, Ersoz A, Turk H, Denizli A (2005) Appl Catal A 286:221–225
67. Bruggemann O (2001) Biomol Eng 18:1–7
68. Visnjevski A, Yilmaz E, Bruggemann O (2004) Appl Catal A 260:169–174
69. Visnjevski A, Schomacker R, Yilmaz E, Bruggemann O (2005) Catal Commun 6:601–606
70. Müller R, Andersson LI, Mosbach K (1993) Macromol Rapid Commun 14:637–641
71. Maddock SC, Pasetto P, Resmini M (2004) Chem Commun 5:536–537
72. Pasetto P, Maddock SC, Resmini M (2005) Anal Chim Acta 542:66–75
73. Wulff G, Chong BO, Kolb U (2006) Angew Chem Int Ed 45:2955–2958
74. Strikovsky A et al (2000) J Am Chem Soc 122(26):6295–6296
75. Chen Z, Hua Z, Wang J, Guan Y, Zhao M, Li Y (2007) Appl Catal A 328:252–258
76. Carboni D, Flavin K, Servant A, Gouverneur V, Resmini M (2008) Chem Eur J 14:7059–7065
77. Katz A, Davis ME (2000) Nature 403:286–289
78. Tada M, Sasaki T, Shido T, Iwasawa Y (2002) Phys Chem Chem Phys 4:5899–5909
79. Tada M, Sasaki T, Iwasawa Y (2002) J Catal 211:496–510
80. Tada M, Sasaki T, Iwasawa Y (2002) Phys Chem Chem Phys 4:4561–4574
81. Tada M, Sasaki T, Iwasawa Y (2004) J Phys Chem B 108:2918–2930
82. Markowitz MA, Kust PR, Deng G, Schoen PE, Dordick JS, Clark DS, Gaber BP (2000) Langmuir 16:1759–1765
83. Markowitz MA, Kust PR, Klaehn J, Deng G, Gaber BP (2001) Anal Chem Acta 435:177–185
84. Markowitz MA, Deng G, Gaber BP (2000) Langmuir 16:6148–6155

Index

343